Measurement
Techniques in
Plant Science

Academic Press Rapid Manuscript Reproduction

Measurement Techniques in Plant Science

Edited by

YASUSHI HASHIMOTO
Department of Biomechanical Systems
College of Agriculture, Ehime University
Tarumi, Matsuyama, Japan

HIROSHI NONAMI
Department of Biomechanical Systems
College of Agriculture, Ehime University
Tarumi, Matsuyama, Japan

PAUL J. KRAMER
Department of Botany
Duke University
Durham, North Carolina

BOYD R. STRAIN
Department of Botany
Duke University
Durham, North Carolina

ACADEMIC PRESS, INC.
Harcourt Brace Jovanovich, Publishers
San Diego   New York   Boston   London   Sydney   Tokyo   Toronto

Copyright © 1990 by Academic Press, Inc.
All Rights Reserved.
No part of this publication may be reproduced or transmitted in any
form or by any means, electronic or mechanical, including photo-
copy, recording, or any information storage and retrieval system,
without permission in writing from the publisher.

Academic Press, Inc.
San Diego, California 92101

United Kingdom Edition published by
Academic Press Limited
24–28 Oval Road, London NW1 7DX

Library of Congress Cataloging-in-Publication Data

Measurement techniques in plant science / edited by Yasushi
Hashimoto
  ... [et al.].
      p. cm.
    Based on a symposium entitled "Instrumentation and Physi-
ological
  Ecology", held in Tokyo in 1985
    ISBN 0-12-330585-3
    1. Plant physiological ecology--Congresses.   2. Growth
(Plants)-
  -Measurement--Congresses.   3. Crops--Growth--Measurement-
  -Congresses.   I. Hashimoto, Yasushi, Date.
  QK905.M43   1990
  581.3'1--dc20
                                                    90-49622
                                                       CIP

Printed in the United States of America
90 91 92 93   9 8 7 6 5 4 3 2 1

CONTENTS

CHAPTER 3:   *Photosynthesis*

CHAPTER 4:   *Translocation*

CHAPTER 5:   *Mineral Nutrition*

CHAPTER 6: *Image Processing*

CONTRIBUTORS

Numbers in parentheses indicate the pages on which the authors' contributions begin.

I. Aiga (343), College of Agriculture, University of Osaka, Prefecture, Sakai, Osaka 591, Japan

J. S. Boyer (101), College of Marine Studies, University of Delaware, Lewes, Delaware 19958

W. Day (207), AFRC Institute of Engineering Research, Wrest Park, Silsoe, Bedford MK45 4HS, United Kingdom

H. Eguchi (361), Biotron Institute, Kyushu University, Fukuoka 812, Japan

E. Epstein (291), Department of Land, Air and Water Resources, University of California, Davis, California 95616

Y. Fares (265), Biosystems Technologies, Inc., Durham, North Carolina 27706-4851

C. B. Field (185), Department of Plant Biology, Carnegie Institution of Washington, 290 Panama Street, Stanford, California 94305-1297

E. L. Fiscus (79), United States Department of Agriculture, Agriculture Research Service, 1701 Center Avenue, Fort Collins, Colorado 80526

J. D. Goeschl (265), Biosystems Technologies, Inc., Durham, North Carolina 27706-4851

Y. Hashimoto (7, 373), Department of Biomechanical Systems, College of Agriculture, Ehime University, Tarumi, Matsuyama 790, Japan

S. E. Hetherington (229), Division of Horticulture, Sydney Laboratories, CSIRO, North Ryde, Sydney 2113, Australia

C. E. Jaeger (265), EPO Biology, University of Colorado, Boulder, Colorado 80309

T. Kaneko (277), Advanced Research Laboratory, Hitachi, Ltd., Kokubunji, Tokyo, Japan

Y. Kano (165), Department of Electrical Engineering, Faculty of Technology, Tokyo University of Agriculture and Technology, Koganei, Tokyo, Japan

M. R. Kaufmann (69), United States Department of Agriculture Forest Service, Rocky Mountain Forest and Range Experiment Station, Fort Collins, Colorado 80521

J. Kondo (343), Science Council of Japan, Roppongi, Tokyo 106, Japan

P. J. Kramer (3, 45, 403), Department of Botany, Duke University, Durham, North Carolina 27706

S. Kuraishi (151), Department of Environmental Sciences, Faculty of Integrated Arts and Sciences, Hiroshima University, Hiroshima 730, Japan

J. S. MacFall (403), School of Forestry and Environmental Studies, Duke University, Durham, North Carolina 27706

C. E. Magnuson (265), Biosystems Technologies, Inc., Durham, North Carolina 27706-4851

H. Miyauchi (151), Department of Environmental Sciences, Faculty of Integrated Arts and Sciences, Hiroshima University, Hiroshima 730, Japan

H. A. Mooney (185), Department of Biological Sciences, Stanford University, Stanford, California 94305

H. Nonami (7, 101), Department of Biomechanical Systems, College of Agriculture, Ehime University, Tarumi, Matsuyama 790, Japan

K. Omasa (343, 387), National Institute for Environmental Studies, Yatabe, Tsukuba, Ibaraki 305, Japan

K. J. Parkinson (207), Analytical Development Company, Ltd., Pindar Road, Hoddesdon, Herts EN1 10AQ, United Kingdom

N. Sakurai (151), Department of Environmental Sciences, Faculty of Integrated Arts and Sciences, Hiroshima University, Hiroshima 730, Japan

K. Shimazaki (387), College of General Education, Kyushu University, Ropponmatsu, Fukuoka 810, Japan

J. N. Siedow (403), Department of Botany, Duke University, Durham, North Carolina 27706

R. M. Smillie (229), Division of Horticulture, Sydney Laboratories, CSIRO, North Ryde, Sydney 2113, Australia

E. Steudle (113), Lehrstuhl für Pflanzenökologie, Universität Bayreuth, 8580 Bayreuth, Federal Republic of Germany

B. R. Strain (265), Department of Botany, Duke University, Durham, North Carolina 27706

K. Supappibul (151), Mangrove Forest Management Unit, Lamngob, Namchew, Trat, Thailand

M. Takatsuji (277), Advanced Research Laboratory, Hitachi, Ltd., Kokubunji, Tokyo, Japan

R. M. Welch (319), United States Department of Agriculture, Agriculture Research Service, U.S. Plant, Soil & Nutrition Laboratory, Ithaca, New York 14853-0331

H. Yamasaki (25), Department of Mathematical Engineering and Information Physics, Faculty of Engineering, University of Tokyo, 7-3-1 Hongo, Bunkyo-ku, Tokyo 113, Japan

PREFACE

Progress in plant science always has been and continues to be dependent on progress in the instrumentation required to measure plant processes and environmental factors. This book has its origins in the international symposium entitled "Instrumentation and Physiological Ecology" held in Tokyo in 1985. The symposium brought together scientists from laboratories in various countries to exchange information about existing instrumentation and provided opportunities to discuss new problems and new apparatus needed to solve them.

The symposium covered the topics of water relations, photosynthesis, translocation, mineral nutrition, and image processing. Since the time of the symposium, there have been important developments and refinements in measurement techniques and technologies, such as intelligent sensing systems and nuclear magnetic resonance (NMR) computer tomography, which are included as topics in this book. In addition, articles have been revised and updated as deemed necessary by their authors.

This book will be of interest not only to specialists in plant science, but also to students who may be interested in reading of the histories of the development of the instrumentations which are included in the reviews. It should also be of interest to engineers working toward practical applications of the techniques for crop management described in the book.

Y. Hashimoto
P. J. Kramer
H. Nonami
B. R. Strain

Chapter 1
Introduction

INSTRUMENTATION AND PROGRESS IN SCIENCE

Paul J. Kramer
Department of Botany
Duke University

Advances in science depend on new ideas and on development of the instrumentation needed to investigate the ideas. Pioneers such as Newton, Darwin, Mendel, and Einstein created new concepts with revolutionary implications for various branches of science, but their concepts are still being explored and expanded as new instrumentation and new research methods permit the acquisition of new information. The importance of instrumentation is well recognized in the scientific world and several Nobel prizes have been awarded to developers of devices such as cyclotrons, transistors, lasers, partition chromatography, and the CAT X-ray scanning technique.

Many good ideas have waited for decades because of the lack of instrumentation to investigate them, and on the other hand development of new instrumentation often results in a notable increase in research activity. The history of the study of photosynthesis provides an example of how progress depends on instrumentation. About 1771 Priestley observed that green plants changed the composition of the air in containers enclosing the plants. However, it was not until improved methods for analyzing the composition of the air and measuring changes in volume of its components were developed early in the 19th century that it was established by de Saussure that green plants in the light absorb CO_2 and release O_2. It was another 60 years before Sachs established that the chloroplasts were involved in this gas exchange and that carbohydrate was produced. It was not until after radioactive isotopes became available in the 1930s that the reductive carbon cycle was worked out, and only after the development of new instrumentation was it possible to study photosynthetic electron flow.

Measurement Techniques in Plant Science
Copyright © 1990 by Academic Press, Inc.
All rights of reproduction in any form reserved.

Few field measurements of photosynthesis were made during the 19th and early 20th century because there was no convenient method of measuring gas exchange. In early studies change in CO_2 concentration of the air passed over leaves was measured by bubbling the air through tubes filled with dilute alkali and titrating the alkali or measuring change in conductivity to determine the amount that had been neutralized by the CO_2. This method was slow, untidy, and not very accurate. During the 1940s infrared gas analyzers began to be used to measure change in concentration of gases and commercially built infrared gas analyzers sensitive to CO_2 became available in the 1950s. However, the first instruments were large and heavy and required mobile laboratories for field measurements of photosynthesis. Improvements in electronics and microprocessors now make it possible to carry the equipment to the field in a suitcase. Infrared gas analyzers proved to be so useful, both in the laboratory and in the field, that there was a great increase in research on gas exchange of plants under various environmental conditions. Today infrared gas exchange measurements are being supplemented by measurements of oxygen production of leaf disks, chlorophyll fluorescence of attached leaves, and carbon isotope discrimination, to provide a broader understanding of photosynthesis. Some of these technologies are discussed in this volume.

There was a similar lag between theory and practice in the field of plant water relations. The idea that the free energy status of water (now termed the water potential) was important was developed early in this century, but several decades passed before practical methods of measuring water potential were developed (see Chapter 2). Good physiological-ecological research often benefits from simultaneous measurement of CO_2 exchange, stomatal conductance, and plant water status, all of which have become possible because reliable, portable gas analyzers, porometers, thermocouple psychrometers, and Scholander pressure chambers are available. As a result of these improvements in instrumentation research in this area has increased many fold in recent years. Improvements in the measurement of plant water status are discussed in Chapter 2.

Improvements in equipment to control the plant environment also have contributed to progress in physiological and ecological research. Technological advances in air conditioning and control systems and improvements in artificial lighting have increased the usefulness of growth chambers and improved the reproducibility of controlled environmental regimes.

New information often indicates new areas of ignorance and the need for more information, and this in turn creates the need for additional instrumentation. Looking to the future we are in a period of rapid development of instrumentation in which the combination of improved electronics and computer capacity combined with existing instrumentation permits measurements that were impossible a decade or two ago. An example is the development of nuclear magnetic resonance spectroscopy and imaging. The physical principles were understood before the technology made it useful in biology and medicine. Only after strong, stable, superconducting magnets and large computers to store and process data became available could the concept be fully exploited. Now improvements in gradient and radio frequency coils and pulse control are resulting in images in the microscopic range (see Chapter 6).

Technology is advancing too rapidly for plant scientists to keep informed concerning new apparatus and new methods that might be useful in their research. An example is the development of a half dozen new scanning probes, following the appearance of the scanning tunneling microscope (STM) in 1981. The STM does not work well on biological material, but it seems possible that some of the new scanning probe microscopes such as the atomic force microscope or the scanning near-field optical microscope may become useful in biological research. Thus there is increasing need for interdisciplinary exchange and collaboration between plant scientists and designers of scientific equipment. Books such as this should increase the transfer of useful technology.

OVERVIEW OF CURRENT MEASUREMENT TECHNIQUES FROM ASPECTS OF PLANT SCIENCE

Yasushi Hashimoto
Hiroshi Nonami

Department of Biomechanical Systems
College of Agriculture, Ehime University
Tarumi, Matsuyama 790, Japan

I. INTRODUCTION

Advancements in sensor designs and developments in measurement techniques have been occurring rapidly in recent years owing to developments in electronics and computer science. Such progress has deeply influenced current measurement techniques in plant science. Agricultural industries, as well as scientists in the field of plant science, have an interest in the recent developments in sensors and measurement technologies, as attempts are being made to automate agricultural production. However, automation of agricultural production has not been widely attained a commercially feasible basis thus far.

Developments in instrumentation have led to smaller and smaller instruments, enabling researchers to measure quantities more precisely. Also, the invention of instrumentation for use with intact living tissues has made it possible to measure physiological information without

Measurement Techniques in Plant Science

7

altering physiological metabolism. Thus, measurement
techniques for determining the physiological properties of
plants now range from molecular analysis at the biochemical
level to macro analysis of the whole plant in ecological
studies. Because this book is written for not only those
with purely scientific interests but also for those
interested in the practical application of measurement
techniques in agricultural production, the present overview
will deal with physicochemical measurement techniques which
are applied to materials larger than the cellular level.
Furthermore, prospects for developing measurement techniques
in application to automation in agricultural production in
greenhouses and plant factories are discussed.

II. WATER STATUS MEASUREMENTS

In order to determine the water status of plants and
their surroundings, water potential is used and is applicable
to the entire soil-plant-atmosphere continuum (see Chapter
2-1 by Kramer in this volume). Thermodynamically, solute and
pressure forces in plants and soils are most exactly
described by water potential, defined as the chemical
potential divided by the partial molal volume of liquid water.
Measurements of water potential are always compared to
reference water potential, which is pure free water at
atmospheric pressure, at a defined gravitational position,
and the same temperature as the system being measured. The
reference is defined as having a water potential of zero.
The water potential value indicates differences between the
amount of free energy in the system and in the reference
state and is defined as the sum of solute forces and local
plus external pressure forces acting on the water as shown in
the following manner;

$$\Psi w = \Psi s + \Psi p$$

$$(1)$$

where Ψw, Ψs and Ψp are water potential, osmotic
potential and pressure potential, respectively. Values for
both Ψw and Ψs are always negative, and Ψp can be
positive or negative depending on whether the pressure is
above or below atmospheric pressure. In the case of plant
cells, Ψp usually represents turgor inside the cell, which
is a positive quantity.

By measuring water potentials, directions of water flow can be determined in not only plants but also in non-living matter surrounding plants. In order to measure water potential, thermocouple psychrometers have been used most frequently because both soil and plant tissues can be used for measurement. There are three types of measurement methods with thermocouple psychrometers, i.e., the Spanner type (1951), the Richards and Ogata type (Richards and Ogata, 1958), and the isopiestic psychrometer described by Boyer and Knipling (1965). Among the three types, the isopiestic psychrometer measures water potential most accurately, because errors caused by water diffusion through the sample in the psychrometer do not affect the determination of water potential (Boyer and Knipling, 1965; Boyer, 1966). Water potential measured with psychrometers indicates the averaged value over the volume of the sample. For example, when a leaf disk is taken as a sample, water potential gradients in the leaf disk will disappear gradually after it is set in the sampling chamber of a psychrometer, and water potential will equilibrate and become equal among cells during measurements. Thus, water potential measured will be the volume-average water potential in tissues of epidermis, mesophyll, phloem, xylem and cell wall spaces in the leaf disk.

Psychrometers can be used to measure water potential in not only excised tissue but also intact plants. Water potential in intact leaves were measured with the Spanner-type psychrometer (Shackel and Brinckmann, 1987) and the isopiestic psychrometer (Boyer, 1968). In order to measure water potential associated with the growth process, Boyer has developed a "guillotine" psychrometer, which uses the isopiestic method, to measure the growth-induced water potential in intact soybean stems (Boyer et al., 1985). He found that excision of tissue in the elongating region caused relaxation of the cell wall and hence, water potential of intact tissue is higher than that of excised tissue in the elongating region (Boyer et al., 1985; see Chapter 2-4 by Boyer and Nonami in this volume). Excision does not cause changes in water potential in the mature region, and therefore, water potential can be measured accurately in excised tissues in the mature region by using conventional psychrometers (Boyer et al., 1985; Nonami and Schulze, 1989). Although excision causes changes in water potential when tissues from the elongating region are used, changes in water potential which are caused by excision are relatively small compared with variations occurring naturally among different individual plants (Nonami and Boyer, 1989), and therefore, for most purposes using the conventional psychrometers is

acceptable for studies in water relations to measure accurate
water potentials.

Psychrometers can be used to measure osmotic potentials
of plant tissues. After measuring water potential of tissue,
cell membranes in the tissue can be broken by freezing and
thawing so that cell turgor is lost (Ehlig 1962). When
measuring water potential in tissue which has lost its turgor
through freezing and thawing, water potential measured can be
considered to be the osmotic potential of the tissue. If
dilution of the protoplast solution by water in the wall can
be assumed, turgor may be calculated by using Eq. 1.

Turgor of individual cells can be measured directly with
a cell pressure probe (Hüsken et al. 1978). Steudle
developed the pressure probe in the 1970's, and the structure
and functions of the pressure probe are described in a
chapter contributed by him in this volume (Chapter 2-5).

Osmotic potential of individual plant cells can be
measured with combinations of the cell pressure probe and
nanoliter osmometer (Nonami and Schulze, 1989). Cell
solution of an individual cell in an intact plant can be
extracted with the pressure probe, and can be placed inside
of the sample holder of the nanoliter osmometer with a
micromanipulator (Nonami and Schulze, 1989). Osmotic
potential of cell solution of volume as small as 10 pl could
be measured reliably with the nanoliter osmometer (Nonami and
Schulze, 1989).

Although Eq. 1 appears in all water-relation textbooks,
the validity of Eq. 1 has not been evident until recently.
Nonami et al. (1987) compared turgor measured with a cell
pressure probe to that measured with an isopiestic
thermocouple psychrometer in the mature regions of soybean
stems. The probe measured turgor directly in cells of intact
stems whereas the psychrometer measured the water potential
and osmotic potential of excised stem segments and turgor was
subsequently obtained by calculating the difference between
the two measurements. Both methods gave similar values of
turgor whether the plants were dehydrating or rehydrating
(Nonami et al., 1987). Boyer and Potter (1973) compared
water potential, osmotic potential and turgor measured with
the pressure chamber and the isopiestic psychrometer in
mature sunflower leaves, and obtained virtually the same
values with both methods.

Turgor and Osmotic potential in cells of well-expanded
leaves of *Tradescantia virginiana* were measured directly by
Nonami and Schulze (1989) while transpiration from the plants
was inhibited by keeping them at 100%RH in the dark, and the
cell water potential was calculated by summing the values

obtained for cell turgor and osmotic potential according to
Eq. 1. The osmotic potential, turgor, and water potential of
epidermal and mesophyll cells measured with the pressure
probe combined together with the nanoliter osmometer were
plotted in relation to the averaged values of osmotic
potential, turgor and water potential measured with the
isopiestic psychrometer in the same leaf (Figures 1A, 1C and
1E). Osmotic potentials of the epidermis were always higher

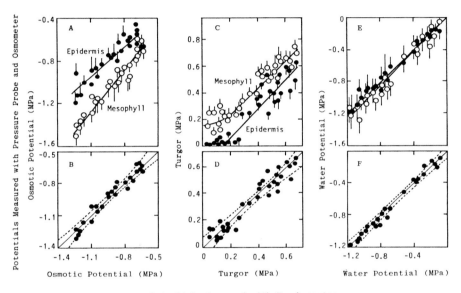

Figure 1. Cell osmotic potential (A, B), turgor (C, D) and
water potential (E, F) of the epidermis (●) and mesophyll
(○) compared with corresponding tissue potential measured
with the isopiestic psychrometer in leaves of *Tradescantia
virginiana* plants kept at 100% RH in the dark. Each point in
A, C and E represents the mean of potentials measured in 7 to
11 cells and vertical bars indicate 95% confidence intervals
calculated from Student's percentage t-distribution. Points
in B, D and F were calculated from the data shown in A, C and
E for tissue osmotic potential, turgor and water potential,
respectively, by averaging cell potentials in proportion to
the volumes occupied by the epidermis (47%) and the mesophyll
(53%) in the leaf. The isopotential position is shown by the
diagonal solid line in B, D and F. The hyperbolic curves in
B, D and F indicate the confidence band at P ≤ 0.01.

than those of the mesophyll (Figure 1A), and mesophyll cell
turgor was always larger than epidermal cell turgor (Figure
1C). Water potentials in both mesophyll and epidermal cells
were equivalent (Figure 1E). In order to make the
cell-potential results more comparable to the
tissue-potential results measured with the isopiestic
psychrometer, potentials of the epidermis and mesophyll cells
were averaged in proportion to their respective volumes in
the leaf (volume percentage of epidermis = 47 ± 6% (average ±
SD; n=21) and that of mesophyll = 53 ± 6% (average ± SD;
n=21)). The volume-averaged cell potentials were plotted
with corresponding psychrometer measurements (Figures 1B, 1D
and 1F, using data from Figures 1A, 1C, and 1E, respectively).
Figures 1B, 1D and 1F show that the isopotential line is
within the calculated confidence band, indicating that the
same tissue potentials can be obtained by using the
isopiestic psychrometer, pressure probe and osmometer.

Because the pressure probe, the nano-liter osmometer,
the psychrometer and the pressure chamber rely on different
principles of measurement, and because different plant
materials were used for the comparison of measured potentials,
it is unlikely that all would be subject to inherent but
undetected inaccuracies that would cause water potential,
osmotic potential and turgor to deviate by the same amount
from the true values of the potentials. Therefore, the
agreement between the methods suggests that Eq. 1 should be
valid in all plant cells.

Measurement methods for soil moisture and soil water
potential have been reviewed by Kano in this volume (Chapter
2-7). Although the psychrometric method can be used to
measure water potentials of both soil and plant roots, the
isolation of materials for measurement is required. Thus,
interactions between soil and roots for water movement cannot
be studied under completely intact conditions. Recently,
direct measurement of water content in soil and roots without
altering growing conditions of plants has become possible by
using magnetic resonance imaging. MacFall et al. (1990) used
magnetic resonance imaging to study a water-depletion region
surrounding loblolly pine roots grown in sand. The sand
phantom information from magnetic resonance imaging could be
interpreted dark regions in sand images as regions of low
water content and bright regions as areas of comparatively
greater water content (MacFall et al., 1990). Images of
loblolly pine seedlings planted in identical sand showed the
formation of a distinct water-depletion region first around
the woody taproot and later showed the region extended and
expanded around the lateral roots and clusters of mycorrhizal

short roots (MacFall et al., 1990). Magnetic resonance
imaging can analyze water distributions in both soil and
plant roots in sequential time course, and its theory and
applications are discussed in more detail by Kramer et al. in
this volume (Chapter 6-5).

III. TRANSPIRATION AND PHOTOSYNTHESIS
MEASUREMENT METHODS

Although gas exchanges of water vapor and CO_2 between
plants and the ambient air has been studied for a long time,
it is not easy to measure transpiration and photosynthesis
accurately in intact plants under natural conditions. Thus,
most techniques to measure transpiration and photosynthesis
accurately have been developed under laboratory conditions.
In laboratories, transpiration rates have usually been
measured together with CO_2 assimilation rates in a leaf
chamber for a single leaf as a part of the gas exchange
system, so that the internal CO_2 concentration can be
determined (Moss and Rawlins, 1963). Recently, some
infrared gas analyzers can measure concentrations of both
water vapor and CO_2 gas, and thus, both transpiration and CO_2
assimilation rates can be determined simply by measuring
concentration differences between inlet and outlet gases of a
leaf cuvette and flow rates of gases through the cuvette.
Such a gas exchange system was used to investigate relations
between cell water status in leaves and gas diffusion through
stomates (Shackel and Brinckmann, 1985; Nonami and Schulze,
1989). Although the system is simple and can be constructed
easily, when ambient CO_2 gas concentration rises to a higher
concentration (more than 1000 ppm), measurement of CO_2
assimilation rates will be difficult, because differential
gas analysis will be disturbed by background noise caused by
the high concentration of CO_2 gas in the ambient air. In
order to measure CO_2 assimilation rates under a high CO_2 gas
environment, a compensating method for measuring carbon
dioxide exchange has been used (Bazzaz and Boyer, 1972;
Martin et al., 1981). A leaf is sealed in a leaf chamber and
a small portion of the air from the chamber is continuously
passed through an infrared gas analyzer to monitor the CO_2
concentration of the leaf chamber atmosphere. The rate of
CO_2 exchange was assayed with a mass flow controller which
regulated the flow of CO_2 which is added to the chamber to
compensate exactly for the CO_2 taken up by the leaf. Thus,

the CO_2 concentration in the leaf chamber atmosphere was
maintained at a constant value regardless of the rate of
photosynthesis. By using a similar compensating method for
water vapor, transpiration from a leaf can be measured
(Bazzaz and Boyer, 1972; Martin et al., 1981). Recent gas
exchange systems and characteristics of various infrared gas
analyzers are discussed by Field and Mooney in this volume
(Chapter 3-1). The design of leaf cuvettes for gas exchange
systems is discussed in detail by Parkinson and Day in this
volume (Chapter 3-2).

More recently, a compact mass spectrometer has been used
to measure gas exchanges between plants and their
surroundings in terms of CO_2, O_2 and H_2O gases to evaluate
photosynthesis, respiration and transpiration simultaneously,
and its system is discussed in more detail in a chapter
written by Kaneko and Takatsuji in this volume (Chapter 4-2).
When isotopes are used in application for research of the
exchange of CO_2 by plants, CO_2 assimilation and translocation
can be measured continuously and non-destructively, and a
method using ^{11}C is described by Strain et al. (Chapter 4-1)
and a method using ^{13}C is described by Kaneko and Takatsuji
in this volume (Chapter 4-2).

When the transpiration rate is measured in a single leaf,
the rate is not necessarily representative of the average
transpiration rate for all leaves. Thus, estimating
transpiration rates for the whole plant is a difficult task
when studying large plants grown in natural habitats.
Kaufmann discusses a method to estimate tree canopy
transpiration from measurements of leaf conductance by using
leaf cuvettes attached to tree branches in this volume
(Chapter 2-2).

Measurements of transpiration can be used for crop
production management. Because plants lose water mostly due
to transpiration, accurate evaluation of transpiration from
crop vegetation helps to ensure adequate and optimum
irrigation for crops in the field. Under field conditions,
however, accurate measurements of transpiration for crop
vegetation are not easily made. Instead of measuring
transpiration, Fiscus suggests the use of porometric
measurements for water stress evaluation in the field for
application to irrigation control, and he discusses his ideas
in more detail in Chapter 2-3.

Transpiration can be measured accurately in an intact
plant without using a leaf cuvette. Because water has a
large specific heat, if water in the xylem is heated, heated
water will move upward along the stem following a
transpiratory flow. First, Huber (1932) paid close attention

to such a physical character of water in plant stems, and
invented a heat pulse method for the measurement of
transpiration. Although the heat pulse method was improved
by several researchers (Huber and Schmidt, 1937; Bloodworth
et al., 1955; Schurer et al., 1979) after Huber invented it,
the heat pulse method contained noise associated with heat
pulses and could not resolve accurately the rate of sap flow.
However, van Zee and Schurer (1983) introduced the
application of system theory and signal processing to the
heat pulse method, and by using computer simulation,
transpiration could thus be measured accurately by
eliminating noise caused by the adverse effects of radial
heat conduction.

Sakuratani (1981) used a similar principle to measure
transpiration, but employed a continuous heating system to
compensate for noise associated with heating. He invented a
heat balance method for measuring water flux in the stem of
intact plants (Sakuratani, 1981). The heat balance method
does not require calibration or computer simulation for
transpiration measurements (Sakuratani, 1981). Kitano and
Eguchi (1989) improved the heat balance method to measure
transpiration rates in stems of cucumber plants with a
resolution of 1×10^{-3} grams per second and a time constant of
1 minute. They found that root water absorption lagged
behind leaf transpiration and affected the dynamics of water
fluxes in plants (Kitano and Eguchi, 1989). Devices for the
heat balance method can be attached to main stems, branches,
and/or petioles to measure water fluxes in course of time at
positions where the devices are attached, and thus,
transpiration can be studied at multiple points
simultaneously in the same plant.

Although the leaf chamber method and the heat balance
method can measure transpiration rates accurately, both
methods cannot resolve very localized stomatal behavior
differences. In order to study localized transpiration,
Hashimoto et al. (1984) measured leaf temperature on various
regions of leaves by using an infrared scanning thermometer.
Because latent heat is taken away from a leaf when water
evaporates from the leaf, the temperature on the leaf surface
becomes lower as transpiration becomes larger. Thus, areas
having lower temperatures indicate the sites where the
greater amount of water evaporation is taking place.
Hashimoto et al. (1984) found that when water stress was
applied to a sunflower plant by pruning its roots, there
were significant differences in temperature and water status
along different areas of a single leaf. Thus, it will be of
interest for future research to determine causes of such

behavior and decide upon which part of a leaf the temperature
and water status should be measured. In this volume,
Hashimoto discusses a thermal image processing system
combined together with a scanning electron microscope in
order to investigate differential transpiration and localized
stomatal behavior (Chapter 6-2).

In order to study photosynthesis, measurements of
chlorophyll fluorescence can be useful and can be made easily.
Thus, recent developments in chlorophyll fluorescence
measurement devices are noteworthy. Smillie and Hetherington
(Chapter 3-3) describe practical applications of chlorophyll
fluorescence measurements, and Omasa and Shimazaki (Chapter
6-4) discuss image processing systems using chlorophyll
fluorescence for studies of environmental stresses.

IV.   MEASUREMENT METHODS
FOR GROWTH PROCESSES

Plant growth can be simply determined by measuring
increases in length or dry weight, using scales and balances.
Such growth measurements can be evaluated in time orders of
days or weeks. When electrical amplifiers are used, length
increases can be measured in time orders of seconds, and
therefore, growth rates can be determined from slopes
obtained with a linear displacement transducer (Cosgrove and
Green, 1981) or a radial displacement transducer (Boyer et
al., 1985). By using a displacement transducer, an
extensiometer can be made in order to measure growth
parameters in intact plants (Nonami and Boyer, 1990).

Kutschera and Schopfer (1986a; 1986b) made an
extensiometer from a displacement to stretch a plant segment
by applying a force to both ends of the segment. Nonami and
Boyer (1990) adapted their method for use with intact growing
plants. When an external force is applied to a plant tissue,
the tissue is extended and deforms. Subsequently, when the
external force applied is removed, the deformation of the
tissue recovers, but not completely. The recovery of length
in the tissue is considered to be an elastic component, and
the length deformed permanently is considered to be a plastic
component of deformation. Nonami and Boyer (1990) found that
under limited external forces applied to intact growing
tissues the elastic and plastic components behave as ideal
elastic (Hookian) and ideal nonelastic (Newtonian) materials,
respectively. Furthermore, the same values for elastic and

plastic components could be obtained repeatedly in the same
elongating plants under identical external forces (Nonami and
Boyer, 1990). The elastic component measured with the
extensiometer was physically related to cell elastic modulus
measured with a cell pressure probe, and not directly related
to the cell expansion process (Nonami and Boyer, 1990). From
strain-stress relations of the plastic components, an average
plastic deformability was calculated, and was related to the
growth process (Nonami and Boyer, 1990).

Using a "guillotine" psychrometer, wall extensibility
can be determined from a relaxation of potentials in the
elongation region after excision of tissue (Boyer et al.,
1985). When values of wall extensibility measured with the
psychrometer and those of plastic deformability measured with
the extensiometer were compared in similarly grown plants,
both quantities could be considered identical (Nonami and
Boyer, 1990). Hence, it is possible to measure wall
extensibility in intact plants with the extensiometer with
simultaneous measurements of growth rates.

Growth can be evaluated by using image processing
systems with the aid of computers, and growth measurements of
plant stems and leaves are discussed by Eguchi in this volume
(Chapter 6-2). Also, Epstein (Chapter 5-1) describes various
growth measurement methods in roots.

When growth is occurring, the physiological status of
expanding cells is drastically different from that of
non-growing mature cells. One distinctive phenomenon
associated with growth is water potential depression in
expanding cells, and measurement methods for water potential
and its components in growing tissues are described in detail
by Boyer and Nonami in this volume (Chapter 2-4). Another
phenomenon associated with growth is the polarized potential
created by growing cells. By using the vibrating probe of
the platinum-black microelectrode (Jaffe and Nuccitelli,
1974), it was found that current consistently entered the
meristematic and elongating tissues of intact growing roots
and mature root regions generated the outward limb of the
current loop (Weisenseel et al., 1979; Behrens et al. 1982;
Miller and Gow, 1989; Rathore et al., 1990). The vibrating
probe directly measures the voltage difference between the
two extreme points of its vibration which are typically 30 μ m
apart (Jaffe and Nuccitelli, 1974). Since the electric field
is nearly constant over this small distance, the electric
field can be approximately calculated by the voltage
difference divided by this distance. The current density in
the direction of the vibration and at the center of vibration
is then given by this field multiplied by the medium's

conductivity. The bulk of the ionic current generated in the
growing roots is considered to be carried by H^+ (Weisenseel
et al., 1979; Miller and Gow, 1989).

Ion-selective microelectrodes have also been used for
studies of root growth (Newman et al., 1987; Kochian et al.,
1989). The K^+- and H^+-selective microelectrodes were placed
approximately 15 μm away from the root surface, and the
simultaneous measurement of net K^+ and H^+ fluxes associated
with individual cells at the root surface were studied
(Newman et al., 1987; Kochian et al., 1989). It was found
that high affinity active K^+ absorption into maize root cells
was not mediated by a K^+/H^+ exchange mechanism, and instead,
was either due to the operation of a K^+-H^+ cotransport system
or mediated by K^+-ATPase (Kochian et al., 1989).

Microelectrodes are also useful in measuring ionic
concentrations in intracellular organelles. The pH in the
cytoplasm and vacuole was measured with a pH-sensitive
microelectrode (Bertl et al., 1984; Felle and Bertl, 1986).
Generally, the cytoplasmic pH seems slightly alkaline
(typically ranged from 7.2 to 7.6), and the vacuolar pH
appears rather acidic (typically ranged from 4.5 to 5.9)
(Felle and Bertl, 1986). Recent advancements in
Ca^{2+}-selective microelectrodes and their application have
been reviewed by Felle (1989).

Studies of root growth and mineral absorption, using
ion-selective electrodes are described by Epstein in Chapter
5-1, and current analysis techniques for nutrient elements
are reviewed by Welch in Chapter 5-2.

V. STRATEGIES FOR DEVELOPMENT OF
PRACTICAL MEASUREMENT METHODS

A. From Viewpoints of
Plant Science

By combining techniques for both tissue level and
cellular level measurements, physiological phenomena can be
analyzed more clearly. Recently, Nonami and Boyer (1989)
measured water potential, osmotic potential and turgor in
elongating tissue with the psychrometer and the pressure
chamber and simultaneously measured turgor of cells in the
same tissue with the pressure probe. Cellular measurements
revealed that cells located near the xylem lost turgor

transiently and the majority of cells in the elongating stem maintained turgor when growth was inhibited by limited water supply in roots (Nonami and Boyer, 1989). Because turgor loss in the tissue level was undetected, they concluded that growth inhibition in growth under water deficient conditions was first caused by changes in the water potential field near the xylem, and hence, water flow into expanding cells was interrupted, resulting in cessation of volume expansion of cells without turgor loss (Nonami and Boyer, 1989). Similarly, the use of a gas exchange system together with a cell pressure probe gave a new interpretation of function of stomatal movement in intact plants (Shackel and Brinckmann, 1985; Nonami and Schulze, 1989). Thus, techniques combining whole plant measurement and cellular measurement in intact plants may lead to new findings in other research areas.

Simulation of cell functions has led to an invention of a new type of osmometer. Steudle (Chapter 2-5) invented a material-selective osmotic cell using artificial membranes. A measurement principle of the osmometer is that membranes attached to the pressure sensor of the osmometer differentially permeate water and soluble materials like living cells, and thus, pressure in the osmometer rises according to permeation of soluble materials through the membranes as turgor increases in plant cells. Steudle describes this mechanism in detail in this volume (Chapter 2-5).

B. From Viewpoints of Information Science

In the field of information science, system theory and signal processing have developed tremendously in recent years owing to advancements in computer technology. If system theory and signal processing are applied appropriately to measurement techniques employed currently, there is a possibility that the accuracy of the techniques may be improved tremendously. As an example, van Zee and Schurer (1983) applied the signal processing theory to remove noises occurring with the heat pulse method, and measured accurate transpiration rates.

More recently, artificial intelligence (AI), fuzzy control and neural network control have been developed in computer science. These control systems are believed to simulate "human thinking" in computers, and have been used for practical applications in recent years (Hashimoto, 1990).

When such new information processing systems are applied to
measurement techniques in plant science, measurement
techniques may be improved for the collection of data under
the field condition.

C. From Viewpoints of
Sensor Instrument Science

Recent developments in electronics have led to new types
of sensors. Combinations of ion selective membrane or metal
layer together with transistors have evolved into ion
sensitive field effect transistors (ISFET) and chemically
sensitive semiconductor devices (CSSD) (Hashimoto, 1990).
Some CSSD's are stable and durable, and thus, they have been
used for nutrient composition management in nutrient solution
in plant factories (Hashimoto et al., 1988). Some ISFET's
can detect enzyme reactions, and thus, they are regarded as
bio-sensors (Hashimoto, 1990). More recently, field-effect
transistors are being combined together with integrated
circuits, forming so-called "intelligent sensors", which are
capable of signal processing within the sensors. Intelligent
sensors are discussed in detail by Yamasaki in this volume
(Chapter 1-3).

D. Application to Agricultural
Production Management

In order to apply measurement techniques to agricultural
production, measurement techniques must be usable in intact
plants under greenhouse or field conditions, and also, must
be automated to be used for a long period of time without
special maintenance care for instruments. As an example,
Fiscus (Chapter 2-3) proposed a porometer method to be used
in the field in order to automate irrigation systems for
optimum watering. Measurement techniques for agricultural
applications can be designed by using already existing
methods.

Practically speaking, under field conditions,
measurement techniques using microelectrodes, pressure probes,
or/and psychrometers are impractical, although they will give
reliable physiological information. Though physiological
information is not measured directly and precisely in intact
plants, automation systems for crop production can be

designed, if measurements of physical quantities affiliated with environmental conditions, such as temperature of plants, images of plants, pH in the nutrient solution, eléctric conductivity of the nutrient solution, voltage of plants, electric current density of plants, etc. can be correlated together with physiological mechanisms of plants grown in the greenhouse or the field. Such physical quantities can be measured reliably for a long period of time without large expense under agricultural conditions. In order to correlate physical quantities affiliated with environment conditions where crops are grown with the plant physiological mechanisms in application to agricultural production, the use of computers has been suggested (Hashimoto, 1989a; Hashimoto, 1989b). Recently, Hatou et al. (1990) have used an AI-computer to manage tomato production in a nutrient culture system by monitoring electric conductivity of the nutrient solution and ion absorption by roots. However, the interrelations between physiological mechanisms and various physical quantities measurable under agricultural conditions have not been well-understood, and further research regarding these matters is required in order to automate agricultural production.

REFERENCES

Bazzaz, F.A. and Boyer, J.S. (1972) A compensating method for measuring carbon dioxide exchange, transpiration, and diffusive resistance of plants under controlled environmental conditions. *Ecology* 53: 343-349

Behrens, H.M., Weisenseel, M.H. and Sievers, A. (1982) Rapid changes in the pattern of electric current around the root tip of *Lepidium sativum* L. following gravistimulation. *Plant Physiol.* 70: 1079-1083

Bertl, A., Felle, H. and Bentrup, F.-W. (1984) Amine transport in *Riccia fluitans*: Cytoplasmic and vacuolar pH recorded by a pH-sensitive microelectrode. *Plant Physiol.* 76: 75-78

Bloodworth, M.E., Page, J.B. and Cowley, W.R. (1955) A thermo-electric method for determining the rate of water movement in plants. *Proc. Soil Sci. Soc. Amer.* 19: 411-414

Boyer, J.S. (1966) Isopiestic technique: Measurement of accurate leaf water potentials. *Science* 154: 1459-1460

Boyer, J.S. (1968) Relationship of water potential to growth of leaves. *Plant Physiol.* 43: 351-364

Boyer, J.S., Cavalieri, A.J. and Schulze, E.-D. (1985) Control of the rate of cell enlargement: Excision, wall relaxation, and growth-induced water potentials. *Planta* 163: 527-543

Boyer, J.S. and Knipling, E.B. (1965) Isopiestic technique for measuring leaf water potentials with a thermocouple psychrometer. *Proc. Natl. Acad. Sci.* 54: 1044-1051

Boyer J.S. and Potter, J.R. (1973) Chloroplast response to low leaf water potentials. I. Role of turgor. *Plant Physiol.* 51: 989-992

Cosgrove, D.J. and Green, P.B. (1981) Rapid suppression of growth by blue light. Biophysical mechanism of action. *Plant Physiol.* 68: 1447-1453

Ehlig, C.F. (1962) Measurement of energy status of water in plants with a thermocouple psychrometer. *Plant Physiol.* 37: 288-290

Felle, H. (1989) Ca^{2+}-selective microelectrodes and their application to plant cells and tissues. *Plant Physiol.* 91: 1239-1242

Felle, H. and Bertl, A. (1986) Light-induced cytoplasmic pH changes and their interrelation to the activity of the electrogenic proton pump in *Riccia fluitans*. *Bioch. Bioph. Acta* 848: 176-182

Hashimoto, Y. (1989a) Dynamical approach to plant factory. *Proc. AGROTIQUE*: 111-118, Bordeaux

Hashimoto, Y. (1989b) Recent strategies of optimal growth regulation by the speaking plant concept. *Acta Horticulturae* 260: 115-121

Hashimoto, Y. (1990) Instrumentation and Information in Bio-systems. (in Japanese) pp. 270 Yokendo Publishing Co. Ltd., Tokyo.

Hashimoto, Y., Ino, T., Kramer, P.J., Naylor, A.W. and Strain, B.R. (1984) Dynamic analysis of water stress of sunflower leaves by means of a thermal image processing system. *Plant Physiol.* 76: 266-269

Hashimoto, Y., Morimoto, T., Fukuyama, T., Watake, H., Yamaguchi, S. and Kikuchi, H. (1988) Identification and control of hydroponic system using ion sensors. *Acta Horticulturae* 245: 490-497

Hatou, K., Nishina, H. and Hashimoto, Y. (1990) Computer integrated agricultural production. Proc. of 11th IFAC World Congress. Pergamon Press, Oxford. in press

Huber, B. (1932) Beobachtung und messung pflanzlicher Saftströme. *Berichte der Deutschen botanischen Gesellschaft* 50: 89-109

Huber, B. and Schmidt, E. (1937) Eine Kompensationmethode zur thermo-elektrischen Messung langsamer Saftströme. *Berichte der Deutschen botanischen Gesellschaft* 55: 514-529

Jaffe, L.F. and Nuccitelli, R. (1974) An ultrasensitive vibrating probe for measuring steady extracellular currents. *J. Cell Biol.* 63: 614-628

Kitano, M. and Eguchi, H. (1989) Quantitative analysis of transpiration stream dynamics in an intact cucumber stem by a heat flux control method. *Plant Physiol.* 89: 643-647

Kochian, L.V., Shaff, J.E. and Lucas, W.J. (1989) High affinity K^+ uptake in maize roots: A lack of coupling with H^+ efflux. *Plant Physiol.* 91: 1202-1211

Kutschera, U. and Schopfer, P. (1986a) Effect of auxin and abscisic acid on cell wall extensibility in maize coleoptiles. *Planta* 167: 527-535

Kutschera, U. and Schopfer, P. (1986b) In-vivo measurement of cell-wall extensibility in maize coleoptiles: Effects of auxin and abscisic acid. *Planta* 169: 437-442

MacFall, J.S., Johnson, G.A. and Kramer, P.J. (1990) Observation of a water-depletion region surrounding loblolly pine roots by magnetic resonance imaging. *Proc. Natl. Acad. Sci. USA* 87: 1203-1207

Miller, A.L. and Gow, N.A. (1989) Correlation between root-generated ionic currents, pH, fusicoccin, indoleacetic acid, and growth of the primary root of *Zea mays*. *Plant Physiol.* 89: 1198-1206

Moss, D.N. and Rawlins, S.L. (1963) Concentration of carbon dioxide inside leaves. *Nature* 197: 1320-1321

Newman, I.A., Kochian, L.V., Grusak, A. and Lucas, W.J. (1987) Fluxes of H^+ and K^+ in corn roots: Characterization and stoichiometries using ion-selective microelectrodes. *Plant Physiol.* 84: 1177-1184

Nonami, H. and Boyer, J.S. (1989) Turgor and growth at low water potentials. *Plant Physiol.* 89: 798-804

Nonami, H. and Boyer, J.S. (1990) Wall extensibility and cell wall hydraulic conductivity decrease in enlarging stem tissues at low water potentials. *Plant Physiol.* in press.

Nonami, H., Boyer, J.S. and Steudle, E. (1987) Pressure probe and isopiestic psychrometer measure similar turgor. *Plant Physiol.* 83: 592-595

Nonami, H. and Schulze, E.-D. (1989) Cell water potential, osmotic potential, and turgor in the epidermis and mesophyll of transpiring leaves: Combined measurements with the cell pressure probe and nanoliter osmometer. *Planta* 177: 35-46

Rathore, K.S., Hotary, K.B. and Robinson, K.R. (1990) A two-dimensional vibrating probe study of currents around lateral roots of *Raphanus sativus* developing in culture. *Plant Physiol.* 92: 543-546

Richards, L.A. and Ogata, G. (1958) Thermocouple for vapor pressure measurement in biological and soil systems at high humidity. *Science* 128: 1089-1090

Sakuratani, T. (1981) A heat balance method for measuring water flux in the stem of intact plants. *J. Agric. Met.* 37: 9-17

Shackel, K. and Brinckmann, E. (1985) *In situ* measurement of epidermal cell turgor, leaf water potential, and gas exchange in *Tradescantia virginiana* L. *Plant Physiol.* 78: 66-70

Spanner, D.C. (1951) The Peltier effect and its use in the measurement of suction pressure. *J. Exp. Bot.* 2: 145-168

van Zee, G.A. and Schurer, K. (1983) On-line estimation of the rate of sap flow in plant stems using stationary thermal response data. *J. Exp. Bot.* 149: 1636-1651

Weisenseel, M.H., Dorn, A. and Jaffe, L.F. 1979 Natural H^+ currents traverse growing roots and root hairs of barley (*Hordeum vulgare* L.). *Plant Physiol.* 64:512-518

FROM SENSOR DEVICES TO INTELLIGENT SENSING SYSTEMS

Hiro Yamasaki

Faculty of Engineering, University of Tokyo
7-3-1, Hongo, Bunkyo-ku, Tokyo, Japan

I. INTRODUCTION

The human sensory system can be considered a highly advanced example of an intelligent sensing system. Its sensitivity and selectivity are flexible in recognizing objects.

We can recognize our names or certain special words which relate to our interests, even if we hear them even in very noisy environments. These characteristics are referred to as the "Cocktail party effect." Similarly, we can notice the abnormal sound from failed components of machines or systems in the ambient operating noise. The mechanism by which we do this has not yet been explained.

Our sophisticated mental recognition system contributes to the function of the sensing system, while the physical level of our auditory system can concentrate its attention on the sound source of interest and improve the signal to noise ratio (S/N). Its sensitivity and selectivity of the system are tuned to the sound-object.

The human sensory system has a hierarchical structure,

Measurement Techniques in Plant Science

and so, a hierarchical system was created to realize advanced sensing functions. Sensor signal integration and sensor signal fusion are adopted as the basic concepts to design the adaptive intelligent sensing system.

Sensors incorporated with dedicated signal processing functions are called "intelligent" or "smart" sensors. The roles of the intelligent sensors are to enhance design flexibility and realize new sensing functions, and additional roles are to reduce loads on central processing units and signal transmission lines by distributed information processing in the lower layer of the system.

II. STRUCTURE OF INTELLIGENT SENSING SYSTEMS

Intelligent sensing systems have a multi-layer hierarchical structure, as shown in Figure 1. On the top layer the most highly intelligent information processing is done. The function of processing is centralized as in the human brain. Processed information is abstract, and it is independent of operating principles and the physical structure of sensors.

```
Upper layer
     TOTAL CONTROL
     Concentrated central processing
     (Digital serial processing)

Middle layer
     INTERMEDIATE CONTROL, TUNING &
     OPTIMIZATION OF LOWER LEVEL
     Distributed processing & control
     (Digital & Analog)

Lower layer
     SENSING & ACTUATING
     [INTELLIGENT SENSORS]
     Distributed parallel processing
     (Analog)
```

Figure 1.   Hierarchical structure of intelligent sensing system

In contrast, information processing in the bottom layer is strongly dependent on the sensor's underlying principles and structures. A group of intelligent sensors on the bottom layer collects information from external objects as do our distributed sensory organs. Signal processing of these intelligent sensors is done in a distributed and parallel manner.

There are intermediate signal processing functions in the middle layer. One function of intermediate signal processing is the integration of signals from multiple intelligent sensors. When the signals come from different types of intelligent sensors, the function is referred to as sensor signal fusion. Another function is tuning of parameters of the sensors to optimize the total system performance.

In general, the information processing done in each layer is more directly oriented to hardware in the lower layer and less hardware oriented in the upper layer. For the same reason, algorithms for information processing are more flexible and need more knowledge in the upper layer and are less flexible and require less knowledge in the lower layer.

If this hierarchical structure of the system is compared with that of the human sensory system, it will be more clearly understood by the mutual correspondence. Our auditory sensing system has multi-layers which have different functions as follows:

Let us imagine that we are in noisy environment and we notice certain interesting words or sounds but we can not understand the meaning easily and correctly from heard words or sounds. We usually react in a sequential pattern (3', 2', and 1').

1) Sensing sound as physical waves (processing in cochlea is included).
2) Selective reception of certain words, sounds, or languages.
3) Recognizing words and phrases semantically.
4) Understanding the situation and decision of necessary reaction.

3') Inference concerning uncertain words or phrases by context.
2') Concentration of auditory power on the sound source.
1') Receiving the target sound with better S/N.

The functions of this adaptive sensing system relate the above functions 1),2),1') and 2'). Refer to Figure 2.

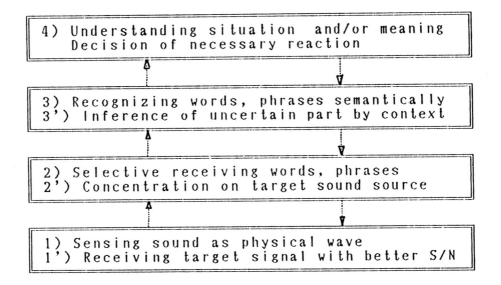

Figure 2. Hierarchy of sound information processing of human in a noisy environment

III. ROLE OF INTELLIGENCE IN SENSORS

The sensor intelligence performs distributed signal processing at the lower layer in the hierarchy of the sensing system.
The role of signal processing function in intelligent sensors can be summarized as follows;

1) Signal enhancement for extraction of useful features of the objects.
2) Reinforcement of inherent characteristics of sensor device.

The important role of the sensor intelligence is to improve the signal selectivity of individual sensors. This includes simple operations of output from multiple sensor devices for feature extraction. However, this does not include optimization of device parameters or signal integration from multiple sensor devices, because this requires knowledge of the sensor devices and the object.

IV. ROLE OF INTELLIGENCE IN MIDDLE LAYER

One of the roles of the intelligence in the middle layer is to extract essential features of the object. The other role is to organize multiple output from the lower layer and to generate intermediate output. In the processing of the middle layer, the output signals from multiple sensors are combined or integrated. The extracted features are utilized for recognition of the situation by upper layer intelligence.

Signals from sensors of different measurands are combined and the results give us new, useful information. Ambiguity or imperfection in the signal of a measurand can be compensated by another measurand. This is sensor signal fusion. The processing creates a new phase of information.

Another function of the middle layer is parameter tuning of the sensors to optimize the total system performance. Optimization is done based upon the extracted feature and knowledge about target signal. The knowledge is given by the upper layer as a form of optimization algorithm.

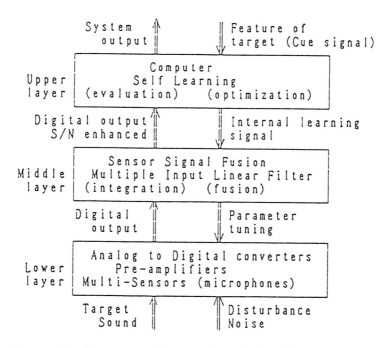

Figure 3. System configuration of intelligent adaptive sound sensing system

An intelligent adaptive sound sensing system was
developed by the authors (Takahashi and Yamasaki, 1989;
Yamasaki and Takahashi, 1990). The system receives the
necessary sound from a signal source out of various noise
environments with improved S/N ratio. The location of the
target sound source is not given but some feature of the
signal is given to discriminate the signal from noise. The
feature is given as a cue signal.

The system consists of one set of multiple microphone
and multiple input linear filters and a self-learning system
for adaptive filtering. The adaptive sensing system has a
three-layered structure as shown in Fig.3. The lower layer
includes multiple microphone and A/D (analog to digital)
converters. The middle layer has a multiple input linear
filter which combines multiple sensor signals and generates
an integrated output. The performance of the filter is tuned
and controlled by a computer in the upper layer. The middle
layer filter plays the role of advanced signal-processing in
the middle layer described here. The upper layer has a
computer which has self-learning functions and optimizes the
filter performance on the basis of knowledge given by an
external source which is in the higher layer (in our case the
source is man).

The output from the optimized filter is used as the final
output of the system. Maximum S/N improvement of 18dB is
obtained experimentally. Possible applications of the system
are detection of abnormal sound from machines malfunctions or
of human voice in a noisy environment.

V. APPROACHES TO REALIZATION OF SENSOR INTELLIGENCE

There are three different approaches to realize sensor
intelligence (Yamasaki, 1984).

1) Use of specific functional materials (Intelligent
   materials).
2) Use of functional mechanical structure (Intelligent
   structure).
3) Integration with computers.

In the 1st and 2nd approach, signal processing is analog
signal discrimination. Only useful signal is selected and
noise or undesirable effects are suppressed. Thus, signal
selectivity is realized by the specific materials or specific

mechanical structure.

In the 1st approach, a unique combination between object material and sensor material contributes to realize the almost ideal signal selectivity. The typical example of the sensor materials is enzyme fixed on the tip of bio-sensors.

In the 2nd approach, propagation of optical and acoustic wave can be controlled by the specific shape of the wave media. Also, reflection of these waves are controlled by the surface shape of the reflectors. A lens is an example; only light which is emitted from a certain point in the object space can be concentrated at a certain point in the image space, and the effect of stray light can be rejected on the image plane.

The hardware for this analog processing is relatively simple and reliable and its processing time is very short due to the intrinsically parallel processing. But the algorithm of analog processing is usually not programmable and it is difficult to modify once it is designed.

The 3rd approach is most popular and it is usually represented as the integration of sensor devices and microprocessors. The algorithm is programmable and it can be changed even after it is designed.

Typical examples of the technical approaches are described in the following sections. These range from single chip sensing devices integrated with microprocessors to big sensor arrays utilizing synthetic aperture techniques, and from two dimensional functional materials to human sensory systems.

VI. APPROACH BY NEW FUNCTIONAL MATERIALS

For obtaining information concerning two or three dimensional objects, sheets of conductive rubber and polyvinylidene fluoride, piezo and pyroelectric plastics are useful as functional material for tactile sensors, and a special simple algorithm has been developed for recognition of an object's shape and calculation of the center of force (Ishikawa and Shimojo, 1982).

Enzymes and microbes have strong selectivity for specific substances. They can even recognize specific molecules. Time for signal processing to reject the effects of coexisting components can be minimized.

Shape memory metals can make sensors unified with actuators. In addition to the unified structure, their memory function may be useful for unique applications.

VII. APPROACH BY FUNCTIONAL MECHANICAL STRUCTURE

If the signal processing function is implemented in the mechanical structure or the form of sensors, processing of the signal is simplified and a rapid response can be expected.

Let us discuss a spatial filter as an example. An object having a two dimensional optical pattern $f(x,y)$ is projected through a spatial filter $g(\xi, \eta)$, and its output is focused and converted into electrical signal $e(t)$ as given by eq.(1):

$$e(t) = \int\int_{D} g(\xi, \eta)[f(x,y,t)*h(\xi, \eta)]d\xi\,d\eta \qquad (1)$$

where $h(\xi, \eta)$ is a point spread function, and * indicates convolution. As seen in eq.(1), two dimensional convolution is carried out in optical configuration. The dominant frequency in $e(t)$ is proportional to the moving velocity of the pattern. Thus typical application of the spatial filter technique is noncontact velocity measurement with simple hardware .

We can locate three dimensional directions of sound sources with two ears. We can also identify the direction of sources even in the median plane. The identification seems to be based on the direction dependency of pinna tragus responses (Bloom, 1977; Mead, 1989). Obtained impulse responses are shown in Fig.4, in which signals are picked up by small electret microphones inserted in the external ear canal, and the spark of an electric discharge is used as the sound source. Differences can be easily observed (Hiranaka and Yamasaki, 1983).

Usually at least three sensors are necessary for identification of three dimensional localization. So pinnae are supposed to act as a kind of signal processing hardware with inherent special shapes. We are studying this mechanism utilizing synthesized sounds which are made by convolutions of impulse responses and natural sound and noise (Hiranaka et al., 1983).

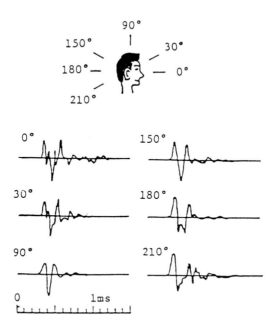

Figure 4. Pinna impulse responses in the median plane.

Not only human ear systems, but other sensory systems of man and animals are good examples of intelligent sensing systems with functional structure. The most important feature of such intelligent sensing systems is integration of multiple functions; sensing and signal processing, sensing and actuating, signal processing and signal transmission. Our fingers are a typical example of the integration of sensors and actuators. Signal processing for noise rejection such as lateral inhibition are carried out in the course of signal transmission in the neural network.

VIII. APPROACH BY INTEGRATION WITH COMPUTER — Part 1

The most popular concept of an intelligent sensor is an integrated monolithic device combining a sensor with a micro-computer within one chip. However, such a device has not been realized as yet. Developmental stages of such

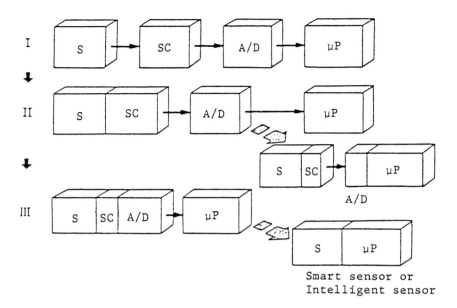

Figure 5. Development trends in integration with microprocessors. S: Sensor; SC: Signal conditioner; A/D: Analog to digital converter; μ P: Microprocessor.

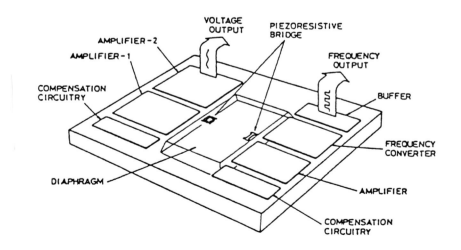

Figure 6. Integrated pressure sensor having voltage and frequency output.

intelligent sensors are illustrated in Fig.5 (Mackintosh
International, 1981). Four separate functional blocks,i.e.,
sensor, signal conditioner, A/D converter, and microprocessor,
are gradually coupled on a single chip then turned into a
direct coupling of sensor and microprocessor.

We are in the second stage in the figure at present.
Let us discuss some examples in the stage. Several results
are reported about research on single chip pressure sensor
devices on which analog circuit for simple signal processing
are integrated (Sugiyama et al., 1983; Yamada et al., 1983).
The circuits consist of amplifier, temperature compensation
circuit, oscillator etc. (Figure 6). No results on single
silicon chip sensor devices integrated with microprocessors
are reported so far. An insulation problem between circuits
and sensors on a common silicon substrate is reported. The
problem may limit the maximum range of signals (Ko et al.,
1983).

Usually sensor devices should tolerate exposure to
severe environmental conditions, but microprocessor devices
are relatively sensitive to ambient condition and
electromagnetic noise induction. Reliability of processors
in severe environments may be another problem. However, in
the case of noncontact measurement or image sensors, early
development is expected.

Figure 7 shows an example for process instrumentation.

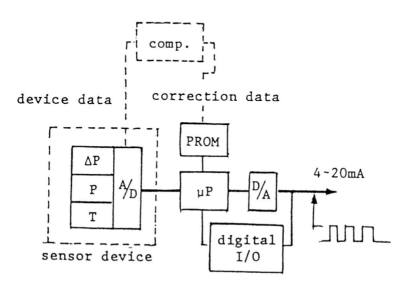

Figure 7.   Digital smart differential pressure
transmitter.

A single chip device including a silicon diaphragm differential pressure sensor, an absolute pressure sensor and a temperature sensor is used. The output signals from these three sensors are applied to a microprocessor via an A/D converter on a separate chip. The processor calculates the output and at the same time compensates for effects of absolute pressure and temperature numerically. The data for compensation of each sensor device is measured in the manufacturing process and loaded in ROM (Read-Only-Memory) of the processor respectively. Thus, the compensation can be precise, and an accuracy of 0.1% is obtained.

The transmitter has a pulse communication ability through a two-wire-analog-signal line and digital communication interface. Remote adjustment of span and zero, remote diagnosis and other maintenance functions can be performed by the use of digital communication means. The range of analog output signal is the International Electrotechnical Commission standard of 4-20mA DC, so total circuits including microprocessor should work within 4mA.

Table 1. Materials and sensed gases of thick film gas sensors.

Sensor	Material	Sensible gas
S_1	ZnO	Organic gases
S_2	ZnO (Pt doped)	Organic gases (Low alcohol sensitivity)
S_3	WO_3	H_2, CO, C_2H_5OH
S_4	WO_3 (Pt doped)	H_2, CO, C_2H_5OH (Low alcohol sensitivity)
S_5	SnO_2	Reducing gases
S_6	SnO_2 (Pd doped)	Reducing gases (High methane sensitivity)

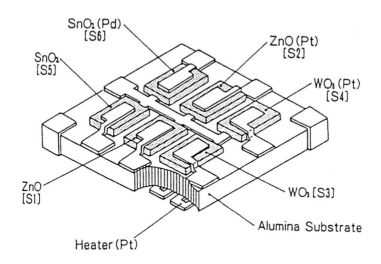

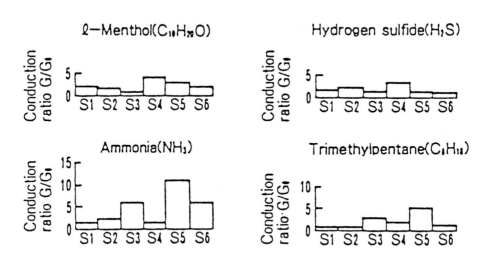

Figure 8. Schematic drawing of integrated sensor, and specific patterns generated by odour-producing materials (i.e., hydrogen sulfide, trimethylpentane, *l*-menthol, and ammonia) when gas concentration of the materials was adjusted to 10 ppm.)

Restriction on the operating current of the microprocessor is overcome by CMOS (Complementary Metal Oxide Semiconductor) circuit approach (Yamatake-Honeywell Inc., 1983). Sensors having frequency output are advantageous in interfacing with microprocessors. Frequency output density sensor and pressure sensor which are compensated and linearized by dedicated processors have been developed (Ohno, 1984).

A new approach to chemical intelligent sensors is proposed. Six thick film gas sensors which have different sensitivity for various gases are mounted on a common substrate, and the sensitivity pattern is recognized by a microcomputer. Sensor materials and sensed gases are shown in Table 1. Figure 8 shows several examples of sensor conductivity patterns for organic and inorganic gases. Typical patterns are memorized and identified by dedicated microprocessor utilizing similarity analysis of patterns. Maximum sensitivity of 1ppm is reported (Ikegami and Kaneyasu, 1983).

IX. APPROACH BY INTEGRATION WITH COMPUTER — Part 2

A more advanced function of coupled systems of sensor and computer is found in synthetic aperture sensing systems. The ultimate resolution of an optical system is determined by the ratio of its aperture size to the wave-length. Because the wave-length of electromagnetic waves and that of ultra sonic waves is much longer than that of visible light, a big aperture is necessary. For example, in a radio telescope the diameter of the parabola antenna must be several tens of kilometers to have a similar resolution to that of optical telescope. Such a big antenna can neither be built nor be driven, even if the design itself is possible. The problem can be solved by the use of synthetic aperture techniques. An array of small aperture sensors coupled with computers can be the substitute for array size aperture, and realize higher resolution.

Figure 9 shows the basic principle of a synthetic aperture system corresponding to a dioptric system using a single lens. Light from the object follows different paths and with different time delays due to lens and then focuses into the image. Each path seems to have a different length, but all of them have the same propagation delay.

In the synthetic aperture system, the outputs of each sensor with different time lags are added, and they are

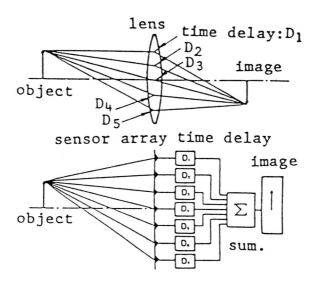

Figure 9. Dioptric system using lens and synthetic
aperture techniques.

summed (Figure 9). The sum makes the constructed image.
Configuration of the two system seems different, but their
functions are the same.
    The direction of the light axis and the focus of the
synthetic aperture system can be driven by adjusting time
delays in a computer, so very rapid scanning is possible.
Even simultaneous focusing into more than one point is
realizable if parallel signal processing is available. Such
a focusing function is never realized in physical systems.
This approach improved the resolution of radar and radio
telescopes substantially.
    Next, we worked on the development of a high resolution
sonar and a measuring system of velocity vector distribution
in the sea water by the use of the synthetic aperture
technique. This is an example of an intelligent sensing
system which has the middle layer for intermediate signal
processing. The processing integrates signals from multiple
sensor devices and reconstruct a total image of object state
(Tamura and Yamasaki, 1982). This technique seems to be a
very powerful approach for higher spatial resolution with
small size point sensors which have poor resolution, thus it
is an effective counter measure to overcome the poor spatial
resolution of present sensor techniques.

X. FUTURE CONCEPTS OF INTELLIGENT SENSING SYSTEMS

Figure 10 shows a concept of an intelligent area image sensing system integrated on single chip device. Research of this device is planned in the R & D Project of Basic Technology for Future Industries which is promoted by the Ministry of International Trade and Industry of the Japanese Government. A concept for future design of intelligent sensors can be seen in the figure (Kataoka, 1982).

The device consists of a multi-layer structure, each layer performing a different function based on the physical properties of the layer materials. A number of light sensing devices are arrayed two-dimensionally on the top layer, signal transmission devices are built in the second layer, memories are in the third, computing devices are in the fourth, and power supplies are in the bottom layer.

A wavelength range of the light sensing devices is determined by suitable selection of the functional materials. A two-dimensional array is an example of the functional mechanical structure. These functional materials and functional mechanical structure are combined on the top layer. They are connected to a computer in the same substrate.

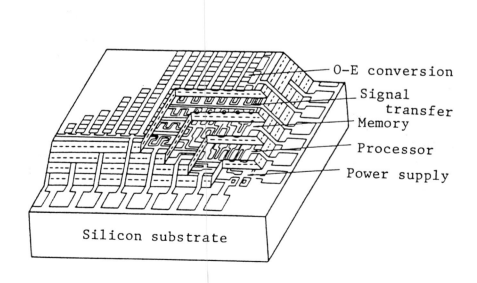

Figure 10. Design of integrated image sensing system with multi-layer structure.

Image processing such as feature extraction and edge enhancement can be performed in the three-dimensional multi-functional structure. This is just like the retina of our eyes. As previously described in this paper, the important feature of sensing systems of man and animals is such integration of multi-functions and distributed signal processing.

It can be said that future concepts or targets of our intelligent sensing systems are the sensory systems of man and advanced animals, in which the three different approaches are combined together.

It is important to note that hierarchical structure is essential for advanced information processing and different roles should be reasonably allocated to each layer.

REFERENCES

Bloom, P.J. (1979) Determination of monaural sensitivity changes due to the pinna by use of minimum-audible-field measurements in the lateral vertical plane. J. Acoustical Society of America 61:820

Hiranaka, Y. et al. (1983) Envelope representation of pinna responses and localization trials by computer simulated sound, Note H-83-31, WG on auditory sense, Acoustical Society of Japan

Hiranaka, Y. and Yamasaki, H. (1983) Envelope representations of pinna impulse responses relating to 3-dimensional localization of sound sources. J. Acoustical Society of America, 73(1): 291

Ikegami, A. and M. Kaneyasu, M. (1983) Olfactory Detection using Integrated Sensor. Technical Digest of Transducers'85: 136

Ishikawa, M. and Shimojo, M. (1982) A Tactile sensor using pressure sensitive conductive rubber. Preprint 2nd Sensor Symposium: 95

Kataoka, H (1982) Sensors and human society. J. IEEJ 102(5): 345

Ko, W.H. et al. (1983) Capacitive pressure transducers with integrated circuits. Sensors & Actuators Vol.4: 403

Mackintosh International, Sensors, Vol.24 (1981)

Mead, C. (1989) Analog VLSI and Neural Systems. pp207-227. Addison-Wesley Pub. Company.

Ohno, I (1984)   Frequency dependent sensors & their
      advantages.   J. SICE 23(3): 327-330
Sugiyama, S. et al.   (1983)   Integrated piezo resistive
      pressure sensor with both voltage and frequency output.
      Abstracts,   Solid State Transducer 83: 115
Takahashi, K. and Yamasaki, H.   (1989)   Self-Adapting
      Multiple Microphone   System,   Abstracts
      TRANSDUCERS'89 C6.5: 198-199
Tamura, Y. and Yamasaki, H.   (1982) Non-contact measurement of
      velocity vector distribution by numerical
      reconstruction of reflected sound field.   Trans. SICE
      18(1): 44
Yamada, K. et al (1983)   A piezo resistive integrated
      pressure sensor.   Abstracts, Solid State Transducer 83:
      113
Yamasaki, H. (1984)   Approaches to Intelligent Sensors,   Proc.
      4th Sensor Symposium: 69-76
Yamasaki, H. and Takahashi, K.   (1990)   An Intelligent
      Adaptive Sensing System.   F. Harashima ed. Proceeding
      of Toyota Conf. '89,   Elsevier Sequoia
Yamatake-Honeywell Inc. (1983)   Digital smart transmitter
      DSTJ 3000,   J.   SICE   22(12): 1054-1055

Chapter 2
Water Relations

A BRIEF HISTORY OF WATER
STRESS MEASUREMENT

Paul J. Kramer

Department of Botany
Duke University
Durham, North Carolina

I.   INTRODUCTION

The history of instrumentation in the field of
plant water relations is related to the development
of new concepts, because most instrumentation arises
from the need to test concepts and theories.
However, there often is a long gap between
development of a concept and development of the
instrumentation required to study it. The injurious
effects of drought must have been recognized by
prehistoric farmers because irrigation systems
existed in Egypt, the Middle East, and China at the
beginning of recorded history, and the first
European explorers found irrigation systems in the
Americas (Kramer, 1983, pp. 107-108). Although
quantitative studies of water absorption and
transpiration were made in the 18th century by
Stephen Hales (1727) and much important work was
done on cell and plant water relations during the
19th century, an organized concept of plant water
balance and water stress did not develop until well
into the 20th century.

Apparently Montfort (1922) first proposed the
analogy between the water economy of plants and
financial bookkeeping in which plant water balance
represents the difference between water absorption
(income) and water loss by transpiration
(expenditures). This concept of plant water balance
developed from earlier observations that on sunny
days when transpiration exceeds absorption there are
significant decreases in the water content of leaves
and other plant organs (Livingston and Brown, 1912;
Krasnoselsky-Maximov, 1917, and others). The
concept of plant water balance as the critical
feature of plant water relations was publicized in
Maximov's book, "The Plant in Relation to Water,"

Measurement Techniques in Plant Science

(1929) and provided a concept for much future
research on plant water relations. However, two
questions had to be answered: (1) what is a good
indicator of the plant water balance or degree of
water stress, and (2) how can it be measured?
Answering these questions required another 40 years
of research, and is still a subject of debate as we
will see in later sections.

II. METHODS OF MEASURING
WATER STRESS

The measurement of water stress has progressed
from simple visual measurements through a variety of
increasingly complex instrumental measurements until
it has now entered the electronic age.

A. Visual Observations

Obviously the oldest indicator of water stress
was the rolling of leaves, and occurrence of
wilting, and such visual indicators are still
useful. Wenkert (1980) found that water stress in
maize is indicated by loss of sheen and development
of a pale color before leaf rolling occurs, and
change in color of bluegrass lawns is said to be a
good indicator of the need for irrigation. O'Toole
and Cruz (1980) also reported that a good
correlation exists between leaf rolling, stomatal
resistance, and leaf water potential in rice.
Oosterhuis et al. (1985) found that increase in
angle of the terminal leaflet of soybean is
correlated with decreasing leaf water potential and
increase in stomatal resistance and might be used
for scheduling irrigation. Thus visual symptoms
sometimes are well correlated with quantitative
measurements of water stress.
Remote sensing with infrared photographic film
was used by Blum et al. (1978) to detect differences
between stressed and unstressed plots of sorghum
because stressed canopies were lighter in color than
unstressed canopies. Idso et al. (1980), Jackson

(1982); and others used remote sensing of the degree of difference between leaf and air temperature as an indicator of water stress and remote sensing may be particularly useful in detecting the development of water stress in crops covering large areas of land. Unfortunately, even the most sophisticated of these visual methods give only qualitative indications of water stress which are useful indicators of the need for irrigation, but not for physiological studies.

B. Water Content

The first quantitative measurements of water stress were based on changes in water content, calculated either on a fresh or a dry weight basis. These were found to be unsatisfactory because fresh weight often varies widely during the day and dry weight also varies measurably because of photosynthesis, translocation, and respiration. One way to minimize those effects is by extracting the soluble constituents and using the residual dry weight, but the residual dry weight of leaves increases during their life because of increase in thickness of cell walls (Weatherley, 1950). Those difficulties resulted in development of other methods of expressing the water content. One method is to express the water content at sampling as a percentage of the water content at full turgor, the water saturation deficit of Stocker (1929) or the relative water content or relative turgor of Weatherley (1950) and other investigators. The problems related to the use of these methods are discussed in Kramer (1983, pp. 376-379) and in Slavik (1974). Relative water content is a good method of monitoring changes in water stress and Sinclair and Ludlow (1985) claim that it is better correlated with physiological activity than water potential. However, measurements of relative water content are incommensurable with measurements of soil or atmospheric water and do not indicate the direction of water movement.

Attempts have been made to monitor changes in leaf water content by measuring electrical conductivity or capacitance. Another method is based on absorption of radiation by the water in leaves (Nakayama and Ehrler, 1964). A source of

beta radiation is placed on one side of the leaf, a
detector on the other side, and changes in radiation
passing through the leaf caused by changes in water
content, are recorded. This method is a good
monitor but difficult to calibrate. All of these
methods were described by Slavik (1974).

Nuclear magnetic resonance (NMR) spectroscopy and
imaging are becoming useful for study of metabolism
and the water status of living plant tissue. The
use of NMR spectroscopy for studies of plant
metabolism in vivo is discussed in the last chapter
of this volume. It also has been used to determine
the amount of bound water in plant tissue (Burke, et
al., 1976), the permeability of cell membranes
(Stout et al., 1978), and water movement out of
cells during freezing of plant tissue (Stout, et.,
1977). NMR imaging is useful for study of root-
soil-water relations and the distribution of roots
in the soil (Bottomley, et al., 1986; Omasa, et al.,
1985a; Rogers and Bottomley, 1987). It also permits
observation of changes in water content and water
binding in living plant tissue (Brown, et al., 1986;
Johnson, et al., 1987). the use of NMR technology
for metabolic studies and imaging is discussed later
in this volume. Improvements in technology probably
will permit study of movement of labeled water at
the cellular level in the near future. However, the
equipment is expensive and requires well trained
technicians to obtain good images.

C.   Osmotic Pressure or
Osmotic Potential

During the first quarter of the 20th century many
measurements were made of the osmotic pressure, now
termed the osmotic potential, of plant cells as an
indicator of water stress and large collections of
data were published (see Miller, 1938, pp. 39-45 for
references). It was found that the osmotic
potential generally is lower in plants of dry
habitats than in those of moist habitats, that
osmotic potential decreases as water stress
develops, and that it usually is lower in woody
plants than in herbaceous plants. The earliest
measurements were mostly made by plasmolysis,

(Fitting, 1911, for example), but the plasmolytic
method was soon superseded by cryoscopic
measurements on expressed sap. Examples are the
data published by Atkins (1916), Harris (1934),
Korstian (1924), and Walter (1960). The freezing
point depression method used in early days has now
been largely superseded by vapor pressure osmometers
which operate rapidly on very small samples. By
1925 or 1930 so many questions were raised
concerning the reliability of osmotic measurements
that they fell into disrepute. Also it began to be
realized that water movement is not controlled by
osmotic potential, but by the free energy of the
water, which we now term the water potential. Since
there were no good methods for measuring plant water
potential, few measurements of plant water stress
were made during the second quarter of the 20th
century.

D.   What Should Be Measured

As the writer pointed out long ago (Kramer,
1963), failure to measure plant water stress
resulted in many inconclusive and contradictory
results in research on the relationship between soil
moisture and plant growth. Many measurements of
soil moisture were made and soil water stress was
evaluated in relation to soil water content at the
permanent wilting percentage and field capacity of
the soil. However, it gradually became clear that
the degree of water stress in a plant cannot be
predicted reliably from measurements of soil water
stress. This is because plants in moist soil often
develop temporary water stress and wilt on hot,
sunny days when transpiration is rapid, but may
exhibit only slight stress on cool, cloudy days,
even when the soil is relatively dry. Thus it was
clear that the only reliable indicator of plant
water stress was a measurement made on the plants
themselves, but there was continuing uncertainty
concerning what to measure and how to measure it.
A satisfactory method of measuring plant water
status and stress should have most of the following
characteristics:

1.   There should be a good correlation between
rates of physiological processes and the degree of
water stress measured by the method.
2.   The degree of water stress measured by the
method should have similar physiological effects in
a wide range of tissues.
3.   The units employed should be applicable to
plants, soil, solutions, and the atmosphere.
4.   The method should be simple, rapid, and
relatively inexpensive.
5.   It should require a relatively small sample
of plant tissue.
Water content and water saturation deficits or
relative water contents of plant tissue cannot be
related quantitatively to soil or atmospheric
moisture and a given osmotic potential has less
physiological significance than a given level of
water potential. Water potential represents the
free energy status of water and can be related
numerically to soil and atmospheric water status.
Furthermore, water movement into and through plants
occurs along gradients of decreasing water
potential. Thus, in spite of the questions raised
by Sinclair and Ludlow (1985), the writer regards
measurements of water potential as the most useful
single measure of plant water status. However,
knowledge of the osmotic and turgor potentials and
cell wall elasticity often are needed in addition to
provide a complete description of cell and tissue
water relations (see chapters by Boyer and Nonami
and by Steudle in this volume and Tyree and Jarvis,
1982).

III.   WATER POTENTIAL MEASUREMENTS

Early in the 20th century a few European
physiologists began to point out that water movement
is controlled by gradients in what was variously
termed suction force, suction tension, or water
absorbing power. This was expressed by a simple
equation, P-T=S, where P represented osmotic
pressure, T turgor pressure or wall pressure, and S
the suction force (today termed water potential) and
was represented schematically by Höfler's (1920)

much-copied diagram (Kramer, 1983, p. 27, for
example). This concept was developed by Renner
(1912), Ursprung and Blum (1916), Thoday (1918) and
others, each using his own terminology, and was
accepted slowly as a better indicator of water
stress than osmotic pressure. The term "diffusion
pressure deficit" publicized in Meyer and Anderson's
textbook (1939) temporarily displaced the confusing
terminology of the 1920's but it in turn was
replaced 20 years later by the more logical
thermodynamic terminology of water potential
(Slatyer and Taylor, 1960, Slatyer, 1967). The
concept of free energy or potential existed early in
the 20th century and it was applied to cell water
relations in an excellent but neglected paper by
Tang and Wang in 1941. It also was used by Owen in
1952, but the intellectual climate was unfavorable
for acceptance of a thermodynamic terminology until
it was again proposed by Slatyer and Taylor in 1960.
Thus a period of over 40 years passed between the
first formulation of the concept of water potential
by Germany physiologists and its general acceptance,
and about the same period of time was required to
develop practical methods of measuring it. These
include liquid, vapor, and pressure equilibration,
and hygrometic methods.

A.   Liquid Equilibration

    The first method of measuring water potential
involved immersing cubes, cylinders, or strips of
plant tissue in a graded series of sucrose or
mannitol solutions of various osmotic potentials for
a few hours. The water potential of the tissue was
considered to be equal to the osmotic potential of
the solution in which no change in weight or length
of the tissue occurred (Ursprung and Blum, 1916;
Ursprung, 1923; Meyer and Wallace, 1941). A
modification of this method, developed in Russia by
Shardakov (1948), is based on change in specific
gravity of the solution in which the tissue is
immersed. It is simple, quick, inexpensive, and can
be used in the field (Brix, 1966, Knipling, 1967).
It was described along with other methods by Barrs
(1968) and Slavik (1974).

B.   Vapor Equilibration

Some of the errors inherent in the liquid
equilibration method can be avoided by allowing
weighed samples of tissue or soil to equilibrate
over a series of solutions of known osmotic
potentials. As with liquid equilibration, the water
potential of the sample is regarded as equal to the
osmotic potential of the solution over which no
change in weight occurs. In a variation of this
method strips of filter paper saturated with
solutions of various osmotic potentials were buried
in the soil to measure its water potential (Hansen,
1926; Gradmann, 1928). The use of the vapor
equilibration method was described by Slatyer
(1958), Barrs (1968), and Slavik (1974). The chief
problems with this method are the large amounts of
tissue used, the long time required to reach an
equilibrium, and the careful control of temperature
required.

C.   Hygrometric Methods

The thermocouple psychrometer is an adaptation of
the vapor equilibration method in which the sample
is enclosed in a small container, usually immersed
in a constant temperature bath, and the relative
humidity of the air in equilibrium with the sample
is measured with a thermocouple psychrometer.
Practical instruments were first described by
Monteith and Owen (1958) and Richards and Ogata
(1958), and soon came into wide use. There are
three principal types of thermocouple psychrometers.
In the Richards and Ogata type a drop of water is
placed on the thermocouple junction and the current
developed by evaporative cooling is measured with a
microvoltmeter. In the Spanner type, described by
Monteith and Owen, Peltier cooling is used to
condense water on the thermocouple junction and the
current generated by evaporative cooling is
measured. Both types must be calibrated with

solutions of known osmotic potentials. The
isopiestic method, described by Boyer and Knipling
(1965) uses a modified Richards psychrometer and two
or three equilibrium readings with tissue samples
are obtained with drops of solution of different
osmotic potentials on the thermocouple junction.
This method eliminates the errors caused by
differences in diffusion resistance of different
samples of tissue, but it cannot be automated.
Various adaptations have been made for in situ
measurements on leaves (Boyer, 1968; Neumann and
Thurtell, 1972; Savage et. al., 1983), roots
(Fiscus, 1972), stems of trees (Wiebe, et al., 1970)
and in the soil (Wiebe and Brown, 1979). It is
difficult to measure water potentials lower than
about -7.5 MPa, but Wiebe (1981) described a method
for measuring water potentials down to about -500
MPa. Psychrometry was covered in detail in Brown
and Van Haveren (1972) and some problems of in-situ
psychrometry were discussed by Shackel (1984).
After measuring the water potential the sample of
plant tissue can be killed by freezing and a new
measurement made that provides the osmotic potential
of the tissue. Some investigators (Markhart et al.,
1981) claim a significant error can result from
dilution of the vacuolar sap with cell wall water
after freezing destroys cytoplasmic membranes. For
example, Markhart and Lin (1985) reported that the
osmotic potential of bean leaves was about 10%
higher (less negative) when measured on expressed
sap with a vapor pressure osmometer than when
calculated from pressure-volume curves. The
importance of dilution by cell wall water varies,
depending on its volume relative to that of vacuolar
water and on its solute concentration. The debate
concerning the solute concentration of cell wall
water is reviewed briefly by Boyer (1985, pp. 499-
500).

D.   Turgor Pressure

The turgor pressure usually is calculated from
the difference between the osmotic potential and the
water potential. However, direct measurements of

turgor pressure can be made by the use of miniature
pressure probes, as described in papers in this
volume by Nonami and Boyer and by Steudle. A
comparison of direct measurements of turgor pressure
with values calculated from psychrometric
measurements was published recently (Nonami, et al.,
1987) and Shackel and Brinckmann (1985) described
simultaneous measurements of leaf water potential
and epidermal cell turgor.

E. Pressure Equilibration

   One of the most widely used methods of measuring
plant water potential is the pressure equilibration
method introduced by Scholander and his colleagues
(1965), but based on a procedure described by Dixon
(1914, Chap. 10). A twig or single leaf is sealed
in a pressure chamber with the cut end protruding
and pressure is applied from a tank of compressed
gas until xylem sap begins to exude from the cut
end. This pressure is taken as equivalent to the
water potential in the xylem sap prior to detaching
the leaf or twig. The method is rapid and usable in
the field, but the results often are more variable
than those obtained with thermocouple psychrometers
(see Kramer, 1983, p. 386). The pressure
equilibration method was described by Slavik
(1974), and Ritchie and Hinckley (1975) discuss
some of its problems. The pressure chamber method
can be used to estimate osmotic potential as well as
xylem pressure potential by the pressure-volume
method described by Tyree and Hammel (1972) and by
Cheung et al. (1975). An application of the
pressure-volume method is described by Roberts and
Knoerr (1977). It avoids the dilution error of
psychrometer measurements, but is time consuming and
requires considerable plant material.
   Various methods of measuring plant water stress
were reviewed by Barrs (1968), Slavik (1974), in a
short but practical article by Turner (1981), and in
Kramer (1983). Those articles provide numerous
additional references.

IV.   MEASUREMENTS OF
STOMATAL APERTURE

Because of their importance in controlling water loss and uptake of CO_2 there has been much interest in measurement of stomatal aperture. Their premature closure also has been used an indicator of water stress (Alvim, 1965; Alvim and Havis, 1954; Shmueli, 1953). Quantitative study of diffusion through stomata seems to have begun with the investigation by Brown and Escombe (1900) of the effect of size and spacing on diffusion through small pores. They concluded that diffusion through small pores is proportional to the diameter rather than the area and that the spacing is very important because of the formation of "diffusion shells" around the pores. Incidentally, the reality of diffusion shells can be demonstrated visually by allowing dye to diffuse through small pores in a thin membrane into a solid medium such as gelatin or agar. Out of experiments of that sort came the concept of boundary layer resistance versus stomatal resistance. Because Brown and Escombe measured diffusion in quiet air where the boundary layer resistance is almost as great as stomatal resistance they concluded that moderate changes in stomatal aperture have little effect on transpiration. Later experiments by Stålfelt (1932), Bange (1953), and others showed that in moving air where the boundary layer resistance is low compared to stomatal resistance, there is a high correlation between stomatal aperture and rate of transpiration.

A.   Visual Observation of
Stomatal Aperture

The earliest observations of stomatal aperture were made by microscopic observation of attached leaves. However, such observations were impractical on most leaves and the method was abandoned in favor of observations made on strips of epidermis peeled off and instantly fixed in absolute alcohol (Lloyd,

1908; Loftfield, 1921). A more recent method is to make impressions of the leaf epidermis in collodion (Clements and Long, 1934), nail polish, or silicone rubber (Zelitch, 1961), and examine them under the microscope. However, Pallardy and Kozlowski (1980) warned that the presence of cuticular stomatal ledges may result in unreliable indications of stomatal aperture. Still another visual method is to infiltrate leaves with liquids of various viscosities. The degree of infiltration indicates whether the stomata are open or closed, but does not permit estimation of stomatal resistance. Literature dealing with infiltration methods is cited by Kramer (1983, p. 328) and by Alvim and Havis (1954).

Modern technology described in this volume has made it possible to return to direct visual observations of stomata and stomatal behavior. For example, scanning electron microscopy reveals details of stomatal structure not previously observed. Another paper by Omasa et al. (1985b) describes use of a remote controlled light microscope combined with a television camera and an image processing system that permits continuous remote observation of the effects of various environmental stresses such as water deficits and air pollutants on stomatal behavior. This system shows that stomata in different parts of a leaf sometimes respond differently to stress, a conclusion supported by more recent observations (Daley, et al., 1989; Downton, et al., 1988) made with other technologies.

B. Porometers

Another method of stomatal observation that has benefited from modern technology is porometry which is the oldest quantitative method of measuring stomatal resistance or conductance (Darwin and Pertz, 1911). Porometers usually measure either viscous flow or diffusion of gas. Recording viscous flow porometers were described by Gregory and Pearse (1934) and Wilson (1947). From these evolved the instrument described by Fiscus in this volume which permits continuous monitoring of stomatal aperture over considerable periods of time in the field and even uses a computer to correct for the effect of cloudy weather. Portable viscous flow porometers also were devised by Alvim (1965) and others for instantaneous measurements in the field. Viscous flow porometers have considerable advantages for long term measurements, but probably should not be used on leaves with stomata on only one surface and numerous bundle extensions that block lateral movement of air.

C. Diffusion Porometers

These instruments measure the approximate rate of diffusion of water vapor, usually by measuring the time required to produce a predetermined change in humidity by means of a sensor enclosed in a small chamber attached to a leaf. Most of them use a humidity sensor connected to an electronic circuit and meter, often with an electronic timer and other accessories that convert the humidity reading into diffusion resistance in s cm^{-1} or its reciprocal, conductance in cm s^{-1} (Kanemasu, et al., 1969; Slavik, 1974). Null-point diffusion porometers also have been introduced in which dry air is introduced to maintain a constant humidity, eliminating some calibration problems and the lag caused by adsorption of water on cuvette walls (Beardsell, et al., 1972). In this volume Kaufmann describes an elaborate cuvette system used to monitor stomatal behavior and estimate transpiration of a forest canopy. This system was made possible by the use of a computer to control the apparatus and record data and electronic sensing equipment to measure environmental factors.

V.   RECENT DEVELOPMENTS

Most of the measurements of plant water status have been made on leaves and twigs. However, some recent research, including experiments in which the soil containing half of a split root system was allowed to dry while the other half was kept moist, suggest that stomatal conductance and photosynthesis sometimes are better correlated with soil and root water status than with leaf water status. This led Davies et al. (1986), Schulze (1986), and Turner (1986) to suggest that roots are the primary sensors of water stress and that biochemical signals from the roots cause the physiological perturbations observed in the leaves of plants subjected to drought.

No doubt the prolonged root stress that develops in drying or flooded soil affects root metabolism and the amounts of hormones and minerals translocated to the shoots. However, shoots of plants growing in moist soil often are subjected to transient midday water stress in hot, sunny weather. Under these conditions it is difficult to conceive of the roots as primary sensors of water stress. This problem was discussed in more detail by Kramer (1988). Apparently, to answer this question we need technology to monitor root and shoot water status simultaneously. This may become possible with on going improvements in NMR imaging (see last chapter).

VI.   RESEARCH NEEDS

In spite of the progress that has been made during the past decade there is continuing need for improvements in existing instrumentation and development of new methods for monitoring various aspects of water stress. A few examples follow:
1.   Improvement in the reliability of in situ psychrometer measurements on plants growing in the field to decrease errors caused by temperature variations.

2. Development of simple, inexpensive methods of monitoring plant water stress to indicate the need for irrigation.

3. Measurement of the water status of plant organs and tissues such as meristematic regions, ovules, and anthers in situ.

4. Methods of studying the effects of mild water stress on membranes and enzymes to understand how metabolic processes are affected by water stress. NMR imaging may assist in such studies.

VII.   SUMMARY

In summary, there has been a gradual development of concepts in the field of plant water relations dating from Hales' work early in the 18th century through the methodical German research in the late 19th century and first quarter of the 20th century. By that time many important concepts had been suggested or even partially developed although few had been generally accepted. The cohesion theory of the ascent of sap was proposed in 1895, the theory of diffusion through stomata was published in 1900, the suction force (water potential) concept of cell water relations was developed by 1915 to 1920, and the treatment of water movement through the soil-plant-atmosphere as analogous to the flow of electricity in a conducting system was first proposed by Huber in 1924. None of these concepts was accepted immediately and few appeared in plant physiology texts until long after their appearance in scientific journals.

There were two important reasons for slow acceptance of the new concepts. One was the natural conservatism of plant scientists, the other was lack of instrumentation to test the concepts. General acceptance of the cohesion theory of the ascent of sap required 50 years because scientists found it difficult to believe or demonstrate that water confined in xylem elements has a very high cohesive force. Although the suction force or water potential concept of cell water relations was cautiously accepted, application of the concept in

research on plants was slow because of lack of
practical methods of measuring it.   However,
development of thermocouple psychrometers and
pressure chambers during the 1960's resulted in
hundreds of research projects and thousands of
measurements of water potential.   Now we even have
miniature pressure probes for direct measurement of
turgor pressure and NMR techniques for studying the
state and distribution of water in cells.   Likewise,
development of new porometer methods has resulted in
a great increase in research based on measurement of
stomatal aperture. The recent outburst of research
activity in plant water relations is analogous to
the outburst of research on CO_2 exchange that
followed the appearance of infrared gas analyzers.
Several of the papers in this volume describe
procedures and apparatus that were made possible
only by the availability of computers and other
electronic devices.   Technology is advancing so
rapidly that one of the major problems for plant
physiologists and ecologists is to learn how the new
technology can be applied to their research.   This
will require increased interdisciplinary exchange
and collaboration between scientists and the
designers and builders of scientific instruments.
These exchanges can be encouraged by workshops such
as the one that produced this volume.

REFERENCES

Alvim P. de T.   (1965).   Stomatal opening as a
       practical indicator of moisture stress in
       cotton.   Physiologia Plant. 19, 308-312.
Alvim P. de T., and Havis, J.R.   (1954).   An
       improved infiltration series for studying
       stomatal opening as illustrated with coffee.
       Plant Physiol. 29:97-98.
Atkins, W.R.G.   (1916).   Some Recent Researches in
       Plant Physiology.   Whitaker and Co., London.
Bange, G.G.J.   (1953).   On the quantitative
       explanation of stomatal transpiration.   Acta
       Bot. Neerl. 2, 255-297.

Barrs, H.D. (1968). Determination of water
    deficits in plant tissues. In T.T. Kozlowski
    (ed.) Water Deficits and Plant Growth, Vol. I,
    pp. 126-368. Academic Press, New York.
Beardsell, M.F., Jarvis, P.G., and Davidson, B.
    (1972). A null-balance diffusion porometer
    suitable for use on leaves of many shapes. J.
    Appl. Ecol. 9:677-690.
Blum, A., Schertz, K.F., Toler, R.W., Welch, R.I.,
    Rosenow, D.T., Robinson, J.W., and Clark, L.E.
    (1978). Selection for drought in sorghum using
    aerial infrared photography. Agron. J. 70:472-
    277.
Bottomley, P.., Rogers, H.H., and Foster, T.H.
    (1986). Nuclear magnetic resonance imaging
    shows water distribution and transport in plant
    root systems in situ. Proc. Nat. Acad. Sci.
    USA. 83:87-89.
Boyer, J.S. (1968). Relationship of water
    potential to growth of leaves. Plant Physiol.
    43:1056-1062.
Boyer, J.S. (1985). Water transport. Ann. Rev.
    Plant Physiol. 36:473-516.
Boyer, J.S., and Knipling, E.B. (1965). Isopiestic
    technique for measuring leaf water potentials
    with a thermocouple psychrometer. Proc. Nat.
    Acad. Sci. USA. 54:1044-1051.
Brix, H. (1966). Errors in measurement of leaf
    water potential of some woody plants with the
    Schardakow dye method. Dept. Pub. No. 1164.
    Forestry Branch, Can. Dept. For. & Rural
    Development.
Brown, H.T., and Escombe, F. (1900). Static
    diffusion of gases and liquids in relation to
    the assimilation of carbon and translocation in
    plants. Phil. Trans. Roy. Soc. (London) B.
    193:223-291.
Brown, J.M., Johnson, G.A., and Kramer, P.J.
    (1986). In vivo magnetic resonance microscopy
    of changing water content in _Pelargonium
    hortorum_ roots. Plant Physiol. 82:1158-1160.
Brown, R.W. and Von Haveren, B.P. (eds.) (1972).
    Psychrometry in Water Relations Research. Utah
    Agr. Exp. Sta., Utah State Univ., Logan, Utah.

Burke, M.J., Gusta, L.V., Quamme, H.A., Weiser,
    C.J., and Li, P.H. (1976). Freezing and
    injury in plants. Ann. Rev. Plant Physiol.
    27:507-528.
Cheung, Y.N.S.., Tyree, M.T., and Dainty, J.
    (1975). Water relations parameters on single
    leaves obtained in a pressure bomb and some
    ecological interpretations. Can. J. Bot.
    53:1342-1346.
Clements, F.E. and Long, F.L. (1934). The method
    of collodion films for stomata. Amer. J. Bot.
    21:7-17.
Daley, P.F., Raschke, P., Ball, J.T. and Berry, J.A.
    (1989). Topography of photosynthetic activity
    of leaves obtained from video images of
    chlorophyll fluorescence. Plant Physiol.
    90:1233-1238.
Darwin, F., and Pertz, D.F.M. (1911). On a new
    method of estimating the aperture of stomata.
    Proc. Roy. Soc. London B 84:136-154.
Davies, W.J., Metcalfe, J., Lodge, T.A. and DaCosta,
    A.R. (1986). Plant growth substances and the
    regulation of growth under drought. Aust.
    Jour. Plant Physiol. 13:105-125.
Dixon, H.H. (1914). Transpiration and the Ascent
    of Sap in Plants. The Macmillan Company, New
    York.
Downton, W.J.S., Loveys, B.R. and Grant, W.J.R.
    (1988). Stomatal closure fully accounts for
    the inhibition of photosynthesis by abscisic
    acid. New Phytol. 108:263-266.
Fiscus, E.L. (1972). In situ measurement of root-
    water potential. Plant Physiol. 50:191-193.
Fitting, H. (1911). Die Wasserversorgung und die
    osmotichen Druckverhältnisse der
    Wüstenpflanzen. Zeitschr. f. Bot. 3:209-375.
Gradmann, H. (1928). Untersuchungen über die
    Wasserverhältnisse des Bodens als Grundlage des
    Pflanzenwachstums. Jahrb. wiss. Bot. 69:1-100.
Gregory, F.G., and Pearse, H.L. (19394). The
    resistance porometer and its application to the
    study of stomatal movement. Proc. Roy. Soc.
    London B 114:477-493.
Hales, S. (1727). Vegetable Staticks. W. & J.
    Innys and T. Woodward, London. London
    Scientific Book guild 1961.

Hansen, C. (1926). The water-retaining power of the soil. J. Ecology 14:111-119.

Harris, J.A. (1934). The Physico-Chemical Properties of Plant saps in Relation to Phytogeography. Univ. Minnesota Press, Minneapolis.

Höfler, K. (1920). Ein Schema für die osmotischen Leistung der Pflanzenzelle. Ber. deut. Bot. Ges. 38:288-298.

Huber, B. (1924). Die Beurteilung des Wasserhaushalts der Pflanze. Jahrb. Wiss. Bot. 64:1-20.

Idso, S.D., Reginato, R.J., Hatfield, J.L., Walker, G.K., Jackson, R.D., and Pinter, P.J., Jr. (1980). A generalization of the stress-degree-day concept of yield prediction to accommodate a diversity of crops. Agr. Meteorol. 21:205-211.

Jackson, R.D. 1982. Canopy temperature and crop water stress. Adv. Irrig. 1:43-85.

Johnson, G.A., Brown, J., and Kramer, P.J. (1987). Magnetic resonance microscopy of changes in water content in stems of transpiring plants. Proc. Nat. Acad. Sci. 84:2752-2755.

Kanemasu, E.T., Thurtell, G.W. and Tanner, C.B. (1969). Design, calibration and field use of a stomatal diffusion porometer. Plant Physiol. 44:881-885.

Knipling, E.B. (1967). Measurement of leaf water potential by the dye method. Ecology 48:1038-1041.

Korstian, C.F. (1924). Density of cell sap in relation to environmental conditions in the Wasatch Mountains of Utah. J. Agr. Res. 28:845-909.

Kramer, P.J. (1963). Water stress and plant growth. Agron, J. 55:31-35.

Kramer, P.J. (1983). Water Relations of Plants. Academic Press, New York.

Kramer, P.J. (1988). Changing concepts regarding plant water relations. Plant, Cell Environ. 11:565-568.

Krasnoselsky-Maximov, T.A. (1917). Daily variations in the water content of leaves. Trav. Jard. Bot. Tiflis 19:1-22. (In Russian).

Livingston, B.E., and Brown, W.H. (1912). Relation
    of the daily march of transpiration to
    variations in the water content of foliage
    leaves. Bot. Gaz. 53:309-330.
Lloyd, F.E. (1908). The physiology of stomata.
    Carnegie Inst. Washington Publ. 82.
Loftfield, J.V.G. (1921). The behavior of stomata.
    Carnegie Inst. Washington Publ. 314.
Markhart, A.H., III., and Lin, T.Y.. (1985). New
    hand operated press for the extraction of
    tissue sap for the measurement of osmotic
    potential. Agron. J. 77:182-185.
Markhart, A.J., III., Sionit, N., and Siedow, J.N.
    (1981). Cell wall water dilution: an
    explanation of apparent negative turgor
    potentials. Can. J. Bot. 59:1722-1725.
Maximov, N.A. (1929). The Plant in Relation to
    Water. Allen and Unwin, Ltd., London.
    (English translation by Yapp.)
Meyer, B.S., and Anderson, D.B. (1939). Plant
    Physiology. 1st ed. D. Van Nostrand Co., New
    York.
Meyer, B.S., and Wallace, A.M. (1941). A
    comparison of two methods of determining the
    diffusion pressure deficits of potato tuber
    tissues. Am. J. Bot. 28:838-843.
Miller, E.C. (1938). Plant Physiology. 2nd ed.
    McGraw-Hill, New York.
Monteith, J.L., and Owen, P.C. (1958). A
    thermocouple method for measuring relative
    humidity in the range 95-100%. J. Sci.
    Instruments 35:443-446.
Montfort, C. (1922). Die Wasserbilanz in
    Nahrlösung, Salzlösung, und Hochmoorwasser.
    Zeitschr. f. Bot. 14:98-172.
Nakayama, F.S., and Ehrler, W.L. (1964). Beta ray
    gauging technique for measuring leaf water
    content changes and moisture status of plants.
    Plant Physiol. 39:95-98.
Neumann, H.H., and Thurtell, G.W. (1972). A
    Peltier cooled thermocouple dewpoint hygrometer
    for in situ measurement of water potentials.
    In R.W. Brown and B.P. Van Haveren (eds.),
    Psychrometry in Water Relations Research. Utah
    Agric. Exp. Sta., Utah State Univ., Logan Utah.

Nonami, H., Boyer, J.S. and Steudle, E. (1987). Pressure probe and isopiestic psychrometer measure similar turgor. Plant Physiol. 83:592-595.

O'Toole, J.C., and Cruz, R.T. (1980). Response of leaf water potential, stomatal resistance, and leaf rolling to water stress. Plant Physiol. 65:428-432.

Omasa, K., Hashimoto, Y., Kramer, P.J., Strain, B.R., Aiga, I., and Kondo, J. (1985). Direct observation of reversible and irreversible stomatal responses of attached sunflower leaves to SO_2. Plant Physiol. 79:153-158.

Omasa, K., Onoe, M., and Yamada. (1985). NMR imaging for measuring root system and soil water content. Environ. Control in Biol. 23:99-102.

Oosterhuis, D.M., Walker, S., and Eastham, J. (1985). Soybean leaflet movements as an indicator of crop water stress. Crop Sci. 25:1101-1106.

Owen, P.C. (1952). The relation of germination of wheat to water potential. J. Exp. Bot. 3:188-203.

Pallardy, S.G. and Kozlowski, T.T. (1980). Cuticle development in the stomatal region of Populus clones. New Phytol. 85:363-368.

Renner, O. (1912). Versuche zur Mechanik der Wasserversorgung. 2 über Wurzeltatigkeit. Ber. deut. Bot. Ges. 30:642-648.

Richards, L.A., and Ogata, G. (1958). Thermocouple for vapor pressure measurement in biological and soil systems at high humidity. Science 128:1089-1090.

Ritchie, G.A., and Hinckley, T.M. (1975). The pressure chamber as an instrument for ecological research. Adv. Ecol. Res. 9:165-254.

Roberts, S.W., and Knoerr, K.R. (1977). Components of water potential estimated from xylem pressure measurements in five tree species. Oecologia 28:191-202.

Rogers, H.H. and Bottomley, P.A. (1987). In situ nuclear magnetic resonance imaging of roots: influence of soil type, ferromagnetic particle content, and soil water. Agron. J. 79:957-965.

Savage, M.J., Wiebe, H.H., and Cass, A. (1983). In situ field measurements of leaf water potentials using thermocouple psychrometers. Plant Physiol. 73:609-613.

Scholander, P.F., Hammel, H.T., Bradstreet, E.D., and Hemmingsen, E.A. (1965). Sap pressure in vascular plants. Science 148:339-346.

Schulze, E.-D. (1986). Whole-plant responses to drought. Aust. Jour. Plant Physiol. 13:127-141.

Shackel, K.A. (1984). Theoretical and experimental errors for in-situ measurements of plant water potential. Plant Physiol. 75:766-772.

Shackel, K.A. and Brinckmann, E. (1985). In situ measurements of epidermal cell turgor, leaf water potential, and gas exchange in Tradescantia virginiana L. Plant Physiol 78:66-70.

Shardakov, V.S. (1948). New field method for the determination of the suction pressure of plants. Doklady Akad, Nauk SSSR 60:169-172. (In Russian.)

Shmueli, E. (1953). Irrigation studies in the Jordan Valley. I. Physiological activity of the banana in relation to soil moisture. Bull. Res. Council Israel 3:228-247.

Sinclair, T.R. and Ludlow, M.M. (1985). Who taught plants thermodynamics? The unfulfilled potential of plant water potential. Aust. J. Plant Physiol. 12:213-217.

Slatyer, R.O. (1958). The measurement of diffusion pressure deficit in plants by a method of vapor-equilibration. Aust. J. Biol. Sci. 11:349-365.

Slatyer, R.O. (1967). Plant Water Relationships. Academic Press, New York.

Slatyer, R.O., and Taylor, S.A. (1960). Terminology in plant and soil-water relations. Nature 187:922.

Slavik, B. (1974). Methods of Studying Plant Water Relations. Springer-Verlag, Berlin and New York.

Stålfelt, M.G. (1932). Der Stomatäre Regulator in der pflanzlichen transpiration. Planta 17:22-85.

Stocker, O. (1929). Das Wasserdefizit von
    Gefässpflanzen in verschiedenen Klimazonen.
    Planta 7:382-387.
Stout, D.G., Cotts, R.M., and Steponkus, P.L.
    (1977). The diffusional water permeability of
    Elodea leaf cells as measured by nuclear
    magnetic resonance. Can. J. Bot. 57:1623-1631.
Stout, D.G., Steponkus, P.L., and Cotts, R.M.
    (1977). Quantitative study of the importance
    of water permeability in plant cold hardiness.
    Plant Physiol. 60:374-378.
Tang, P.S., and Wang, J.S. (1941). A thermodynamic
    formulation of the water relations in an
    isolated living cell. J. Physical Chem.
    45:443-453.
Thoday, D. (1918). On turgescence and the
    absorption of water by the cells of plants.
    New Phytol. 17:108-113.
Turner, N.C. (1981). Techniques and experimental
    approaches for the measurement of plant water
    status. Plant Soil 58:333-366.
Turner, N.C. (1986). Crop water deficits: a decade
    of progress. Advances in Agronomy 39:1-51.
Tyree, M.T. and Jarvis, P.G. (1982). Water in
    tissues and cells. In O.L. Lange, P.S. Nobel,
    C.B. Osmond, and H. Ziegler (eds.).
    Encyclopedia of Plant Physiology. n.s. 12B:35-
    75.
Tyree, M.T., and Hammmel, H.T. (1972). The
    measurement of the turgor pressure and the
    water relations of plants by the pressure-bomb
    technique. J. Exp. Bot. 23:267-282.
Ursprung, A. (1923). Zur Kenntnis der Saugkraft.
    VII. Eine neue vereinfachte Methode Zur
    Messung der Saugkraft. Ber. deut. Bot. Ges.
    41:338-343.
Ursprung, A., and Blum, G. 91916). Zur Methode der
    Saugkraftmessung. Ber. deut. Bot. Ges. 34:525-
    539.
Walter, H. (1960). Einfuhrung in die Phytologie.
    Vol. 3, Part 1. 2nd ed. E. Ulmer, Stuttgart.
Weatherley, P.E. (1950). Studies in the water
    relations of the cotton plant I. The field
    measurements of water deficits in leaves. New
    Phytol. 49:81-97.

Wenkert, W. (1980). Measurement of tissue osmotic pressure. Plant Physiol. 65:614-617.

Wiebe, H.H. (1981). Measuring water potential (activity) from free water to oven dryness. Plant Physiol. 68:1218-1221.

Wiebe, H.H., and Brown, R.W. (1979). Temperature gradient effects on in situ hygrometer measurements of soil water potential. II. Water movement. Agron. J. 71:397-401.

Wiebe, H.H., Brown, R.W., Daniel, T.W., and Campbell, E. (1970). Water potential measurement in trees. BioScience 20:225-226.

Wilson, C.C. (1948). The effect of some environmental factors on the movements of guard cells. Plant Physiol. 23:5-37.

Zelitch, I. (1961). Biochemical control of stomatal opening in leaves. Proc. Nat. Acad. sci. USA 47, 1423-1433.

ESTIMATING TREE CANOPY TRANSPIRATION
FROM MEASUREMENTS OF LEAF CONDUCTANCE

Merrill R. Kaufmann

USDA Forest Service
Rocky Mountain Forest and Range
Experiment Station
Fort Collins, CO 80526 USA

I. INTRODUCTION

Stomata play a crucial role in controlling water vapor and CO_2 exchange between plants and their environment, and their influence on transpiration and photosynthesis depends upon boundary layer conditions of the foliage. Stomata exert maximum control over gas exchange when air movement around the foliage is high (boundary layer resistance is low). Jarvis and McNaughton (1986) use the term "omega factor," a dimensionless number between 0 and 1, to refer to the degree to which transpiration is uncoupled from the atmospheric water vapor saturation deficit. When an atmospheric saturation deficit exists at the leaf surface, transpiration is directly influenced by the vapor gradient and by surface conductance of the foliage, and omega approaches 0. This condition exists when atmospheric mixing is adequate. However, when mixing is inadequate -- for example, over the surface of very large leaves, with a dense canopy, or when air movement is quite low -- the atmospheric water vapor deficit does not exist at the leaf surface. Then stomatal regulation of water vapor movement is reduced, as observed by Bange (1953) and others, and transpiration is more directly related to radiation input (Jarvis, 1985). Under these conditions, omega approaches 1. Generally, transpiration is the result of both aerodynamic and radiation-driven processes occurring simultaneously, and the omega factor lies somewhere between 0 and 1.

The relative importance of stomatal and boundary layer effects on transpiration is one factor determining the choice of methods for estimating transpiration. For many crop plants, conditions often are such that aerodynamic mixing is low; this results in an increase in temperature and ambient water vapor pressure in the most dense part of the canopy. Transpiration is controlled primarily by radiation, and measurements of stomatal behavior provide somewhat limited insight into the transpiration process. Fiscus (this volume) discusses the continuous measurement of stomatal opening in corn. It is noteworthy that his measurements are used primarily as a guide to the development of water stress rather than for estimating conductance or transpiration.

In forest canopies more than a few meters tall, aerodynamic mixing is generally quite high, with omega values of 0.2 or less (Jarvis, 1985). Vertical temperature and humidity variation through the canopy is typically low, and transpiration is controlled directly by stomata and the vapor pressure differences from leaf to air. These conditions have led various researchers to base estimates of forest transpiration on measurements or estimates of stomatal conductance (Jarvis, 1976; Jarvis et al., 1985; Kaufmann, 1984; Riha and Campbell, 1985; Running, 1984; and Tan et al., 1978).

Recently Kaufmann and Kelliher (1990) reviewed and evaluated techniques for estimating transpiration in forest stands. Various approaches are available, including the estimation of forest evapotranspiration using the water balance of catchments or plots or using micrometeorological techniques, direct estimation on individual trees, or modelling canopy transpiration rates. Each approach has advantages and disadvantages that determine its appropriateness for given situations.

This paper considers those situations in which stomatal conductance data are needed to estimate transpiration or photosynthesis of individual tree crowns and forest canopies. This approach is appropriate when information is required for individual species in mixed forests, for stands not large enough to employ eddy correlation or other stand-level techniques, or for studies concerned with crown or canopy processes. In process studies, conductance data are needed in certain mechanistic ecosystem models to parameterize relationships between gas exchange and environmental conditions.

At the outset, it must be noted that the use of stomatal conductance measurements made on portions of tree crowns or of estimates of average conductances for entire crowns is a difficult and uncertain approach for estimating transpiration of trees and stands. The problems arise when small-scale measurements are extrapolated to large-scale systems without a thorough understanding of stomatal response and of specific environmental conditions throughout the system. While the temperature and humidity conditions in many forest stands are rather uniform throughout the canopy, the radiation environment is extremely variable. This variability introduces considerable complexity into the estimation of stomatal conductance. A number of radiation absorption models are being developed or improved for forest canopies and individual tree crowns, but their application in transpiration calculations is difficult, and in most cases researchers resort to "big leaf" or layer approaches that treat the canopy as a single layer or several homogeneous layers.

While stomatal behavior under steady-state conditions is reasonably well understood, there is considerable uncertainty associated with predicting stomatal behavior under conditions of variable light intensity involving high-frequency sun flecks or lower frequency shifts associated with changing sun angles or clouds. For forests having low omega factors, this is perhaps the single largest obstacle in using conductance measurements or estimates obtained at the level of leaf or branch elements for total crown estimates of transpiration. With this disclaimer, the following sections examine the evaluation of stomatal response in forests.

II. STOMATAL BEHAVIOR UNDER FOREST CONDITIONS

A. Measurement of Stomatal Conductance and Interpretation of Data

(1) Selection and Use of a Measurement System. While stomatal behavior in controlled environment studies is often of interest, the most pressing problems regarding stomatal function presently exist in attempting to integrate physiological function of trees into models of ecosystem behavior. For using stomatal conductance information predictively, the critical needs are a suitable data base that includes conductance measurements and all appropriate environmental and plant measurements, and a suitable model, parameterized from these data, that interprets or predicts stomatal conductance for the range of environmental conditions encountered.

Most available data come from transpiration or photosynthesis studies involving the use of porometers or gas exchange cuvettes, primarily the former. There seems to be little technical difficulty in measuring conductance by porometry, but from a modelling standpoint it is logistically difficult to make a sufficient number of measurements under field conditions to characterize stomatal response to the environment. Furthermore, porometer measurements are made on relatively small samples of foliage, and this contributes to variability by increasing the likelihood of sampling tissue having conductances well above or below the average conductance for a larger sample. An added complication is that porometry provides spot measurements with no record of the stability of environmental conditions in the time immediately preceding the measurement, including light as discussed earlier. Nonetheless, porometry may be the only feasible method for collecting conductance data, and it is more capable than cuvette systems of capturing spatial variability within tree crowns or the canopy. The use of porometers to measure conductance is carefully reviewed by Smith and Hollinger (1990).

Gas exchange cuvettes offer the opportunity to collect leaf conductance and environmental data for extended periods of time on the same foliage. The volume of foliage enclosed in cuvettes often can be larger than that used in porometry, resulting in an integration of self-shading and leaf age effects. It is much more difficult, however, to evaluate spatial variation in conductance among branches within the crown envelope or among crowns or leaf layers in the canopy. A major strength of cuvette measurements is that conductance data may be collected over a broad range of conditions on the same foliage. This is particularly valuable for developing coefficients in models relating conductance to environmental conditions, provided, of course, that the sampled foliage is representative.

Most research suggests that four factors are important in regulating stomatal response under forest conditions (Kaufmann, 1982a and b; Landsberg, 1986). The two primary factors are photosynthetic photon flux density and the vapor pressure deficit or absolute humidity difference from leaf to air. The other two factors, temperature and plant water stress, are considered to be secondary because their effects are intermittent. Most temperature effects on conductance and transpiration occur through effects on the vapor gradient, but temperature extremes may reduce conductance.

Similarly, stomata are unaffected by moderate levels of plant water stress, but below threshold plant water potentials (in the leaves and perhaps also in the roots), conductance is reduced in proportion to the level of water stress.

For instantaneous conductance measurements made simply to determine conductance in a given situation, little additional data may be needed. However, for developing mechanistic or phenomenological relationships between conductance and the important environmental and plant factors indicated above, additional data are required. A number of porometer systems presently available are equipped for measurements of light intensity, humidity, and temperature. Similarly, cuvette systems installed under field conditions also require these measurements, and for both systems water stress measurements are needed, particularly during periods when water stress affects leaf conductance.

Cuvette systems capable of continuous or frequent measurements must be capable of being left in place without unduly modifying the environment around the sampled foliage. This is a difficult task. In many cases the researcher requires both water vapor flux and CO_2 flux data. Systems capable of these measurements include a CO_2 analyzing system and are generally much more complex because of the need for careful control of the cuvette environment.

A simplified cuvette system capable of measuring water vapor flux from tree branches was devised by Kaufmann (1981). This system is based on earlier porometer principles involving the measurement of humidity transients during a short time interval when the cuvette is sealed. While theoretically not as free of potential errors as the null-balance approach, this system is far easier to construct and operate and less expensive than would be possible if a climatized cuvette or null balance approach were used, particularly when multiple cuvettes are used in the system. The cuvettes are designed to have outside air circulated through them continuously except during a brief measurement period when trap doors seal the chambers and changes in relative humidity and temperature are measured (Figure 1). Measurements also include ambient light intensity, temperature, and relative humidity.

This system meets several requirements. First, it operates relatively trouble-free for long periods, even under harsh field conditions including snow. Second, conditions within the cuvettes remain sufficiently close to ambient that there is a minimal chamber effect on the results. And third, the cuvettes are large enough to enclose branches up to 30 cm long, much larger than can be sampled with porometers. During one six-month period, a series of eight cuvettes were used for about 35,000 conductance measurements, with a missing data rate of 2.7 percent (chiefly caused by lightning damage).

(2) Leaf Conductance Models. The cuvette system briefly described above was used to acquire data on four subalpine tree species: Engelmann spruce (*Picea engelmanni* Parry ex Engelm.), subalpine fir (*Abies lasiocarpa* [Hook.] Nutt.), lodgepole pine (*Pinus contorta* var. *latifolia* Engelm.), and aspen (*Populus tremuloides* Michx.). Measurements were made on branches located near the midpoint of the crown, about 6 m above the ground. Data were screened to exclude measurements made when relative humidity was above 90 percent, temperatures were below freezing, or light intensity during the 20-minute period preceding conductance measurements varied by more than one-third.

Initial analyses of these data were presented by Kaufmann (1982a and b). More recently, Massman and Kaufmann (1990) re-examined these data and analyzed them using a number of models available in the literature. These include an adaptation of a linear model presented by Jarvis (1976), a model from Lohammer et al. (1980), a feed-forward model from Farquhar

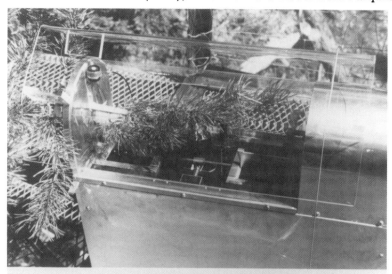

Fig. 1. Transient humidity cuvette for measuring leaf conductance: (a) upper enclosure shown with a lodgepole pine branch; (b) interior of lower enclosure.

(1978), an additional model provided by Farquhar through personal communication, and a model from Ball et al. (1987).

These models differ primarily in the way they treat humidity and temperature. The models have the general form as follows:

$$g = g_0 + g_1 f_Q f_D f_T \qquad (1)$$

where g is leaf conductance, g_0 is the minimum (nighttime) conductance and is a fitting parameter, g_1 is also a fitting parameter related to maximum conductance, and the remaining terms are modifying functions related to photosynthetic photon flux density (Q), leaf-to-air vapor pressure difference (D), and temperature (T). The modifier f_Q is similar for all models and is equal to Q/(Q + b), where b is a fitted parameter for each model.

The modifiers f_D and f_T depend on the specific model under consideration. For the Jarvis (1976) model,

$$f_D = 1 - b' D/P \qquad (2)$$

where b' is a fitting parameter and P is ambient pressure. For the Lohammer et al. (1980) model,

$$f_D = 1/(1 + b'' D/P) \qquad (3)$$

where b'' is another fitting parameter. For the Farquhar (1978) model,

$$f_D = (1 - b' D/P)/(1 + b'' D/P) \qquad (4)$$

For an adaptation of the Farquhar model (G. D. Farquhar, personal communication),

$$f_D = [1 - b' (D/P)^{1/2}]/(D/P)^{1/2} \qquad (5)$$

For each of these models, Massman and Kaufmann (1990) used an adaptation from Jarvis (1976) for f_T,

$$f_T = [(T-T_l)/(T_0-T_l)][((T_h-T)/(T_h-T_0))^a] \qquad (6)$$

where T is the current temperature, T_l is the lowest temperature for stomatal functioning, T_h is the highest temperature for stomatal functioning, T_0 is a fitted parameter representing the optimal temperature, and the exponent, a, is a constant. For the final model, an adaptation from Ball et al. (1987), the f_D and f_T terms are combined into a single term, h, which is the leaf surface relative humidity expressed as a fraction.

$$f_D f_T = h \qquad (7)$$

Kaufmann's (1982a) model is given by

$$g = b_1(Q^{1/2}/D^{1/2}) + b_2(Q^{1/2}/D) + b_3(Q^{1/2}/D^2) \qquad (8)$$

Massman and Kaufmann's (1990) analyses indicate moderate improvements in R^2 values over those presented by Kaufmann (1982a) for the four species (Table 1). Coefficients for the models and standard errors of the means are given in Massman and Kaufmann. The stepwise analyses demonstrated clearly the importance of light intensity as a dominant factor regulating leaf conductance. Significant improvements in fit were found by including humidity terms, and in some cases, particularly for shade-tolerant species, the improvement was substantial. These new analyses suggest that shade-tolerant species may be more strongly coupled to ambient leaf-to-air-vapor differences than shade-intolerant species. Temperature was the least important of the tested variables, as observed by Kaufmann (1982a and b). These analyses are valuable in that they provide comparisons among models representing stomatal behavior, and they provide an evaluation of stomatal response to environmental conditions for several species.

B. Canopy Conductance and Transpiration

Eddy correlation and Bowen ratio energy balance techniques are useful approaches for measuring canopy evapotranspiration, and they do not require estimates of leaf or canopy conductance. However, these techniques have not been proven to be reliable in hilly or mountainous terrain. The most common approach for estimating canopy evapotranspiration under these conditions is to use modelling techniques employing estimates of leaf conductance. The Penman-Monteith equation is such a model, and Kaufmann and Kelliher (1990) review its application in forest situations.

In most forests, canopies are sufficiently open and tall that aerodynamic mixing is high, and under these conditions conductance exerts considerable control over evapotranspiration. The Penman-Monteith equation, also known as the combination method, includes a radiation-driven component and a canopy conductance component, with the proportioning of evapotranspiration between these components represented by the omega factor discussed at the beginning of this paper (Jarvis and McNaughton, 1986). The combination method assumes that a single canopy resistance (the inverse of canopy conductance) can be used to represent the canopy as a single "big leaf". The method has been applied successfully in a number of situations for estimating canopy evapotranspiration and for examining understory evapotranspiration as well (Kaufmann and Kelliher, 1990), but the caution noted earlier regarding the difficulty of representing leaf conductances in

Table 1. R^2 values obtained by fitting observed leaf conductances to models. See text for model description by equation number.

Species	\multicolumn{6}{c}{R^2 for Equations}
	2+6	3+6	4+6	5+6	7	8
Aspen	0.60	0.71	0.71	0.71	0.66	0.67
Lodgepole Pine	0.71	0.74	0.74	0.73	0.72	0.74
Engelmann Spruce	0.64	0.76	0.76	0.76	0.73	0.72
Subalpine Fir	0.57	0.72	0.73	0.71	0.69	0.67

canopies where light intensity varies in space and time should be kept in mind.

Until better methods are available for treating the effects of variable environmental conditions on leaf conductance, leaf conductance data and models discussed in the previous section relating leaf conductance to environmental conditions are useful in estimating transpiration from forest canopies. The approach presently used is to estimate conductance for various layers of the canopy and to sum those estimates to obtain a canopy conductance. Kaufmann and Kelliher (1990) discuss procedures for species in which stomata are found on both leaf surfaces (amphistomatous leaves) or only on one surface (hypostomatous leaves).

III. SUMMARY

The collection of leaf conductance data and the subsequent development of appropriate leaf conductance models accounting for variable conditions throughout the forest canopy are difficult and large tasks. Nonetheless, this research is particularly valuable for determining explicit effects of environmental and plant conditions on transpiration, and for assessing differences among species. There is considerable uncertainty associated with the modelled conductances, and reducing this variability requires a more thorough understanding of stomatal behavior than presently in our grasp. While it is possible to expand on models discussed in this paper to include effects of such factors as low nocturnal temperatures and low leaf water potentials, the effects of variable environmental conditions in the canopy, age differences encountered through the season, and morphological effects associated with position of the foliage in the canopy will not be adequately understood until more research is done.

REFERENCES

Ball, J. T., I. E. Woodrow, and J. A. Berry. (1987). A model predicting stomatal conductance and its contribution to the control of photosynthesis under different environmental conditions. *In*: J. Biggins (ed.), Progress in Photosynthesis Research Vol. IV. Martinus Nijhoff Publishers, Dordrecht, The Netherlands, p. 221-224.

Bange, G. G. J. (1953). On the quantitative explanation of stomatal transpiration. *Acta Botanica Neerlandica* 2:255-297.

Farquhar, G. D. (1978). Feedforward responses of stomata to humidity. *Australian J. Plant Physiology* 5:787-800.

Jarvis, P. G. (1976). The interpretation of the variations in leaf water potential and stomatal conductance found in canopies in the field. *Phil. Trans. Royal Soc. Lond. B.* 273:593-610.

Jarvis, P. G. (1985). Transpiration and assimilation of tree and agricultural crops: the 'omega' factor. *In* M. G. R. Cannell and J. E. Jackson (eds.), Attributes of Trees as Crop Plants, Inst. of Terrestrial Ecol., Huntington, England, p. 460-480.

Jarvis, P. G. and K. G. McNaughton. (1986). Stomatal control of transpiration: scaling up from leaf to region. *Adv. Ecol. Res.* 15:1-49.

Jarvis, P. G., H. S. Miranda, and R. I. Muetzelfeldt. (1985). Modeling canopy exchanges of water vapor and carbon dioxide in coniferous forest plantations. *In* B. A. Hutchison and B. B. Hicks (eds.), The Forest-Atmosphere Interaction. Reidel Publishing Co., Boston, MA. p. 521-542.

Kaufmann, M. R. (1981). Automatic determination of conductance, transpiration, and environmental conditions in forest trees. *Forest Sci.* 27:817-827.

Kaufmann, M. R. (1982a). Leaf conductance as a function of photosynthetic photon flux density and absolute humidity difference from leaf to air. *Plant Physiol.* 69:1018-1022.

Kaufmann, M. R. (1982b). Evaluation of season, temperature, and water stress effects on stomata using a leaf conductance model. *Plant Physiol.* 69:1023-1026.

Kaufmann, M. R. (1984). A canopy model (RM-CWU) for determining transpiration of subalpine forests. I. Model development. *Can. J. For. Res.* 14:218-226.

Kaufmann, M. R. and F. M. Kelliher. (1990). Estimating tree transpiration rates in forest stands. *In* J. P. Lassoie and T. M. Hinckley (eds.), Techniques and Approaches in Forest Tree Ecophysiology. CRC Press, Boca Raton, FL. (In press.)

Landsberg, J. J. (1986). Physiological Ecology of Forest Production. Academic Press, New York. 198 p.

Lohammer, T., S. Larsson, S. Linder, and O. Falk. (1980). FAST - simulation models of gaseous exchange in Scots pine. *Ecological Bulletin (Stockholm)* 32:505-523.

Massman, W. J. and M. R. Kaufmann. (1990). Stomatal response to certain environmental factors: a comparison of models for subalpine trees in the Rocky Mountains. *Agric. For. Meteorol.* (In press.)

Riha, S. J. and G. S. Campbell. (1985). Estimating water fluxes in Douglas-fir plantations. *Can. J. For. Res.* 15:701-707.

Running, S. W. (1984). Documentation and preliminary validation of H2OTRANS and DAYTRANS, two models for predicting transpiration and water stress in western coniferous forests. *USDA Forest Service Rocky Mtn. For. and Range Exp. Sta. Res. Pap.* RM-252, 45 p.

Smith, W. K. and B. Y. Hollinger. (1990). Stomatal behavior. *In* J. P. Lassoie and T. M. Hinckley (eds.), Techniques and Approaches in Forest Tree Ecophysiology. CRC Press, Boca Raton, FL. (In press.)

Tan, C. S., T. A. Black, and J. U. Nnyamah. (1978). A simple diffusion model of transpiration applied to a thinned Douglas-fir stand. *Ecology* 59:1221-1229.

FIELD USE OF RECORDING
VISCOUS FLOW POROMETERS

Edwin L. Fiscus

United States Department
of Agriculture
Agricultural Research Service
Fort Collins, Colorado

I. INTRODUCTION

The fundamental idea behind mass flow (or viscous flow)
porometry is quite simple. We reason that if we try to force
a given quantity of air, or any other gas, through a leaf,
entering through one surface and exiting through the other,
it will be easier to accomplish this if the stomata are open
than if they are closed. There are several basic ways in
which this concept may be used to estimate the degree of
stomatal opening. To obtain measurements in real resistance
(or conductance) terms it is necessary to have some measure
of the flux of gas through the leaf as well as the pressure
head driving that flux.

Over the years a wide array of ingenious devices has been
developed for that purpose and correlations demonstrated
between mass flow resistance, diffusive resistance, and
direct measurements of stomatal aperture. Most of these
devices are adequately reviewed by Hsiao and Fischer (1975),
Meidner (1981) and Slavik (1974) and we will say no more
about them except that generally they are not well suited for
both continuous monitoring of stomatal aperture and use in
field situations.

If it is desired to obtain only relative measurements of
stomatal opening then it is necessary to measure only the
flux or the pressure head or the rate of change of either.
The requirements for a device to make such measurements are

Measurement Techniques in Plant Science

much less stringent. Therefore such devices are more
adaptable to continuous monitoring in a field situation.
This is in fact the approach we have used for monitoring
stomatal opening as a means of detecting plant water stress.

   In this paper we will describe equipment and the data
analysis used to relate the porometer measurements to stress
levels, and compare this technique with more conventional
measures of water stress.

II. EXPERIMENTAL PROCEDURES

   The porometer cup (Figure 1) was a permanently attached
type which consisted of a soft rubber cushion appressed to
the abaxial surface of the leaf. The cushion was connected
to an air supply through a very fine metering valve and the
gas pressure at the lower surface of the leaf was sensed with
an electronic pressure transducer. The cup was held in place
by a Plexiglas plate placed on the upper surface of the leaf
with an aperture cut in it to allow the free passage of gas.
A CO_2-free mixture of nitrogen and oxygen was used as the gas
supply to the porometers. The pressure was regulated to a
nominal head of 0.7 kPa and the supply was fed simultaneously
to an array of porometers (Figure 2). Pressures measured at
the lower surface of the leaf ranged from nearly zero, when
the stomata were fully open, to the line pressure when they
were closed to the maximum extent possible. The pressure
readings from each leaf were normalized to the maximum
pressure read during a 24-hr period to allow for variability
in installation and individual leaf differences, although
these differences tended to be small. The photosynthetic
photon flux density (PPFD) was the only other environmental
parameter necessary for using the porometers as stress
detectors since we were primarily interested in the stomatal
response to light. A more detailed description of the system
is given by Fiscus et al., (1984a).

   Typical data are shown in Figures 3 and 4 where the
normalized pressure readings are expressed as relative
stomatal closure, the higher pressures indicating tighter
closure. Figure 3 includes data taken on the same day from
plants under 3 different irrigation treatments. Figure 3A
shows the response from a well watered plant, 3B from a plant
in the intermediate stages of water stress, and 3C from a
plant that is stressed to the point of complete stomatal
closure. The plant represented in Figure 3C still is capable

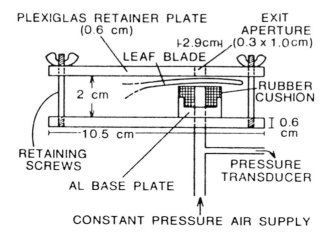

Figure 1. Detail of the mass flow porometer.

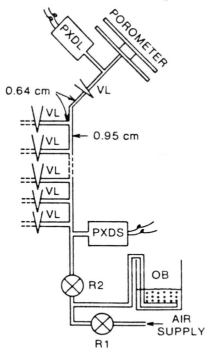

Figure 2.   Manifold system for driving several
porometers from the same air supply.   The needle valves are
set so that this resistance to flow is much larger than the
leaf resistance over most of the range of opening.

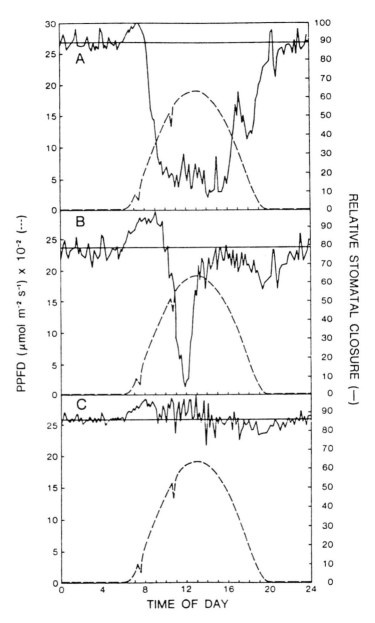

Figure 3. Typical data from three irrigation treatments on the same sunny day. Solid curves are the normalized porometer pressure readings and the dashed curve is the light intensity. Solid straight lines are the baselines.

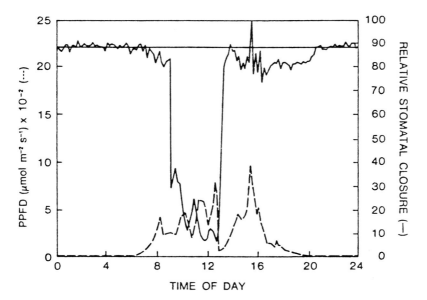

Figure 4. Typical data from a well-watered plant on an overcast day.

of recovery but obviously it is unable to fix much CO_2 under these conditions and will be damaged irreversibly if the stress is allowed to persist.

Figure 4 shows the response from a well watered plant on an overcast day and illustrates the reason why we have to deal with the stomatal response to light as the stress indicator and not just the degree of stomatal opening on any particular day. In this instance the stomata do not stay open during the day, not because the plant is stressed but because there is not enough light to keep them open. Simply monitoring relative stomatal opening therefore is not enough since data from cloudy days might be interpreted as advanced levels of stress. Also, as can be seen in Figures 3A and 3B, single measurements of the more conventional water status indicators, as well as mass flow porometry, taken at the wrong time of day (noon for example) might lead one to believe that both of those plants (Figures 3A and 3B) were equally well off. It is for these reasons that we go to the trouble of integrating the light and the relative stomatal opening each day. The integration of light is performed very simply by summing the intervals under the PPFD curve for each day. This sum is called the integrated light intensity (ILI).

The relative stomatal opening is integrated by taking the
average of the readings between midnight and dawn and using
that value as a baseline. The area below the baseline
(Figures 3 and 4) is then summed over the 24 hr period and
expressed as the integrated stomatal opening (ISO) for that
day. Values of ISO are plotted against the corresponding
values for ILI (Figure 5) and a boundary line drawn to
represent the maximum possible values for ISO at any measured
level of ILI.

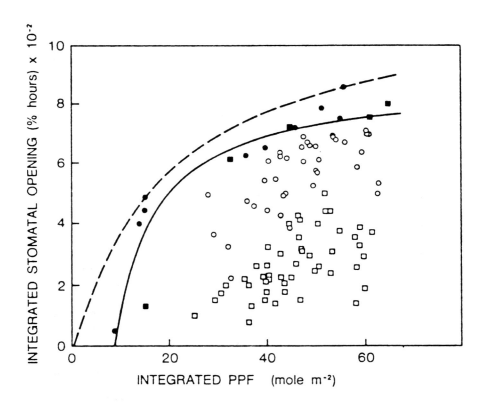

Figure 5. Plot of ISO vs. ILI. Solid line is the
boundary line calculated from a hyperbolic regression of
points near the maximum. Dashed line is the boundary , line
calculated from the double reciprocal plot (Figure 6).

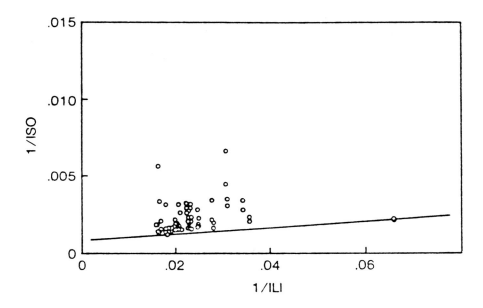

Figure 6. Double reciprocal plot of ILI and ISO. The straight line is drawn between two data points such that all other data have ordinate valves greater than or equal to the line. The equation for this line is the dashed boundary line in Figure 5.

Determining the boundary line for the data has presented something of a problem in the past but the use of a double reciprocal plot of the data (1/ISO vs 1/ILI) seems to take much of the guesswork out of the process. The data are plotted as in Figure 6 and a straight line drawn between two data points picked such that all other data have ordinate values greater than or equal to the line. The equation for this line is $1/ISO_{max} = a + b/ILI$, where a and b are the intercept and slope respectively of the double reciprocal line. Solving this equation for ISO_{max} which represents the maximum possible value for ISO at the measured ILI, yields the envelope shown as the dashed line in Figure 5. Three years of field data have shown that this procedure determines a true boundary line more accurately than our original method

of using a least squares hyperbola determined from a number
of points close to the boundary (solid line in Figure 5).
Particularly, the double reciprocal method seems to be more
realistic at the lowest levels of ILI encountered.

As an indicator of stress, the actual ISO for any day was
calculated from the data and compared to the ISO_{max}
corresponding to the ILI for that same day. The single
number thus resulting from a days data was called the
fractional integrated stomatal opening (FISO) and provides a
measure of plant water stress. The question of how FISO, as
a stress indicator, compares with the more conventional
methods of estimating stress is the subject of the next
section.

III. COMPARISON OF FISO WITH
CONVENTIONAL STRESS INDICATORS

The most widely used measures of plant-water stress
consist of various ways of assessing the water content of
leaves, the energy level of the water in the leaves, or the
degree of stomatal opening. Of these, water content is
little used for field work anymore and we are left with the
other two. The most widely used field method for assessing
the energy content of leaf water is the measurement of leaf
water potential with the Scholander pressure chamber.
Measures of the degree of stomatal opening are now largely
confined to measurement of the diffusion resistance of leaves
with some type of diffusion porometer, preferably a steady
state device. In this section of the paper we will compare
the results from mass flow porometry to measurements of leaf
water potential and diffusion resistance taken on the same
crop over the same period of time as the mass flow data.

A. Leaf Water Potential

Midday (solar noon $\pm$ 1 hr) leaf water potentials taken
on two plots of *Zea mays* in 1982 are shown in Figure 7. One
plot was irrigated regularly in an attempt to maximize
productivity. The other plot was subjected to two cycles of
stress for comparison. The stress cycles were induced by

withholding irrigation water. The first cycle was relieved
by irrigation and precipitation on day 207, while the second
stress period was allowed to progress through the rest of the
season.   Leaf water potentials were measured in each plot
with a pressure chamber.
    Leaf water potentials for the irrigated plot generally
ranged between -1 and -1.4 MPa. While the values for the
stressed plot overlapped that range considerably, they
reached values as low as -1.8 MPa. But, there was never

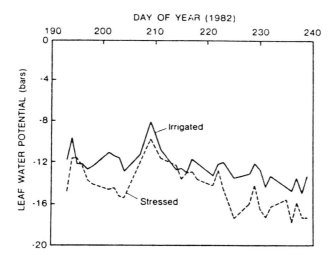

Figure 7.   Midday leaf water potentials from stressed
and irrigated plots.

more than a 0.4 MPa difference between them even under what
appeared to be severe stress conditions, characterized by
stunted vegetative growth, severe leaf curl, and
discoloration.
    For comparison we plotted the difference between the
irrigated and stressed plot leaf water potentials, along with
the fractional integrated stomatal closure (1-FISO) in Figure
8.   In this instance stomatal closure was plotted so the
trends would go in the same direction and make comparison
easier.   The first point of interest in Figure 8 is that
before there was any noticeable difference in the leaf water
potentials the mass flow porometers were already showing a
20% closure. Following this trend, stomatal closure reached
about 50% before there was as much as a 0.2 MPa difference in
leaf water potentials. After the irrigation on day 207 both

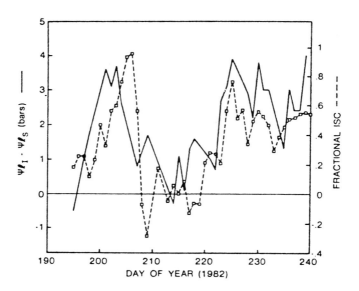

Figure 8.   Comparison of FISO with the difference in
midday leaf water potentials between the stressed and
irrigated plots.

the potential difference and the stomatal closure decreased.
Unfortunately there was a two day gap in the potential data
immediately preceding the irrigation but despite the gap the
trends are clear.

Also, during the first stress cycle the water potential
difference appears to have reached its maximum several days
before complete stomatal closure. Then after irrigation the
potential difference was much slower to reach its minimum
than was the stomatal closure. After the recovery period
there were several days (about 215-220) when the water
potential difference rose by 0.1-0.15 MPa while the stomata
showed full opening. Perhaps this was an indication of some
osmotic adjustment by the stressed plants. We will see other
indications of adjustment when we examine the diffusion
resistances for these same plants.

During the rest of the 1982 season the water potential
difference and the stomatal opening seemed to track each
other, with coincidental peaks and valleys. The same
severity of stress was not repeated during that year,
possibly because the weather was unusually cool and overcast.

B. Diffusion Resistance

Measurements of midday leaf diffusion resistance on both plots are shown with FISO in Figure 9. As with the water potential difference FISO seems already to have dropped to about 75% by the time there was any apparent difference in the diffusion resistances between the stressed and irrigated plots. Unlike the potential data, however, the diffusion resistance recovery follows very closely the recovery in FISO.

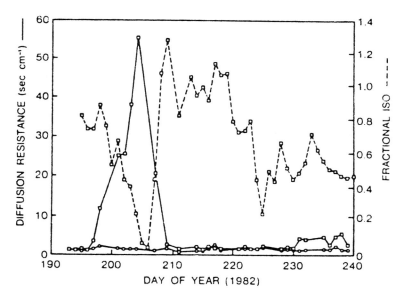

Figure 9. Comparison of FISO with midday leaf diffusion resistance determined with a steady-state porometer.

During the second stress cycle the diffusion resistance remained low and similar to the irrigated plot until 11 days after FISO reached levels that indicated moderate stress. The reasons for this apparent shift in the relationship between mass flow resistance and diffusion resistance are unclear. Perhaps the adjustment of plant water relations due to the first stress cycle involved more than simple osmotic responses. It is possible that the adjustment resulted in an increased internal mass flow resistance or an inability of the stomata to shut down as tightly at night. There are several other alternatives such as a shortening of the

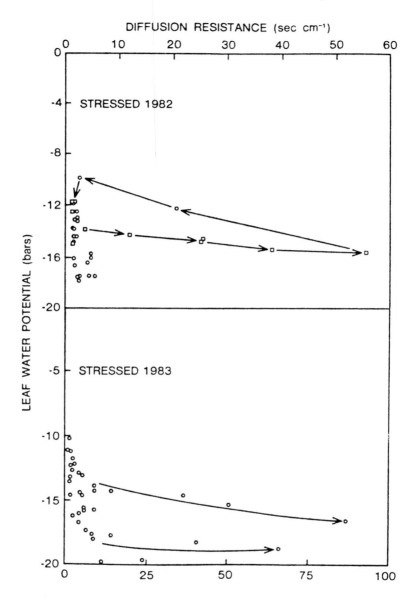

Figure 10. Relationship between midday leaf water potential and midday diffusion resistance for the stressed plots of 1982 (A) and 1983 (B). The second stress cycle in both cases shows apparent osmotic adjustment brought about by the first stress cycle.

diffusion path, a decrease in the stomatal response to light, perhaps involving an increase in the opening thereshold intensity, or a difference in response of the adaxial and abaxial stomata. Further experimentation and analysis is necessary before it will be possible to choose between these alternatives.

The expected adjustment of the relationship between diffusion resistance and leaf water potential occurred as a result of the first stress cycle. Figure 10A shows the data from 1982 when it was not possible to increase the resistance very much during the second stress cycle. Figure 10B from 1983 is included to show that when conditions are favorable for inducing stress during the latter part of the growing season the stomata are still capable of a closing response and that this closure is shifted to lower water potentials after the first stress cycle.

IV.   FISO AS AN IRRIGATION
CONTROL PARAMETER

Because FISO is as good an indicator of plant water stress as either diffusion resistance or leaf water potential measurements, we decided to try using it as the feedback element in an automatic irrigation control system (Fiscus et al., 1984b). The system was implemented as in Figures 11 and 12 with a small microcomputer system with data acquisition and output control capabilities. The plots where the water application was automatically controlled were irrigated with a drip (trickle) system in order to control precisely the application rates and quantities. Although there are distinct advantages to using trickle irrigation the control system should work equally well with any other type of system which could be remotely activated such as electrically operated ditch gates, center pivot systems, or automatic gated pipe. The other plots in the field (Figure 12) were furrow irrigated on an arbitrary schedule. As in previous years the data consisted of porometer pressure and PPFD measurements acquired at 10 minute intervals throughout the growing season. The integrations and calculation of FISO for each of the plots were carried out by the computer immediately after midnight each day. The flow diagram for the control program is shown in Figure 13.

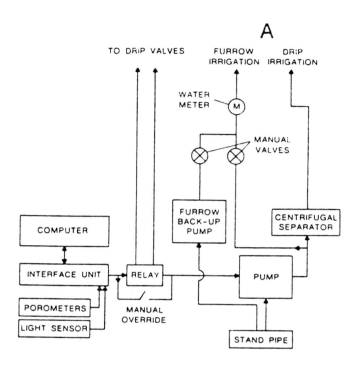

Figure 11.   Diagram of computer interface and control system.

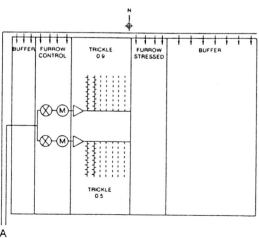

Figure 12.   Layout of field plots for 1983.   Ms are water meters and valves are activated by the computer.

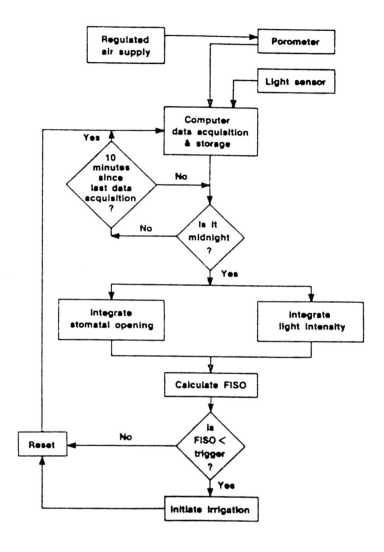

Figure 13. Flow diagram of irrigation control system
program.

After calculation of FISO the computer simply checks the
predetermined irrigation trigger level (that FISO below
which irrigation should be initiated) for each controlled
plot. If the FISO for that day was below the trigger level
then the pump and appropriate valves are activated to carry

out the irrigation.

In addition to the obvious advantages of such a system in terms of water conservation and irrigation efficiency it should be a very useful experimental tool in the study of stress effects in the field. Its usefulness derives from the fact that for the first time it is possible to control the level of stress in a plant population at a wide range of arbitrary levels instead of simply putting the plants through extreme drying cycles by withholding irrigation water. In effect the way the system works is simply to decrease the amplitude of the drying cycles such that variations in the degree and duration of stomatal opening are minimized. This technique should allow more detailed studies of stress adaptive processes as well as the effects of water stress on a wide range of physiological processes.

An indication of the effectiveness of the control system may be seen in Figure 14. Here, we have accumulated (1-FISO) for each day of the active growing season for each plot and expressed it as accumulated stomatal stress days (SSD). One SSD is therefore equivalent to a single 24 hr period during which the stomata do not open at all. For example, four days with FISOs equal to 0.75 would add up to 1 SSD (4 x 1-0.75). Also shown in Figure 14 are the accumulated SSDs from the

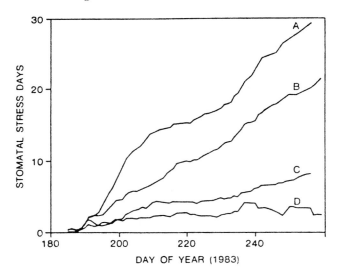

Figure 14. Accumulated stomatal stress days for the four plots of 1983. A and C are furrow irrigated stressed and control plots respectively. B and D are the drip irrigated stressed and control plots.

furrow irrigated plots. The major point to be made here is
that in the furrow irrigated stressed plot the stress cycles
are clearly in evidence. Also, the accumulation of stress in
the well irrigated plot is irregular and somewhat
unpredictable. The computer controlled plots, however, show
both a more regular and predictable accumulation of stress
days.

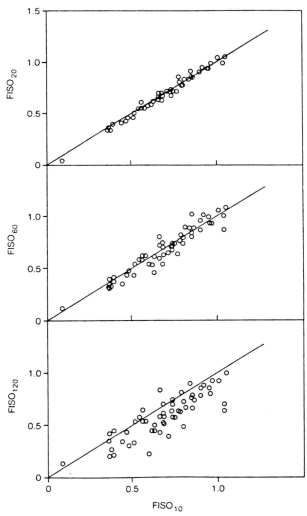

Figure 15. Scatter diagram for FISOs calculated from 3
different timing intervals compared to FISOs calculated at 10
minute intervals. The lines are the 1:1 relationship.

Finally, as a matter of practical significance in
developing an irrigation control system or simply for
monitoring plant stress, it is necessary to know the timing
intervals required to obtain reliable data. It is also an
important consideration in the development of battery
operated instrumentation where power requirements and memory
costs may limit the feasibility of such systems. For this
purpose we analyzed the data from our 1983 field plots at
different time intervals. Figure 15 shows how the scatter in
the data increases as the time interval between readings is
extended from 10 to 120 minutes. Figure 16 shows the r^2
values for both stressed and irrigated plots as functions of
the timing interval. In both cases the r^2 declined
relatively slowly up until the 60 minute interval after which
the rate of decline accelerated indicating decreasing
reliability of the average FISO for intervals longer than 60
minutes.

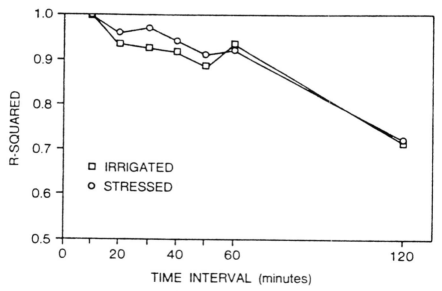

Figure 16. Decreasing data reliability with increasing
sampling intervals for the stressed and irrigated plots of
1983.

The mean FISOs for 1983 are shown in Figure 17. Clearly,
it makes little difference in this instance if the data
acquisition interval is increased from 10 to 60 minutes.
Beyond 60 minutes, however, the mean FISO shows a significant

decline.

Although the mean FISOs for the whole season appear relatively insensitive to the timing interval up to 60 minutes the degree and nature of the increased data scatter dictate caution when using the system for irrigation control, especially if the objective is to minimize water use and irrigation costs. This fact is illustrated by Figure 18 which shows the amount of irrigation water which would have been applied based on the FISOs calculated at different timing intervals. Maximum conservation cannot be achieved in the well irrigated plot with timing intervals greater than about 40 minutes. In fact, if the interval had been extended to 2 hours the system would have applied more than twice the amount of water necessary to achieve good production. The reason for this apparently is that the increased data scatter resulting from the longer timing intervals increases the probability that the measured FISO will fall below the trigger point.

The conclusion to be drawn from the timing interval data is that the interval used must be determined to a very large extent by what objective is desired. If one wishes to study transient stomatal responses then readings must be taken many times per minute. If monitoring accumulated stress is all that is desired then reasonable accuracy may be achieved with intervals as great as 60 minutes. However, good irrigation control requires intervals of 40 minutes or less.

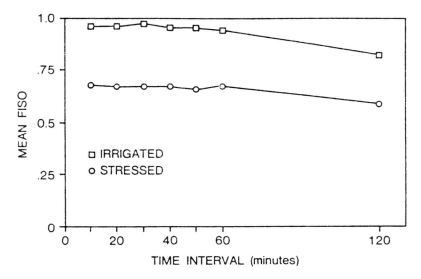

Figure 17.   Mean seasonal FISOs for 1983.

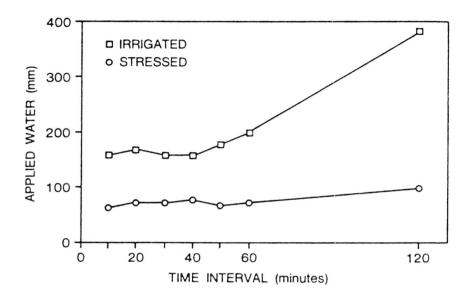

Figure 18. Effect of sampling interval on the quantity
of water which would have been applied by the automatic
irrigation control system. Grain yields on the two plots
were 146 bushels/acre (irrigated) and 110 bushels/acre
(stressed).

V.   SUMMARY

    In summary, the mass flow porometer is a useful technique
for the detection and monitoring of water stress in the field.
It compares favorably with the more traditional measures of
stress such as leaf water potential and stomatal diffusion
resistance. The instruments described here are well suited
for continuous monitoring of stomatal aperture and therefore
can be used as the feedback elements in a totally automated
irrigation control system. Such a control system allows, for
the first time, the growth of experimental field plants under
controlled steady levels of water stress.

REFERENCES

Fiscus, E.L., Wullschleger, S.D., and Duke, H.R. (1984a).
    Integrated stomatal opening as an indicator of water
    stress in *Zea*. *Crop Sci*. 24:245-249.
Fiscus, E.L., Wullschleger, S.D., and Duke, H.R. (1984b).
    Stomatal sensors control the water supply to *Zea mays.In*:
    Agricultural Electronics - 1983 and Beyond - Vl. I.
    Proc. Natl. Conf. on Agric. Elec. Appl., Amer. Soc. Ag.
    Eng., St. Joseph, pp. 278-285.
Hsaio, T.C. and Fischer, R.A. (1975). Mass flow porometers.
    In: Measurement of stomatal aperture and diffusive
    resistance. Bull. 809, Coll. Agric. Res. Ctr.,
    Washington State Univ. pp. 5-11.
Meidner, H. (1981). Measurements of stomatal aperture and
    responses to stimuli. In: Stomatal Physiology. P.G.
    Jarvis and T.A. Mansfield (eds). Cambridge Univ. Press,
    Cambridge, London, New York. pp. 26-49.
Slavik, B. (1974). Methods of Studying Plant Water
    Relations. Springer-Verlag, New York, Heidelberg, Berlin.
    pp. 314-325.

APPENDIX

RSC:	Relative Stomatal Closure = $(P/Pmax)100$, where P is the measured pressure at the lower surface of the leaf, Pmax is the maximum pressure observed during a 24 hr period indicating the greatest closure during that period. See figures 3 and 4.
RSO:	Relative Stomatal Opening = $100 - RSC$
ILI:	Integrated Light Intensity = $\int_{0}^{24\ hr} PPFD$
ISO:	Integrated Stomatal Opening = $\int_{0}^{24\ hr} RSO - RSOref$
RSOref:	RSO reference or baseline = mean of RSO between midnight and dawn.
ISOmax:	Maximum possible ISO for any ILI. This is the boundary line.
FISO:	Fractional Integrated Stomatal Opening = $ISO/ISOmax$
FISC:	Fractional Integrated Stomatal Closure = $1 - FISO$

SSD: Stomatal Stress Day; equivalent to one day when
                  FISC = 1. Fractions of FISC may be accumulted
                  over time and the sum is an indicator of
                  accumluted stress.

WATER POTENTIAL AND ITS COMPONENTS IN GROWING TISSUES

John S. Boyer
Hiroshi Nonami[1]

College of Marine Studies and Department of Plant Sciences
University of Delaware
Lewes, DE 19958

I. INTRODUCTION

Thermodynamically sound methods are available for measuring the water status of plant cells and tissues (Scholander et al., 1965; Boyer and Knipling, 1965; Hüsken et al., 1978). Rather than measuring water content, they indicate the chemical potential of water and have the advantage that the water status is defined in physical terms not affected by the plant or its history. This allows experiments to be compared, which is of the utmost importance for studies involving water whose supply can vary over even short times. Other methods do not generally allow this kind of comparison because they are defined in biological terms that can differ between experiments.

The methods of measuring chemical potential rely on equilibrating the pressure or vapor pressure of a sensor with that in the cell or tissue. No net transport occurs in the measured system or between the system and sensor. Thermodynamically, this means that the sensor has a chemical potential that is the same as that in the measured system and, because the sensor potential is known, the system potential is known. With this approach, very accurate measurements are possible.

---

[1]Current Address: Department of Biomechanical Systems, College of Agriculture, Ehime University, Tarumi, Matsuyama 790, Japan

Measurement Techniques in Plant Science

When plant tissue is growing, however, the measurements become more complex (Boyer, 1985). Intact tissues can grow during the measurement and thus water is absorbed, which indicates that potential gradients are present in the tissue and a true equilibrium cannot be achieved. When these tissues are excised, no water can enter and growth stops. Although internal gradients then can equilibrate, the potential of the tissue may change as a result of the excision. Moreover, when growing and nongrowing tissues are excised together, water is transferred toward the growing region from the nongrowing tissue.

These effects, while complicating the measurement of water status in growing tissues, can be used with good effect not only to determine the water status of growing tissue but also to study the growth process itself. From the growth work, we have learned that there are growth-induced potentials and that cell enlargement depends on them to a considerable degree. It also has been possible to monitor the physicochemical properties of cell walls that are important for cell enlargement. The principles emerging from this work form the basis of our understanding of cell enlargement but also suggest ways to improve the accuracy of measurements of the potential of growing tissue. The following review provides a brief summary of these principles and of the measurement methods.

II. THE PROCESS OF CELL ENLARGEMENT

Plants increase in size mostly because cell water content increases (see Cleland, 1971; Taiz, 1984; Boyer, 1985 for reviews). The effect results from the enlargement of individual cells as water enters, driven by osmotic forces. Enlargement continues until the cells reach a maximum size whereupon enlargement ceases. As a consequence, sustained growth requires a steady supply of new cells and they are produced in localized meristems where cell division takes place. In these regions, modest cell enlargement occurs because the daughter cells must increase in size before they can divide again. Some of the daughter cells do not redivide and instead continue to enlarge until the cells become many times larger than the meristematic cells. The resulting enlargement displaces the cells from the meristematic region into a region where only enlargement is occurring. As a result, most growing plant organs have zones of meristematic activity and zones of enlarging activity, and the most rapid growth occurs in the enlarging zone.

The osmotic force that drives enlargement is caused by the concentration difference across the plasma membrane, the solution inside the cell being more concentrated than that outside. Water enters in response to the concentration difference and generates a hydrostatic pressure, the turgor, internally as the constraining effect of the cell wall begins to operate. The turgor counters the tendency for further water to enter but it also stretches the wall in a direction determined by the molecular orientation of the wall polymers (Green et al., 1971; Taiz, 1984). With large turgors, the wall stretches irreversibly to a larger size as the wall yields and the polymers slip past each other.

The yielding of the wall keeps turgor from building as high as it otherwise would (Boyer, 1968; Nonami and Boyer, 1987). As a consequence, the turgor does not completely oppose the tendency for water to enter and a difference in chemical potential is generated across the plasmalemma during the growth process. For simplicity, we will express the chemical potential and its components in terms of the pressure equivalent, the water potential (Ψ_w), so that the difference across the plasmalemma is shown by:

$$\Delta \Psi_w = \Delta \Psi_p + \sigma \Delta \Psi_s \tag{1}$$

where Ψ_p is the pressure (inside the cell, the pressure is the turgor; outside, the pressure often is a tension), Ψ_s is the osmotic potential, σ is the reflection coefficient which approaches unity for most solutes involved in metabolism, and Δ indicates the difference across the membrane. The $\Delta \Psi_w$ is termed a growth-induced Ψ_w because it is generated by the yielding of the walls during growth (Molz and Boyer, 1978).

For isolated cells, the $\Delta \Psi_w$ is very small but for complex tissues, $\Delta \Psi_w$ across the tissue may be quite large because water must traverse many cells as it travels from the vascular supply to the farthest cells during growth (Molz and Boyer, 1978).

As the cells enlarge, the incoming water dilutes the solution in the cell interior. The solutes are mostly in the form of small substrate molecules such as sugars and amino acids together with some inorganic salts, and they are used extensively for respiration and for synthesizing new polymers. These activities remove solute from the cell solution and the osmotic force quickly runs down. As a consequence, new solutes must be imported continuously.

The requirement for solute and water uptake as well as a yielding cell wall indicate that growth relies on interlinked processes that must occur at balanced rates (Boyer, 1985). The failure of any one leads to a slowdown or cessation of growth even if the others could support rapid growth. Because growth will not occur without osmotic forces sufficient for water entry, the Ψ_w and its components are the fundamental driving forces and growth cannot be understood without them.

III. MEASURING THE WATER STATUS OF CELLS AND TISSUES WITHOUT THE COMPLICATIONS OF GROWTH

When growth is not occurring, the measurement of plant water status is relatively straightforward. Three instruments are available that can provide the measurements: the thermocouple psychrometer, the pressure chamber, and the pressure probe. They are based on developing a vapor pressure (psychrometer) or pressure (pressure chamber and pressure probe) that equals the vapor pressure or pressure displayed by water in the cells. Thus they are equilibrium methods (for reviews, see Boyer, 1969; Zimmermann and Steudle, 1978). Because no net transfer occurs between the sensor and the cells, the conductance for water transport does not affect the measurement. For vapor pressure methods, this eliminates the effect of the epidermis on the diffusion of water vapor between the sensor and the tissue. For pressure methods, it eliminates pressure differences in various parts of the plant and measurement system.

The psychrometer is the most versatile of the methods because it can measure the water status of virtually any plant part or soil in terms of the Ψ_w or any of its components (Eq. 1). It detects the vapor pressure of the sample by exposing a solution of known vapor pressure to an atmosphere in a chamber into which the sample is sealed (Boyer and Knipling, 1965). The solution that neither loses nor gains water from the atmosphere has a vapor pressure and thus a potential identical to that of the sample. This method, termed isopiestic psychrometry (Boyer and Knipling, 1965), typically employs a thermocouple to measure the temperature of the solution. The solution that neither heats nor cools the thermocouple is neither gaining nor losing water. Knowing the potential of the solution thus identifies the potential of the water in the sample.

The psychrometer can be used as an osmometer to measure the osmotic potential of solutions extracted from the cells. With the Ψ_w data, the relation $\Psi_w = \Psi_p + \Psi_s$ allows the turgor to be calculated.

There are other variants of this method that use water on the thermocouple instead of solutions, and calibrate the rate of water transfer (Spanner, 1951) or dewpoint (Neumann and Thurtell, 1972). These are not equilibrium techniques and suffer from errors caused by diffusive conductances to water vapor movement (Boyer and Knipling, 1965; Shackel, 1984). In a few cases, the errors are small enough to be tolerated (Boyer and Knipling, 1965) but in others they may be large. Especially in tissues with thick cuticles and low conductances for water vapor, the errors should not be ignored. In addition, when tissue is growing, small differences in potential are common and the higher accuracy (Boyer, 1966) of equilibrium (isopiestic) psychrometry is essential.

Enclosing nongrowing tissue in a vapor chamber inevitably eliminates transpiration and collapses Ψ_w gradients that had been moving water through the tissue to replace the losses from transpiration. When the tissues remain intact and attached to the plant, the cells equilibrate with the Ψ_w of the xylem solution. When the tissues are excised, the cells equilibrate with themselves because they are disconnected from the xylem. This means that excised tissue will display a Ψ_w that differs from that of the intact tissue, although the difference is usually not large. Interpretations of the data need to reflect the differences, however.

The second method uses a pressure chamber and is the best method for field conditions although it requires excised tissue (Scholander et al., 1965; Boyer, 1967). The plant part, usually a leaf or branch, is placed into the chamber and sealed into the top so that the cut end extends outside the chamber. Pressure is applied, and water moves out of the cells and into the xylem until liquid appears at the cut surface. When the pressure is adjusted so that no liquid moves in or out of the tissue, the xylem solution has returned to the same position it held in the intact plant, and the pressure counteracts the tension normally present in the xylem. For simplification, solute effects often can be ignored because the concentrations are usually small in the xylem solution (Scholander et al., 1965; Boyer, 1967). The counteracting pressure then can be considered to be the Ψ_w of the xylem solution. Because the measurement is made with a pressure that causes no flow, it is an equilibrium measurement and the Ψ_w of the xylem solution equals the Ψ_w of the surrounding tissues (Boyer, 1967). However, the most accurate measurements require the osmotic potential of the xylem solution to be determined and added to the equilibrium pressure to give a truer Ψ_w of the xylem solution and surrounding tissues (Boyer, 1967).

A variant of this method (Waring and Cleary, 1967) uses the pressure chamber to detect the first appearance of xylem solution at the cut surface when pressure rises in the chamber at a controlled rate. This approach suffers from the lack of equilibrium between the tissue and the applied pressure. As a consequence, the pressure reading at the endpoint is often considerably different from the balancing or equilibrium pressure, and errors may result.

The third instrument for measuring plant water status is the pressure probe, an oil-filled capillary whose tip can be inserted into individual cells of intact plants or plant parts (Hüsken et al., 1978). It detects the turgor in the cells by pressurizing the oil until the meniscus between the oil and cell solution is returned to its original position before inserting the tip into the cell. The pressure inside the capillary is then equal to that in the cell. Although the cell wall and membranes must be penetrated to make the measurements, there appears to be little effect on the turgor readings (Nonami et al., 1987). It is the only single-cell method available and because it measures turgor directly, it is especially important for growth studies (e.g., Cosgrove and Steudle, 1981; Nonami and Boyer, 1987; 1989).

The pressure probe has also been used recently (Shackel, 1987; Nonami and Schulze, 1989) for microsampling the solution inside single cells. The sample is placed under oil and observed while it is frozen. From the freezing point, the osmotic potential of the solution can be calculated. Thus, measurements of the turgor and osmotic potential are possible from single cells, and the Ψ_w can be calculated from $\Psi_p + \Psi_s$.

IV. MEASURING THE WATER STATUS
OF GROWING CELLS AND TISSUES

Several early attempts were made to measure the water status of growing plant parts by using osmotica (Bennett-Clark, 1956; Brouwer, 1963; Burström, 1953; Burström et al., 1967; Cleland, 1959; 1967; Ordin et al., 1956), but the methods gave only indirect estimates and were subject to the side effects of solute exchanges with the tissues. The first direct measurements were made in intact tissues using a thermocouple psychrometer (Boyer, 1968). Leaf Ψ_w were 0.15 to 0.25 MPa below those in the xylem and were associated with water uptake in support of expansion. These Ψ_w did not appear in older leaves that were no longer expanding (Boyer, 1974). It was proposed that Ψ_w below those of the xylem were involved in transporting water into the enlarging cells and were generated by the yielding of the cell walls during the growth process (Boyer, 1968).

The Ψ_w associated with growth raised the question of why such low Ψ_w were required when the same leaves could move water many times faster during transpiration and display Ψ_w that were only slightly lower. The answer appeared to be that the transpirational water bypassed many of the leaf cells probably by evaporating into the intercellular spaces (Boyer, 1974; Boyer, 1985; Nonami and Schulze, 1989). In contrast, water had to enter all the cells for growth. Thus, the frictional resistances were less for the transpiration path than for the growth path and smaller $\Delta\Psi_w$ were required to move a unit of water for transpiration than for growth.

A model of water transport in stems supported this idea (Molz and Boyer, 1978). Using published values for the hydraulic conductivity of plant cells, the kinetics of water movement during growth could be predicted for the whole tissue. The model and kinetics predicted Ψ_w that averaged about 0.2 MPa below that of the xylem in growing tissue, which was confirmed by direct measurement with a psychrometer.

Measurements with a pressure probe also supported the concept of a consistently lower turgor in growing cells than would be expected if the same cells were not growing (Cosgrove and Steudle, 1981). In a related study, however, large amounts of solute were found (Cosgrove and Cleland, 1983) in exudates that were considered to be from the cell wall solution (the apoplast). These investigators used excised tissue pieces to which pressure or centrifugal force was applied to collect exudates. Also seedlings were infiltrated with water to extract solutes from the cell walls. It was proposed (Cosgrove and Cleland, 1983) that this solute (to 120 mM or an osmotic potential of -0.3 MPa) decreased the apoplast Ψ_s sufficiently to account for the low Ψ_w of the cells.

Results with the pressure chamber were not in agreement with this conclusion (Scholander et al., 1965; Boyer, 1967). The pressure chamber required large pressures to balance the negative pressures in the xylem, and these would not be expected if the apoplast contained a concentrated solution. Moreover, the Ψ_s of the xylem solution was near zero in many plants and conditions (Scholander et al., 1965; Boyer, 1967). If large concentrations of solute were present in the apoplast, they should also be present in the xylem solution. A recent detailed analysis of the solutes released to the xylem by the apoplast showed that the concentration was indeed low (about 8mM, Jachetta et al., 1986).

When the Cosgrove and Cleland (1983) experiments were repeated but with intact noninfiltrated seedlings, the high solute concentrations were not observed (Nonami and Boyer, 1987). Instead, concentrations were about 15mM (about -0.04 MPa) in agreement with the measurements in leaves (Scholander et al., 1965; Boyer, 1967; Jachetta et al., 1986). Thus, most of the solute seen by Cosgrove and Cleland (1983) appeared to have been released from cut cells or leaky membranes.

If the growth-induced Ψ_w of the cell is not balanced by solute in the cell wall, what other factors might be involved? The answer was provided by pressure chamber measurements which indicated that most of the force acting on water in the apoplast was a tension (Nonami and Boyer, 1987). This conclusion was based on the fact that the pressure in the chamber balanced tensions rather than osmotic potentials, as described above (Section III). Evidently, the growth-induced Ψ_w was transmitted to the wall solution as a tension. As a consequence, water would be drawn out of the xylem whenever the tension in the surrounding cell walls exceeded that in the xylem, and this water would feed cell enlargement.

The only exception to this concept seems to be the high apoplast concentrations at the base of developing seeds. Here, the embryo and storage tissues are not connected to the vascular system of the maternal plant. The phloem releases its solute, and concentrations can be quite high. Concentrations of 500mM have been reported (Maness and McBee, 1986).

From these concepts, it might be expected that the turgor would always be lower in growing cells than in nongrowing cells. This was not the case, however, when turgor was compared in the two kinds of cells (Cavalieri and Boyer, 1982). The turgor was the same even though growth-induced Ψ_w were present in the enlarging cells but not in the mature cells. However, the accuracy of these measurements was questioned because of the use of excised tissues (Cosgrove et al., 1984). There was concern that the turgor might change after excision. Indeed, the theory of cell enlargement predicts that turgor should decrease if the water supply is removed (Cosgrove et al., 1984; Boyer et al., 1985). The cell wall would relax as it yielded without water entry, and the turgor would diminish to the yield threshold where yielding would stop and the turgor would become stable.

Tests with a pressure probe indeed showed that turgor dropped when growing tissue was excised (Cosgrove et al., 1984; Cosgrove, 1985; 1987). The decrease was slow and large. A new stable value was reached after about 5 hours and was considered to be the yield threshold for the cell walls, i.e., the turgor below which wall yielding would not occur. A theory was developed to describe the kinetics of the decrease (Cosgrove, 1985; Ortega, 1985).

The implication of this work was that the large changes in potential brought about by excision would make excised tissue unsuitable for measuring the water status of plant growing regions. However, in earlier work, the water status of intact and excised tissues had frequently been compared and found to be similar (Boyer, 1968; Boyer and Wu, 1978; Cavalieri and Boyer, 1982). This puzzling contradiction was made more curious by markedly different psychrometer measurements that were reported (Boyer et al., 1985) soon after the first pressure probe work (Cosgrove et al., 1984). There was little change in turgor when the tissue was excised and the change was completed in only 5 min.

This contradiction led to careful comparisons of the turgor measured with the two instruments. The comparisons gave very similar results (Nonami et al., 1987). Thus, the source of the discrepancy was sought elsewhere.

Eventually, it was found (Matyssek et al., 1988) that the large slow relaxations observed by Cosgrove et al. (1984), and Cosgrove (1985, 1987) were measured in tissue having mature or slowly growing tissue attached. The relaxation was delayed because water was transported from the attached tissues to the enlarging tissues. In effect, although the tissue was excised, the water supply had not been totally removed and cell wall relaxation did not immediately occur. After 5 hours, the yield threshold had itself undergone a change and when relaxation did occur, it was much larger than in the freshly excised growing tissues. The main evidence to support this concept was that removal of the mature or slowly growing tissue removed the delayed relaxation, which then became rapid and small. The subsequent results agreed: turgor decreased rapidly but only by a small amount (Matyssek et al., 1988).

A detailed comparison of Ψ_w and turgor measurements recently was reported for growing tissue (Nonami and Boyer, 1989). Four techniques were used involving thermocouple psychrometry with excised tissue, psychrometry with intact tissue, a pressure chamber with excised tissue, and a pressure probe with intact tissue. The methods gave similar results regardless of whether excised or intact tissue was used. The variation between samples was larger than the changes caused by excision. Thus, although cell wall relaxation caused the Ψ_w and turgor to decrease to the yield threshold, the effects so far have been too small (less than 0.1 MPa) to affect most interpretations of the data.

V. RECOMMENDATIONS FOR MEASURING THE WATER STATUS OF GROWING CELLS AND TISSUES

These findings provide some guidelines for evaluating the water status of growing cells based on principles that are consistent with our theoretical understanding of how growth occurs. Predictions made from theory thus help design appropriate measurements.

A. Mature Tissue

The main factor affecting the measurements is the rate at which water moves through the tissue to replace transpirational losses. If movement is negligible, the tissue Ψ_w in intact plants reflects that of the water supply in the xylem. If tissue is excised from the intact plant in this condition, the Ψ_w of the excised tissue is the same as that of the intact tissue because the Ψ_w was equilibrated in the intact plant before sampling (Cavalieri and Boyer, 1982; Boyer et al., 1985; Matyssek et al., 1988). On the other hand, if transpiration is occurring in the intact plant, Ψ_w gradients are present. These equilibrate internally when the tissue is excised provided that one prevents further transpiration. The Ψ_w

obtained is then a volume-averaged Ψ_w for the gradient included in the sample. The isopiestic psychrometer, pressure probe and pressure chamber give comparable values with this tissue (Boyer, 1967; Nonami et al., 1987; DeRoo, 1969). For intact transpiring tissue, the pressure probe probably gives the most reliable measurements because it does not perturb the potential gradients. When the psychrometer encloses this tissue, it prevents transpiration and collapses the potential gradients. The water status then reflects that in the xylem.

For the greatest accuracy, care should be taken to make the measurements at equilibrium with each technique. Calibrated psychrometers, or pressure probes not having balancing pressures, or pressure chambers based on the first appearance of xylem solution will give values subject to considerable error.

B. Enlarging Tissue

The Ψ_w of enlarging tissue is lower than it would be if growth were not occurring. This is because the walls yield rapidly enough to prevent turgor from developing fully, and the Ψ_w is thus kept below that in the xylem. The low Ψ_w is usually not caused by concentrated solutions in the apoplast although small amounts of solute exist there. The exception appears to be in immature seeds where the contents of the phloem are released to the apoplast and concentrations can be quite high. In other enlarging tissues, the low Ψ_w in the cell walls mostly represents a tension that forms as the low Ψ_w of the protoplast pulls water from the wall into the protoplast. The faster the enlargement, the lower is Ψ_w and the greater is the pull.

The preferred measurements are in intact tissue where the water status reflects that actually present during growth. The psychrometer and pressure probe can give these kinds of data (Boyer et al., 1985; Nonami and Boyer, 1989; Shackel et al., 1987). The pressure chamber cannot and, if the excised tissue consists only of enlarging cells, some decrease in Ψ_w and turgor should be expected. In many cases, the decrease is small enough to have little effect on the data but in others the decrease will need to be taken into account.

C. Excised Tissue With Mature or Slowly Growing Tissue Attached

The principles enumerated above often make this approach useful for measuring the water status of excised tissue. Mature and rapidly growing tissues are excised together and, because the Ψ_w of the enlarging tissue is below that of the attached surrounding tissues, water moves from the surrounding tissues to the enlarging tissue and delays wall relaxation. Thus, for a time (usually at least an hour), the Ψ_w and turgor of the enlarging tissue are the same as in the intact plant.

This is particularly important for measurements with a pressure chamber, which must use excised tissues. The measurement is done by exposing only the enlarging tissue to the pressure and using the mature tissue to form the seal in the top of the chamber. The Ψ_w and turgor then are close to those in the intact plant for at least an hour (Nonami and Boyer, 1987).

Slowly growing or mature tissue can help measurements with a psychrometer as well. The slowly growing or mature tissue is included with the enlarging tissue and slowly transfers water to the enlarging tissue during the measurement. Accordingly, the water status indicates an average for these tissues (both the mature and enlarging tissues are exposed to the vapor atmosphere in this case), but the water status of the tissue remains reasonably stable for long enough to complete the measurement. The average value that is obtained should be close to the average for the same tissues when intact, although there will be some tendency for the average to be lower because of the gradual dehydration of the mature tissue. This effect tends to cause the water status to approach that of the enlarging tissue. After long times, the potentials in the mature and enlarging tissues will decrease further as a slow wall relaxation begins after the water in the mature tissue is depleted (Baughn and Tanner, 1976; Cosgrove et al., 1984). The measurements should be completed before this relaxation occurs

REFERENCES

Baughn, J.W., Tanner, C.B. (1976). Excision effects on leaf water potential of five herbaceous species. Crop Sci. 16:184-190.

Bennett-Clark, T.A. (1956). Salt accumulation and mode of action of auxin. A preliminary hypothesis. In: The Chemistry and Mode of Action of Plant Growth Substances, R.L. Wain and F. Wightman, Eds., Academic Press, Inc., New York, pp. 284-291.

Boyer, J.S. (1966). Isopiestic technique: measurement of accurate leaf water potentials. Science 154:1459-1460.

Boyer, J.S. (1967). Leaf water potentials measured with a pressure chamber. Plant Physiol. 42:133-137.

Boyer, J.S. (1968). Relationship of water potential to growth of leaves. Plant Physiol. 43:1056-1062.

Boyer, J.S. (1969). Measurement of the water status of plants. Ann. Rev. Plant Physiol. 20:351-364.

Boyer, J.S. (1974). Water transport in plants: mechanism of apparent changes in resistance during absorption. Planta 117:187-207.

Boyer, J.S. (1985). Water transport in plants. Ann. Rev. Plant Physiol. 36:473-516.

Boyer, J.S., Cavalieri, A.R., Schulze, E.D. (1985). Control of cell enlargement: effects of excision, wall relaxation, and growth-induced water potentials. Planta 163:527-543.

Boyer, J.S., Knipling, E.B. (1965). Isopiestic technique for measuring leaf water potentials with a thermocouple psychrometer. Proc. Natl. Acad. Sci. 54:1044-1051.

Boyer, J.S., Wu, G. (1978). Auxin increases the hydraulic conductivity of auxin-sensitive hypocotyl tissue. Planta 139:227-237.

Brouwer, R. (1963). The influence of the suction tension of the nutrient solutions on growth, transpiration and diffusion pressure deficit of bean leaves (Phaseolus vulgaris). Acta Bot. Neerl. 12:248-261.

Burström, H.G. (1953). Studies on growth and metabolism of roots. IX. Cell elongation and water absorption. Physiol. Plant. 6:262-276.

Burström, H.G., Uhrström, I., Wurscher, R. (1967). Growth, turgor, water potential, and Young's modulus in pea internodes. Physiol. Plant. 20:213-231.

Cavalieri, A.J., Boyer, J.S. (1982). Water potentials induced by growth in soybean hypocotyls. Plant Physiol. 69:492-496.

Cleland, R. (1959). Effect of osmotic concentration on auxin-action and on irreversible and reversible expansion of the Avena coleoptile. Physiol. Plant. 12:809-825.

Cleland, R. (1967). A dual role of turgor pressure in auxin-induced cell elongation in Avena coleoptile. Planta 77:182-191.

Cleland, R.E. (1971). Cell wall extension. Annu. Rev. Plant Physiol. 22:197-222.

Cosgrove, D.J. (1985). Cell wall yield properties of growing tissue: evaluation by in vivo stress relaxation. Plant Physiol. 78:347-356.

Cosgrove, D.J. (1987). Wall relaxation in growing stems: comparison of four species and assessment of measurement techniques. Planta 171:266-278.

Cosgrove, D.J., Cleland, R.E. (1983). Solutes in the free space of growing stem tissues. Plant Physiol. 72:326-331.

Cosgrove, D.J., Steudle, E. (1981). Water relations of growing pea epicotyl segments. Planta 153:343-350.

Cosgrove, D.J., Van Volkenburgh, E., Cleland, R.E. (1984). Stress relaxation of cell walls and the yield threshold for growth: demonstration and measurement by micropressure probe and psychrometer techniques. Planta 162:46-54.

DeRoo, H.C. (1969). Leaf water potentials of sorghum and corn, estimated with the pressure chamber. Agron. J. 61:969-970.

Green, P.B., Erickson, R.O., Buggy, J. (1971). Metabolic and physical control of cell elongation rate. In vivo studies in Nitella. Plant Physiol. 47:423-430.

Hüsken, D., Steudle, E., Zimmermann, U. (1978). Pressure probe technique for measuring water relations of cells in higher plants. Plant Physiol. 61:158-163.

Jachetta, J.J., Appleby, A.P., Boersma, L. (1986). Use of the pressure vessel to measure concentrations of solutes in apoplastic and membrane-filtered symplastic sap in sunflower leaves. Plant Physiol. 82:995-999.

Maness, N.O., McBee, G.G. (1986). Role of placental sac in endosperm carbohydrate import in sorghum caryopses. Crop Science 26:1201-1207.

Matyssek, R., Maruyama, S., Boyer, J.S. (1988). Rapid wall relaxation in elongating tissues. Plant Physiol. 86:1163-1167.

Molz, F.J., Boyer, J.S. (1978). Growth-induced water potentials in plant cells and tissues. Plant Physiol. 62:423-429.

Neumann, H.H., Thurtell, G.W. (1972). A Peltier cooled thermocouple dewpoint hygrometer for in situ measurements of water potentials. In: Psychrometry in Water Relations Research, ed. R.W. Brown, B.P. Van Haveren, pp. 103-112, Logan: Utah State University, Utah Agricultural Experiment Station, 342 pp.

Nonami, H., Boyer, J.S. (1987). Origin of growth-induced water potential: solute concentration is low in apoplast of enlarging tissues. Plant Physiol. 83:596-601.

Nonami, H., Boyer, J.S. (1989). Turgor and growth at low water potentials. Plant Physiol. 89:798-804.

Nonami, H., Boyer, J.S., Steudle, E.S. (1987). Pressure probe and isopiestic psychrometer measure similar turgor. Plant Physiol. 83:592-595.

Nonami, H., Schulze, E.-D. (1989). Cell water potential, osmotic potential, and turgor in the epidermis and mesophyll of transpiring leaves. Planta 177:35-46.

Ordin, L., Applewhite, T.H., Bonner, J. (1956). Auxin-induced water uptake by Avena coleoptile sections. Plant Physiol., 31:44-53.

Ortega, J.K.E. (1985). Augmented growth equation for cell wall expansion. Plant Physiol. 79:318-320.

Scholander, P.F., Hammel, H.T., Bradstreet, E.D., Hemmingsen, E.A. (1965). Sap pressure in vascular plants. Science 148:339-346.

Shackel, K.A. (1984). Theoretical and experimental errors for in situ measurements of plant water potential. Plant Physiol. 75:766-772.

Shackel, K.A. (1987) Direct measurement of turgor and osmotic potential in individual epidermal cells: Independent confirmation of leaf water potential as determined by in situ psychrometry. Plant Physiol. 83:719-722.

Shackel, K. A., Matthews, M.A. & Morrison, J.C. (1987). Dynamic relation between expansion and cellular turgor in growing grape (Vitis vinifera L.) leaves. Plant Physiol. 84:1166-1171.

Spanner, D.C. (1951). The Peltier effect and its use in the measurement of suction pressure. J. Exptl. Botany 2:145-168.

Taiz, L. (1984). Plant cell expansion: regulation of cell wall mechanical properties. Annu. Rev. Plant Physiol. 35:585-657.

Waring, R.H., Cleary, B.D. (1967). Plant moisture stress: evaluation by pressure bomb. Science 155:1248-1254.

Zimmermann, U., Steudle, E. (1978). Physical aspects of water relations of plant cells. Adv. Bot. Res. 6:45-117.

METHODS FOR STUDYING WATER RELATIONS OF PLANT CELLS AND TISSUES

Ernst Steudle

Lehrstuhl Pflanzenökologie
Universität Bayreuth
Federal Republic of Germany

I. INTRODUCTION

In many physiologically and ecologically important processes water transport at the cell and tissue levels plays an important role. For example, the radial uptake of water into roots requires a transport across several tissues, i.e. the rhizodermis, cortex, endodermis, and stele which form hydraulic resistances in series (Steudle, 1989b,c). It is generally accepted that, in the root, water moves primarily within the apoplast. In the endodermis, the Casparian band interrupts this movement and requires a cellular transport step. The uptake of water may be coupled to the active uptake of nutrients which creates an osmotic gradient across the root (Steudle, 1989a). As in the root, the hydraulic properties of cells and of the apoplast of the different tissues contribute to the overall hydraulic properties of leaves. These properties determine the relative importance of the different pathways for water from the vessels to the stomata and play an important role in the regulation of gas exchange. For example, the evaporation of liquid water from the leaf mesophyll into the substomatal cavity rather than the peristomatal evaporation may be the dominating path for the water flow in the leaf (Nonami et al., 1990). However, at high flow rates, the path along the epidermis may become important as well.

Another example is the extension (volume) growth of plants which requires large amounts of water. It has been proposed that the supply of enlarging cells with water could become a limiting or co-limiting factor during growth (Boyer, 1985; Matyssek et al., 1988; Nonami and Boyer, 1989). This hypothesis has been questioned (Cosgrove, 1986; 1987). To

Measurement Techniques in Plant Science

113

decide whether or not the water supply plays a limiting role in growing tissue, it is necessary to measure the hydraulic resistances and to compare them with the mechanical resistance to growth (irreversible cell elongation) and the "resistances" due to solute uptake (Steudle, 1985, 1989a).

The adaptation of plants to arid conditions has resulted in various mechanisms that reduce water loss and increase water storage. It is obvious that cell and tissue water relations play an important role in these processes, but quantitative data for the water relations parameters are lacking in most of the cases. During the osmoregulation of plants, the cell and tissue water relations determine the rate at which rapid changes in cell turgor occur. The rates in changes of cell turgor due to active or passive solute flow are much slower. It has been proposed that turgor directly affects active and passive solute transport across the plasmamembrane (Coster and Zimmermann, 1976). There is also evidence that changes in water potential cause metabolic signals such as the level of abscisic acid (ABA), but the interactions between hydraulic and metabolic properties (e.g. the sensing of water potentials) are not understood (Schulze et al., 1988). A better understanding will require a detailed knowledge of water relations at the levels of cells and tissues.

This list of processes would be incomplete without mentioning the flow of solution in the phloem. The flow of water and sugars in the sieve tube elements is a typical coupled flow which could, in principle, be quantitatively described (Tyree and Dainty, 1975). The problem with the processes given above is that the absolute values of the hydraulic resistances, water storage capacities, and solute flows in the tissues involved are not known in detail. If they were known, it should be possible to completely describe the movement of water and solutes in terms of mathematical models (Philip, 1958, Molz and Ikenberry, 1974; Molz and Ferrier, 1982; Steudle, 1989a; Zhu and Steudle, 1990). In this context, it is essential that the contributions of the different pathways for water and solute transport in a tissue (i.e. the aploplasmic, symplasmic, and transcellular components) are also known. This, in turn, requires techniques for the measurement of the parameters at the cell and tissue levels. This chapter summarizes some technical progress in the field. Emphasis is given to recent developments of pressure probes for measuring the water and solute relations of cells and roots and for techniques to study phenomena of negative pressure which are important for the long distance transport in the xylem.

II. CELL WATER RELATIONS

Water relations of higher plants at the cell level have been studied in the past by various techniques. With the plasmolytic method (e.g. Lee-Stadelmann and Stadelmann, 1989) the rate of shrinking or swelling of plasmolyzed protoplasts has been measured under the microscope in order to evaluate the hydraulic conductivity of the cell membrane (Lp). Tracer flux experiments using tritiated water as a marker for diffusional water flow have also been employed (see reviews: Dainty, 1976; Zimmermann and Steudle, 1978; Steudle, 1989a) as well as NMR-techniques (Stout et al., 1977). These techniques suffer from unstirred layer effects and from the fact that plasmolyzed cells may exhibit hydraulic properties different from those of turgid cells (Steudle, 1989a).

With the cell pressure probe (Hüsken et al., 1978), these difficulties can be avoided. The technique permits the direct measurement of water relations parameters of turgid cells in intact tissue. Unstirred layer effects are negligible under the conditions of the measurement because only small amounts of water are moved across the membrane of an individual cell while the water status of the surrounding tissue remains unchanged. Furthermore, solute transport parameters of cells (permeability, P_S, and reflection, σ_S, coefficients) can be determined in osmotic experiments under certain conditions.

A. Cell Pressure Probe

1. Hydrostatic pressure relaxations. For the measurement of water relations parameters with the cell pressure probe (Figure 1), a glass microcapillary is introduced into a tissue cell under the microscope. The microcapillary (tip diameter: 2 to 7 μm) is filled with silicone oil and connected to a small pressure chamber which also contains silicone oil. A pressure transducer within the chamber measures the pressure in the system as a proportional voltage. The volume of the pressure chamber (and the hydrostatic pressure) can be artificially changed by moving a metal rod into the chamber or out of it by means of a micrometer screw.

When the tip of the probe is introduced into a cell, the cell turgor suddenly moves the oil in the tip of the probe backwards so that a meniscus is formed in the tip between oil and cell sap. This meniscus serves as a reference point during turgor measurements. With the aid of the movable rod, the meniscus is positioned close to the cell so that the volume of the cell sap in the tip is very small compared to the cell volume. Under these conditions, the cell turgor can be measured directly when the meniscus is fixed at a certain position. In most of the cells measured up to now, a stationary cell turgor can be recorded a few minutes after puncturing the cell. The size of the microcapillary has to be adapted to the size of the cell to minimize damages. At present, cells with diameters of down to 20 μm can be measured.

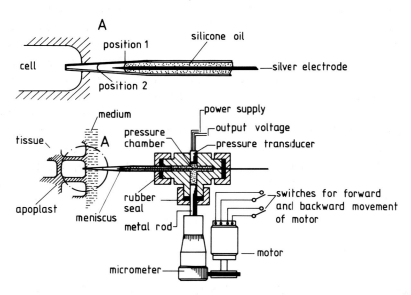

Figure 1. Pressure probe for measuring water relations of plant cells (schematical). Cell turgor is determined by introducing a microcapillary into a cell under the microscope. Turgor is balanced by the silicone oil within the pressure chamber. A meniscus is formed in the tip of the capillary which serves as a reference point during the measurements. By moving the meniscus with a metal rod, the cell volume can be changed in a defined way and the corresponding changes in turgor can be measured. For further explanations, see text.

In order to determine the elastic coefficient of an individual cell, the meniscus is first fixed at a certain position (e.g. position 1 in Figure 1) and is then rapidly moved to a second position, whereby the distance of the movement is measured under the microscope. If the inner diameter of the capillary is also determined, the changes in cell volume (ΔV) can be calculated. The corresponding changes in cell turgor (ΔP) are also recorded and, therefore, the volumetric elastic modulus of the cell (ϵ) can be evaluated:

$$\epsilon = V \frac{dP}{dV} \approx V \frac{\Delta P}{\Delta V} \, , \tag{1}$$

if the cell volume (V) is evaluated from the dimensions and shape of the cell.

The half-time of water exchange, $T^{w}_{1/2}$, of an individual cell is determined by pressure relaxation experiments. After water flux equilibrium has been achieved and the meniscus has been fixed at a position 1, the meniscus is rapidly moved to a position 2 and is fixed there again with the aid of the movable rod. Under these conditions, a water flow is induced across the cell membrane which tends to decline exponentially with time. A "pressure relaxation curve" is obtained. From the rate constant of the relaxation (k_{w}), the hydraulic conductivity (Lp) can be evaluated provided that ϵ is known, since the half-time of the relaxation is given by:

$$T^{w}_{1/2} = \frac{\ln(2)}{k_{w}} = \frac{V}{A} \frac{\ln(2)}{Lp(\epsilon + \pi^{i})} \, . \tag{2}$$

V/A is the volume to surface area ratio of the cell and π^{i} the osmotic pressure. π^{i} can be estimated from the stationary (equilibrium) turgor pressure of the cell and the osmotic pressure of the medium. $T^{w}_{1/2}$ as well as ϵ are ecologically important parameters. $T^{w}_{1/2}$ is directly related to the rate at which changes in water potential (or cell volume or turgor) are propagated across a tissue (see below). To some extent, the volumetric elastic modulus determines $T^{w}_{1/2}$ (Eq. (2)). ϵ also controls the stationary changes of cell and tissue volume in response to a change of the water potential. It can be easily shown that if $\epsilon >> \pi^i$, water potential changes will result in changes of turgor pressure rather than in changes of V or π^i, whereas for low ϵ (i.e. for $\epsilon \approx \pi^i$) changes in V and π^i are also important (Steudle, 1989a). Under these conditions, a cell may keep its turgor more constant. $\epsilon >> \pi^i$ may hold for higher plants at high turgidity. However, since ϵ is usually a strong function of cell turgor, the other situation ($\epsilon \approx \pi^i$) could be also met a low turgor pressure and, thus, the pressure dependence of ϵ may result in a protection of tissues against plasmolysis. This fact could be important under certain environmental conditions (water stress, osmotic stress, drought, etc.; Steudle, 1989a).

The pressure clamp technique represents a variation of the usual pressure probe technique and provides a method for measuring cell water relations as well as the cell volume (Wendler and Zimmermann, 1982; Zimmermann, 1989). In the technique, the stationary turgor, P, is changed instantaneously with the aid of the probe and is then clamped at a new constant value P + ΔP by moving the meniscus in the tip of the capillary to compensate for the water flow across the membrane. The movement of the meniscus is recorded as a function of time which results in a "volume relaxation curve" from which Lp can be evaluated without knowing ϵ. The half-time of volume relaxations is usually much larger than that of pressure relaxations, and this can be an experimental advantage. On the other hand, the amounts of water moved across the cell membrane are much larger than during pressure relaxations and, therefore, care must be taken about unstirred layer effects.

2. Osmotic processes of plant cells. Evaluation of solute transport parameters. The pressure probe technique not only permits the measurement of water relations parameters of individual plant cells, but also allows to determine solute flows. Transport of osmotically active solutes across the cell membrane changes the osmotic gradient across the membrane. If the changes of the internal concentration are sufficiently large, measurable changes in turgor will result. For solutes with reflection coefficients (σ_s) close to unity, a change in turgor by one bar is brought about by a concentration change of about 40 mOsmol. Since the sensitivity of the pressure probe is, at present, 0.02 to 0.03 bar, concentration changes of the order of a few mOsmol are measurable with sufficient accuracy. Passive as well as active solute uptake can be measured. For some solutes which are taken up by

active transport mechanisms, the net changes in the internal osmolarity with time will, perhaps, be too small. Therefore, the technique is, at present, mainly used to determine the passive permeability of rapidly permeating solutes such as non-electrolytes of low molecular weight (alcohols, amides, ketones, etc.).

It is possibe to measure Lp, the permeability (P_s) and the reflection (σ_s) coefficient in a single experiment. Thus, there is an independent second experiment ("osmotic pressure relaxation") for determining Lp besides the hydrostatic pressure relaxations (see above). As compared with tracer experiments, the technique is much more rapid so that a large number of substances can be measured in a short time in order determine relations between the chemical structure of solutes and the permeation rate etc. The technique could be useful to investigate the influence of inhibitors, hormones, and pollutants on general permeation properties of plant cell membranes, i.e. changes in the structure and composition of the plasmamembrane. During osmoregulation, the rapid uptake of solutes (e.g. NaCl) can be also monitored and changes in the solute transport can be measured. As already mentioned, the main disadvantage of the technique is that the concentration changes have to be in a concentration range of the order of mOsmol. A complication may arise from unstirred layers, if the method is appplied to tissue cells (Steudle and Tyerman, 1983; Tyerman and Steudle, 1982, 1984).

Figure 2A shows the pressure/time courses (pressure relaxations) in a plant cell in hydrostatic experiments or when a non-permeating solute is added to the medium. In this case, the responses are monophasic and half-times will be given by Eq.(2) as for the "hydrostatic relaxation" which is also shown in Figure 2A. In the presence of permeating solutes, however, biphasic responses are obtained, i.e. after a (usually short) "water phase" and a pressure minimum (P_{min}), turgor is again increasing and returning to the original value ("solute phase"; Figure 2B,C). It can be seen from the figure that the osmotic effects are completely reversible as it is theoretically expected.

The second "solute phase" of biphasic pressure responses is due to an equilibration of the solute across the cell membrane and is strictly exponential. From the "water phase", $T^W_{1/2}$ and Lp can be calculated and from the "solute phase" the permeability coefficient (P_s). For solutes which permeate at a high rate a correction for solute flow is necessary when Lp is calculated from $T^W_{1/2}$ (Steudle and Tyerman, 1983). The permeability coefficient of the solute is evaluated from the rate constant of the solute phase, k_S, since:

$$k_S = \frac{\ln(2)}{T^S_{1/2}} = P_S \cdot \frac{A}{V} \quad . \quad (3)$$

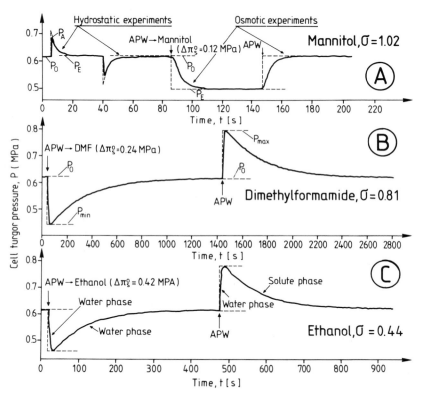

Figure 2. Pressure relaxation experiments on an isolated plant cell (giant internode of Chara corallina). In (A) "hydrostatic experiments" are shown and "osmotic experiments" with a non-permeating solute ($P_S = 0$; $\sigma_S = 1$). In (B) and (C), biphasic responses with permeating solutes are given (slowly permeating solute: dimethylformamide; rapidly permeating solute: ethanol) from which permeability coefficients were evaluated besides σ_S and Lp (Eqs.(2) to (4). Note the different time scales in (A), (B), and (C). APW = "artificial pond water" (= 1 mM NaCl and 0.1 mM KCl, CaCl$_2$, and MgCl$_2$, respectively).

An analytical expression can be given for the pressure/time course of biphasic curves (Tyerman and Steudle, 1982; Steudle and Tyerman, 1983). The reflection coefficient (σ_s) can be calculated from the change of the osmotic pressure of the medium ($\Delta\pi_s^0$) and the measured change in turgor at the minimum ($P_0 - P_{min}$), i.e.:

$$\sigma_s = \frac{P_0 - P_{min}}{\Delta\pi_s^0} \; \frac{\epsilon + \pi^i}{\epsilon} \; \exp(k_s \cdot t_{min}) \quad , \qquad (4)$$

where t_{min} is the time interval between addition of solute and the minimum. Thus, it is possible to obtain the three coefficients (Lp, P_s, σ_s) that govern the shrinking and swelling properties of plant cells and to completely describe the transport properties of the system water, solute, and membrane in the absence of active transport. Up to now, the technique has been applied to the measurement of transport coefficients of internodes of Chara cells, but also of cells in the isolated epidermis of leaves (Tyerman and Steudle, 1982, 1984; Steudle and Tyermann, 1983). A modified version of the probe has been used to measure water and solute relations of roots ("root pressure probe"; see below).

B. Water Relations Parameters of Higher Plant Cells

1. Absolute values of elastic moduli. So far, absolute values of ϵ have been determined for about 20 higher plant species using the cell pressure probe (see reviews: Zimmermann and Steudle, 1978, Steudle, 1989a). In most of the cases, ϵ exhibits a strong pressure dependence. In some cases, also a volume dependence has been observed. The increase of ϵ with both turgor and volume (cell diameter) is due to a stress-hardening effect. High pressure as well as an increase in cell dimensions increase the tension within the cell wall which results in an increase of the Young's moduli of the wall material (Steudle, 1980; Steudle et al., 1977; 1982).

Elastic moduli of higher plant cells range between a few bars and some hundred bars depending on the type of tissue and the turgidity (see reviews: Dainty, 1976; Zimmermann and Steudle, 1978; Steudle,

1989a). This range is similar to that obtained by other techniques such as the Scholander bomb, psychrometry, or the external-force method. However, some caution is necessary when comparing pressure probe data with those obtained with these techniques. With the pressure bomb as well as with psychrometric methods, volume changes of tissues are measured as a function of equilibrium water potentials in order to determine pressure-volume curves, average turgor values, and average ϵ. Very often, the average values refer to plant parts or entire organs which consist of different types of tissues which may have cells with quite different ϵ as the pressure probe data show.

The meaning of average ϵ values is not clear. There are neither physical nor physiological reasons for averaging the parameters of quite different tissues. The pressure bomb concept has been questioned because the pressure dependence of the mean ϵ determined from pressure volume curves may not reflect the real situation at the cell level (Cheung et. al., 1976). Furthermore, Tyree and Richter (1981; 1982) have shown that the basic assumption of the pressure bomb concept that the water content of the apoplast does not influence the measurements may not hold. If this is true, the whole concept would have to be reconsidered.

The external-force method introduced by Ferrier and Dainty (1977; 1978) should be mentioned. By this technique, elastic properties of plant tissues are determined by measuring the tissue compression in dependence on an applied external force (weight). The assumption is made that changes in tissue thickness are mainly due to an elastic extension of the cell walls and not to a deformation of cells. At least for isolated internodes of <u>Chara</u>, the application of the external-force method (in combination with the cell pressure probe) showed that this assumption may be questioned (Steudle et al., 1982). If this is also true for tissues, the average ϵ values obtained by the external-force technique may be substantially underestimated.

<u>2. Half-times of water exchange $(T^{W}_{1/2})$ and hydraulic conductivity (Lp)</u>. The half-times of water exchange measured with the pressure probe for higher plant cells are on average short and of an order of some seconds to some ten seconds. Cells of growing tissue exhibit fairly short half-times ($T^{W}_{1/2}$ = 0.3 - 1.2 s for the epicotyl of pea and $T^{W}_{1/2}$ = 0.3 - 15.1 s for the soybean hypocotyl; Cosgrove and Steudle, 1981;

Steudle and Boyer, 1985) which are due to high Lp values of the membrane. This finding is of interest with respect to a possible limitation of growth by water transport (Boyer, 1985; Cosgrove, 1986; Steudle, 1985).

Most of the Lp values found so far for higher plant cells with the pressure probe range between 10^{-6} and 10^{-8} m s^{-1} MPa^{-1} and are, thus, in the upper part of the range given in the literature (Steudle, 1989a). Extremely low Lp values have been found in tracer-flux experiments (e.g. House and Jarvis, 1968; Glinka and Reinhold, 1972). It is very likely that they are underestimates because of unstirred layer effects (Dainty, 1976). The results for $T^W_{1/2}$ and Lp of higher plant cells suggest that $T^W_{1/2}$ is rather small and Lp fairly large. If this trend holds in general, it should have consequences for our understanding of plant water relations because it should also affect tissue water relations and the transport of water in the entire plant (see below).

III. WATER TRANSPORT IN TISSUES

In a quantitative description of water transport in tissues, three possible pathways of water movement have to be considered, i.e. (1) the apoplasmic path around the cell protoplasts, (2) the symplasmic pathway via plasmodesmata, and (3) the transcellular (vacuolar) pathway across cell membranes and the cell wall between adjacent cells (see Figure 5 and Steudle and Jeschke, 1983).

The relative importance of the three pathways will depend on their hydraulic conductivity, their relative cross-sectional area, and on factors such as the number and size of plasmodesmata etc. To date, it is not possible to separate the symplasmic and transcellular components and, therefore, it is reasonable to summarize both components as a "cell-to-cell" component. For example, the hydraulic conductivity of cell membranes (Lp) as measured with the pressure probe will also include a symplasmic component.

Although the cross-sectional area for the apoplasmic component is much smaller than that for the cell-to-cell component, it is commonly agreed that the contribution of this component is much larger than that of the cell-to-cell component (see reviews: Läuchli, 1976; Weatherley, 1982). This point of view is based on the assumption that the cell Lp is fairly low and on the fact that in qualitative experiments in which dyes and other material have been introduced into the transpiration stream, these substances moved quickly in the apoplast (Strugger, 1938; Tanton and Crowdy, 1972). However, dyes are by no means good markers for water transport and quantitive data collected up to now with the pressure probe show that the cell Lp can be high (see above). This suggests that a reasonable part of water may move from cell to cell. The unkown figure is the hydraulic conductance of the apoplast which is difficult to be measured directly. Estimates of the hydraulic conductivity of cell wall material are rare (Table 1).

Recent experiments in which the pressure probe technique has been combined with techniques at the tissue level suggest that there could be a substantial cell-to-cell component of water flow in tissues. These results are briefly summarized and discussed in the following. However, before this is done a few remarks are necessary which refer to the mathematical description of tissue water transport.

1. Stationary water transport in tissues. When a water potential gradient is set up across a tissue, the water transport through the tissue will become steady after some time. The water flow per m^2 of tissue cross section (J_{vr}) will depend on the tissue thickness (l) and on factors such as the fractional mean cross-sectional areas of the two pathways (γ_{cw} and γ_{cc}; $\gamma_{cc} >> \gamma_{cw}$) and on the hydraulic conductivities of the apoplast (Lp_{cw}) and the cell membranes (Lp). The hydraulic conductance of the tissue (Lp_r) will be the sum of the conductances in the two parallel pathways, i.e.:

$$Lp_r = \gamma_{cw} \frac{Lp_{cw}}{l} + \gamma_{cc} \frac{Lp}{2 \, l/\Delta x} \quad . \tag{5}$$

Δx is the mean thickness of the cell layers in the direction of the water flow and $l/\Delta x$ represents the number of cell layers which have to be crossed. The factor of two in Eq. (5) arises because two membranes have be crossed per cell layer.

2. Non-stationary conditions ("Dynamic water relations"). Under non-stationary conditions, the situation is more complex. Besides the hydraulic conductances of the pathways, the storage capacities of the

cell-to-cell (C_c) and apoplasmic (C_{cw}) path have to be taken into account. The per cell capacities of the pathways are given by:

$$C_c = \frac{dV}{d\psi} = \frac{V}{(\epsilon + \pi^i)} \quad \text{and} \quad C_{cw} = \frac{dV_{cw}}{d\psi} \quad , \quad (6)$$

respectively. V_{cw} is the per cell water content of the apoplast.

Table 1. Some estimates of the hydraulic conductivity of cell walls (Lp_{cw}). Note: Lp_{cw} is given in units different from those of the membrane Lp, i.e. per unit cross section and length. Thus, the values of Lp_{cw} and Lp are not directly comparable.

Plant Species	Hydraulic conductivity, $Lp_{cw}(m^2s^{-1}MPa^{-1})$	Ref.
Nitella fexilis wall of internode (wall thickness: 7-10 μm)	5×10^{-11} 1.4×10^{-10} $(7\text{-}10)\times10^{-11}$	A B C
Pinus sp. lignified wall	5.6×10^{-10}	D
Glycine max. cortex of growing hypocotyl	8×10^{-8}	E
Zea mays root apoplast	$(0.3\text{-}6)\times10^{-9}$	F

(A) Kamiya et al., 1962, (B) Tyree, 1969, (C) Zimmermann and Steudle 1975, (D) Briggs, 1967, (E) Steudle and Boyer, 1985, (F) Zhu and Steudle, 1990.

If a tissue is held at a certain water potential (ψ) and is then transferred to an evironment with a different ψ, the change in the water potential ($\Delta\psi$) is first sensed by the cells at the tissue surface, and they will shrink or swell. The $\Delta\psi$ is then sensed by the next cell layer and so on. The change in water potential (or volume or turgor) will propagate into the tissue. The propagation can be described in terms of a diffusion type of kinetics, since (analogous to Fick's first law of diffusion) the rate of water exchange between adjacent cell layers is proportional to the difference of water potentials.

As early as 1958, Philip derived an expression for the "diffusivity" of the process. He neglected the apoplasmic water movement and for the cell-to-cell path the diffusivity, D_c, was:

$$D_c = \frac{\alpha}{2} \, Lp \cdot \Delta x \, (\epsilon + \pi^i) \, \backsim \, \frac{1}{T^W_{1/2}} \quad , \tag{7}$$

where α is a shape factor, approximately equal to one. It can be seen from Eqs. (2) and (7) that D_c is inversely proportional to $T^W_{1/2}$ and that the factor of proportionality is given only by the dimensions and shape of the cells. Thus, pressure probe data of $T^W_{1/2}$ directly yield D_c.

Molz and co-workers (e.g. Molz and Ikenberry, 1974) have extended the Philip approach and have proposed a theory in which the apoplasmic water movement is included. In their theory, they assumed local water flux equilibrium between cell-to-cell and apoplasmic path, an assumption which is reasonable because (as we know now from pressure probe data) $T^W_{1/2}$ is generally short in higher plants. For the tissue diffusivity (D_t), Molz and Ikenberry derived the following expression (a_{cw} and a_{cc} are the cross-sectional areas for the two pathways in m^2):

$$D_t = \frac{Lp_{cw} \cdot a_{cw} \cdot \Delta x + Lp \cdot a_{cc} \cdot \Delta x^2 / 2 \cdot}{C_c + C_{cw}} \quad , \tag{8}$$

which contains the sum of the hydraulic conductances in the numerator and the sum of the per cell water capacities in the denominator. An increase of the hydraulic conductances of the pathways increases D_t, whereas the rate of propagation of volume (water potential) changes in the tissue is damped as the storage capacities increases.

Equations (5), (7), and (8) allow us to quantify tissue water transport and to evaluate effects of different environmental conditions on plant water transport (e.g. water stress). They also allow to check transport models for water movement in tissues. For example, the comparison of D_t values (obtained by overall shrinking or swelling measurements of tissues) with D_c values (from $T^W{}_{1/2}$ values) should give an estimate about the contribution of apoplasmic transport. Tissue diffusivity (D_t) should be larger than the cell-to-cell diffusivity (D_c), and, if $D_t \gg D_c$, the apoplasmic component should dominate. On the other hand, if $D_t \approx D_c$ is valid, the apoplasmic contribution should be small.

Another possibility of working out the relative importance of pathways is the comparison of the hydraulic conductance of tissue (Lp_r) and conductance of cell membranes (Lp) (Eq. (5)). Lp_r can be estimated by the cell-to-cell component in Eq. (5). This value may be compared with the measured Lp_r and so on (Steudle and Boyer, 1985; Westgate and Steudle, 1985).

For determining D_t, conventional techniques can be used. D_t can be obtained from the osmotic swelling or shrinking of tissue slices, where the uptake or loss of water is measured by weighing. The rate of hydration or dehydration in these experiments will depend on D_t and on tissue shape and dimensions. A constant D_t can be expected only, if the cells are fairly homogenous in shape and size and if their Lp values are similar (Eq.(8)). Another possibility to determine D_t is to dehydrate the tissue or plant part first in dry air and then to rehydrate in distilled water (sorption experiments). In this type of experiment, a rehydration via the xylem system is possible. As compared with osmotic experiments, the sorption technique avoids complications due to unstirred layer effects and due to the diffusion of solutes into the wall space.

Sorption kinetics of tissues have been described in the literature by different authors (e.g. Molz and Klepper, 1972; Molz and Boyer, 1978). Steudle and Boyer (1985) and Westgate and Steudle (1985) have modified the sorption technique for the use in growing segments of soybean seedlings and in maize leaf tissue. They rehydrated tissues via the xylem system to obtain D_t. They were also able to determine D_t in a different type of experiment where they changed the hydrostatic pressure in the xylem and measured the pressure propagation across the tissue using the cell pressure probe. Furthermore, they determined

Lp_r of the same tissue by a pressure perfusion technique as well as water relations parameters at the cell level.

In the experiments with segments of soybean seedlings, the water flow was controlled by the cuticle which could be removed by abrasion. For sufficiently small gradients of hydrostatic pressure, the radial volume flow depended linearly on the hydrostatic gradient and yielded Lp_r. Lp_r values were rather large and even larger than the cell Lp. This result is not compatible with a cell-to-cell transport and points to a dominating apoplasmic transport. It has to be stated that under the conditions of perfusion, the cellular spaces were flooded, although progressive flooding did not increase the absolute value of Lp_r.

The comparison between the D_t and D_c values (sorption kinetics and pressure probe data for $T^W_{1/2}$) yielded a quite different result. Although there was quite a range of the absolute values of D_c and D_t due to inhomogeneities of the tissues, D_t was of the same order as D_c and, therefore, under the conditions of sorption, the cell-to-cell path should have contributed substantially to the overall transport across the tissue. From the results, it has been concluded that in the case of pressure perfusion an external hydrostatic gradient is set up across the tissue which drives a large water flow across the apoplast which has a high potential hydraulic conductivity, Lp_{cw}. Perhaps, the system of intercellular spaces plays an important role during hydrostatic water flow.

Under the conditions of sorption (which should refer to most of the "natural conditions"), the situation is different. Since a gradient in hydrostatic pressure is missing, water flow in the apoplast can be only driven by osmotic and/or matric gradients. Matric gradients should have a similar effect as hydrostatic, but osmotic gradients ($\Delta\pi_{cw}$) should have a much smaller effect, since the cell walls should have a low reflection coefficient and, therefore, the effective driving force for water in the apoplast ($\sigma_{cw} \cdot \Delta\pi_{cw}$) should be low. This may result in an apparently higher resistance of the apoplasmic path and in a relatively high contribution of the cell-to-cell component. The comparison between D_t and D_c values for tissues other than the soybean hypocotyl shows the same trend. This finding suggests that the hypothesis of a dominant apoplasmic transport only holds in the presence of an external hydrostatic pressure gradient. Measurement on roots indicate that this may be also valid for other tissue (see below). It should have consequences for our understanding of plant water transport and its regulation at the level of the whole plant. For example, a membrane-controlled cell-to-cell transport may result in considerable hydraulic resistances in tissues, depending on the absolute values of Lp, Lp_{cw}, γ_{cc}, γ_{cw} etc., as it has been proposed for growing tissue (Molz and Boyer, 1978). Lp may be affected by internal (e.g. growth hormones) or external factors (low water potential, pollutants, etc.).

The results for the midrib tissue of the maize leaf were similar to those obtained with the soybean hypocotyl in the presence and absence of hydrostatic gradients (Westgate and Steudle, 1985). For this tissue, an additional pressure propagation experiment has been performed in which a step change in the hydrostatic pressure (water potential) in the xylem was propagated across the tissue and sensed by a pressure probe inserted into cells at certain distances from the vessels. As expected, the propagation of pressure changes in the tissue followed a diffusion type of kinetics, but the D_t was by more than one order of magnitude larger than for the sorption experiment. This means that again under "hydrostatic conditions", the apoplast was the dominant path as postulated in the interpretation given above for the different results obtained in the presence and absence of hydrostatic pressure gradients.

In principle, pressure perfusion and propagation techniques as applied to the soybean hypocytl and maize leaf tissue can be also used for other tissues or plant organs whenever it is possible to connect the xylem system tightly to a pressure chamber and to measure water flows across the tissue. For the calculation of D_t and Lp_r, it would be of advantage if the tissues were fairly homogenous and regularly shaped.

Further experiments with different pressure methods at the cell and tissue level may help to improve the accuracy of the water relations parameters and to get a better quantitative basis for modeling water transport in plant tissues. In these experiments, it is also necessary to apply negative pressures (tensions) to the xylem in order to simulate the conditions naturally occuring in transpiring plants and to see whether the absolute values of Lp_r and of the hydrostatic D_t are different under these conditions (see section V).

IV. WATER TRANSPORT IN ROOTS

For the measurement of the hydraulic conductance of roots, Steudle and Jeschke (1983) developed a root pressure probe which continuously measures the root pressure of excised roots (Figure 3).

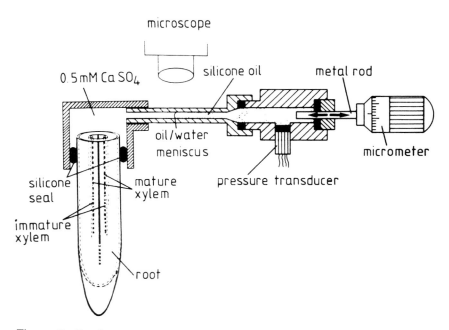

<u>Figure 3</u>. Root pressure probe for measuring water and solute flows in roots (schematical). The excised root or root system is tightly connected to the apparatus by seals so that the root pressure is built up in the system which can be then changed by either moving the metal rod or by changing the osmotic pressure of the medium ("hydrostatic" and "osmotic" experiments; see also Figure 1).

With the equipment, root pressure can be also manipulated and radial water flows across the root (J_{vr}) can be induced. The technique is analogous to that for the measurement of $T^{w}_{1/2}$, ϵ, and Lp of cells, but there are differences in the application and in the theory used for the evaluation of water transport parameters (Steudle et al., 1987). As for the cell pressure probe, a meniscus between the silicone oil in a pressure chamber and water serves as a reference point (Figure 3). The elastic extensibility of the apparatus plays an important role during the

Table 2. Hydraulic conductivities of roots (Lp_r) and of individual root cells (Lp) for some plant species. Results refer to osmotic (osm.) and hydrostatic (hydr.) experiments.

Plant Species	Root $Lp_r \times 10^8$ $(ms^{-1}MPa^{-1})$	Cell $Lp \times 10^8$	Techniques; Ref.
Hordeum distichon(a)			
osm.	0.5-4.3	-	
hydr.	0.3-4.0	12	Root
			and
Zea mays			cell
osm./hyd.(b)	1.4/10	-/24	press.
osm.(c)	1.6-2.8	-	probe
hydr.(c)	12-23	19-140	(a-d)
Phaseolus coccineus(d)			
osm./hydr.	(2-8)/(3-7)	-/190	
Triticum aestivum(e)			Osmot. stop
osm./hydr.	(1.6-5.5)/-	-/12	flow;
			cell
Zea mays(e)			press.
osm./hydr.	(0.9-5)/-	-/12	probe
Gossypium hirsutum(f)			Press. flow;
hydr.	23	12	cell press. probe

(a) Steudle and Jeschke, 1983; (b) Steudle et al., 1987; (c) Zhu and Steudle, 1990; (d) Steudle and Brinckmann, 1989; (e) Jones et al., 1988; (f) Radin and Matthews, 1989.

measurement and for the calculation of Lp_r. At the present state of development, the volumetric elastic modulus of the xylem (ϵ_x) cannot be measured with the root pressure probe because this would require an extremely large elastic modulus (extremely low extensibility) of the equipment.

When the excised root is carefully attached to the apparatus (Figure 3), it behaves like an osmometer and zero J_{vr} is obtained when the osmotic pressure difference (or the $\sigma_{sr} \cdot \Delta\pi$, where σ_{sr} = reflection coefficient of the root) between the root xylem and the medium ($\Delta\pi$) equals the root pressure. For young roots of barley and maize grown in nutrient solution, the stationary root pressure varied between 1 and 3 bar, whereas for roots of bean (Phaseolus coccineus) 0.5 to 1.5 bar have been measured. Root pressures of excised roots can be recorded for more than one day.

With the root pressure probe, there are three different ways to determine the root Lp_r, i.e. by hydrostatic and osmotic pressure relaxations (as for cells) and by a constant flow experiment which is analogous to the pressure clamp experiment for cells (see section II, A; Wendler and Zimmermann, 1982). The root pressure is increased stepwise and then kept constant in order to induce a constant efflux of water from the root. The rate of water flow can be evaluated from the movement of the meniscus in the calibrated capillary while P_r is adjusted by moving the metal rod introduced into the probe.

For roots of barley and bean, the different types of experiments yielded similar results for Lp_r (Table 2). Large differences between osmotic and hydrostatic experiments (as for the other tissues; see section III) were not observed. Furthermore, it was found that the cell Lp of cortex and rhizodermis cells was much larger than the root Lp_r (by one to two orders of magnitude). The Lp_r calculated from the cell Lp assuming a cell-to-cell transport and also assuming that endodermis and stelar cells had a similar Lp as cells in the rhizodermis and cortex, was similar to the measured Lp_r. This result suggested that there was at least a substantial cell-to-cell transport in the root as it has been also proposed for other tissues in this review.

However, it has to be pointed out that there is an alternative to this model, if the membranes of the endodermis have a much smaller Lp than cortex and rhizodermis. Under these conditions, the measured Lpr may represent mainly the hydraulic resistance of this structure while the other resistances in series (rhizodermis, cortex, stele) do not play an important role. To decide this question, direct measurements of the endodermal Lp are necessary which can be performed with the probe. However, I think that the alternative is unlikely to be true because it would require that the membranes of the endodermis exhibit transport properties which are quite different from those of the cortex and rhizodermis cells.

The finding that osmotic and hydrostatic Lp_r were similar in barley and bean seems to contradict the results obtained from the other tissues (see section III). However, it has to be taken into account that in the root the Casparian band should block off the continuous apoplasmic path and interrupt the intercellular system so that the hydrostatic gradient cannot be very effective provided that the Casparian strip is a barrier with a sufficiently large hydraulic resistance. In this case, the differences between the two types of experiments should level off and the cell-to-cell component should dominate. If this is true, one should expect that in roots, where the Casparian strip still has a reasonable hydraulic conductance, the differences between osmotic and hydrostatic experiments should also become evident. This seems to be the case in maize roots, where large differences between hydrostatic and osmotic Lp_r have been found (Steudle et al., 1987; Steudle and Frensch, 1989; Zhu and Steudle, 1990). Furthermore, one may expect that in the upper (older) parts of the root where the Casparian strip becomes more permeable for water, differences between the two types of experiments may occur. With the root pressure probe, it is possible to analyze different parts of the root. This type of experiments is currently performed.

The absolute values of root Lp_r found with the root pressure probe for different roots were similar to those found with other techniques (see Table 2 and reviews: Steudle, 1989a,b,c). A dependence of the root Lp_r on the absolute value of J_{Vr} as it has been reported in the literature has not yet been observed with the root pressure probe, but it has to be noted that Lp_r was determined only in a rather small range of J_{Vr} ($J_{Vr} = 0$ to 1×10^{-7} m s^{-1}). Larger values of Lp_r may be obtained, if larger forces are applied across the root cylinder. These conditions would be met in transpiring plants, whereas positive root pressures only occur during guttation when J_{Vr} is close to zero.

As with cells, osmotic experiments can be performed with roots while the root pressure is continuously monitored with the probe. In experiments with permeating solutes, the relaxations are biphasic (Figure 4) and allow to evaluate permeability (P_{sr}) and reflection (σ_{sr})

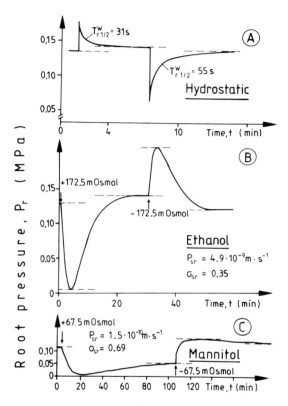

<u>Figure 4</u>. Root pressure relaxations as obtained with the root pressure probe (Figure 3) on roots of <u>Phaseolus coccineus</u>. Hydrostatic experiments as well as osmotic experiments with two different solutes are shown. The experiments are analogous to those with cells (see Figure 2). From the relaxations, transport coefficients (Lp_r, P_{sr}, and σ_{sr}) have been calculated (see Tables 2 and 3). The rapidly permeating solute (ethanol) exhibited a much shorter half-time of solute exchange and a larger P_{sr} and smaller σ_{sr} than the slowly permeating solute (mannitol).

coefficients. Compared with cells, complications may occur in the root because of the assumption made of a simple two compartment model and of a membranelike structure which acts as an "osmotic barrier" between the compartments. This structure is usually identified by the endodermis. However, the barrier may be much more complicated. In the root cylinder, there should be membranelike elements arranged in series such as the exodermis, cortex, and endodermis. Others, such as

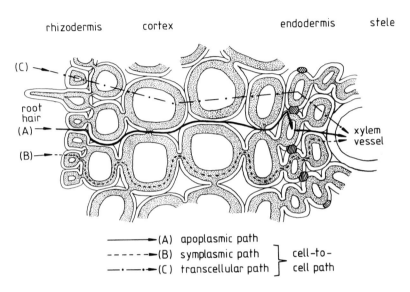

Figure 5. Pathways for water and solutes in roots. The apoplasmic path (A) refers to flows in the cell wall around cell protoplasts, whereas the symplasmic pathway (B) is defined by a transport from cell to cell via plasmodesmata. On the third transcellular or vacuolar path (C), water or solutes have to cross two cell membranes per cell layer as well as the wall between adjacent cells. At least for water, pathways (B) and (C) cannot be separated experimentally to date, and, therefore, it is reasonable to define a cell-to-cell path which comprises both components. It is usually assumed that in the root, the Casparian band in the endodermis interrupts the apoplasmic transport of water and solutes and requires a cell-to-cell transport step.

the apoplasmic, symplasmic, transcellular pathways, are arranged in parallel (Figure 5). A parallel arrangement of different elements may also result from the fact that the root is developing along the axis. Therefore, the root should exhibit properties of a "composite membrane" rather than of a single, homogeneous membrane (Steudle, 1989a). The latter model is the classical way of looking at roots (e.g. Dainty, 1985; Fiscus, 1986; Passioura, 1988).

The situation would be furthermore complicated by unstirred layers (Steudle and Frensch, 1989). In the hydrostatic experiments, these effects are small, but in osmotic experiments, the entire cortex may act as an unstirred layer. However, a thorough analysis performed on maize roots showed that the effects of unstirred layers on the absolute values of the transport coefficients should not be larger than 10% (Steudle and Frensch, 1989). Effects of osmoregulatory responses of roots during the osmotic experiments should be even lower (Steudle and Brinckmann, 1989).

Despite of these complications, the experimental results (see below) indicate that the simple two compartment model of the root is basically correct, although the composite structure of the root has to be considered when discussing the absolute values of transport coefficients.

Some results of Lp_r, P_{sr}, and σ_{sr} as obtained with the root pressure probe are given in Tables 2 and 3 together with some results from the literature. Table 2 only contains data of the hydraulic conductivity from experiments in which measurements were performed at both the cell and root level. The differences between the hydrostatic and osmotic Lp_r for maize are obvious. They have been explained by different transport mechanisms (see above and section III). Work from other laboratories using different techniques supports the view of differences in the mechanisms. Jones et al. (1988) also interpreted their Lp_r and Lp values in terms of a considerable apoplasmic water transport for wheat and maize. The same was true for cotton roots (Radin and Matthews, 1989). This picture suggests that, at certain stages of development, the Casparian band could be fairly permeable for water or that the initials of secondary roots provide the apoplasmic path (Peterson et al., 1981).

Table 3 indicates that reflection coefficients of roots are low even for solutes for which cells exhibit a $\sigma_s \approx 1$. Effects of unstirred layers and/or active transport cannot completely explain the effect. The result has been interpreted by the composite structure of the osmotic barrier in the root (see above). Under certain conditions, composite membranes could exhibit low values of σ_{sr}, although P_{sr} is fairly low, and the roots are by no means "leaky" despite of a low σ_{sr} (for a detailed discussion see Steudle et al.; 1987; Steudle and Frensch, 1989; Steudle and Brinckmann, 1989; Steudle, 1989a,b,c). It is interesting that low reflection coefficients of roots are also indicated from results which have been obtained by other techniques (Table 3).

Table 3. Root permeability (P_{sr}) and reflection (σ_{sr}) coefficients as obtained from root pressure probe experiments (refs. e-i) and by other techniques (refs. a-d).

Species	Osmoticum	Root σ_{sr} (1)	Root $P_{sr} \times 10^9$ (ms^{-1})
Lycopersicon esculentum(a)	nutrients	0.76	--
Glycine max(b)	nutrients	0.90	--
Phaseolus vulgaris(c)	nutrients	0.98	2.2
Zea mays, excised roots (e-g)	nutrients(d)	0.85	--
	ethanol	0.3	6-19
	mannitol	0.4-0.7	--
	sucrose	0.54	3
	PEG 6000	0.64-0.82	--
	NaCl	0.5-0.6	6-14
	KCl	0.53	--
	KNO3	0.5-0.7	<1-8
	NH4NO3	0.5	3-32
Hordeum distichon(h)	mannitol	0.5	--
Phaseolus coccineus(i)	ethanol	0.31	6.3
	methanol	0.25	4.5
	urea	0.46	0.8
	mannitol	0.68	0.15
	KCl	0.49	0.82
	NaCl	0.59	0.21
	NaNO3	0.54	0.37

(a) Mees and Weatherley, 1957; (b) Fiscus, 1977; (c) Fiscus, 1986; (d) Miller, 1985; (e) Steudle et. al., 1987; (f) Steudle and Frensch, 1989; (g) Zhu and Steudle, 1990; (h) Steudle and Jeschke, 1983; (i) Steudle and Brinckmann, 1989.

Measurements with the root pressure probe have been combined with those with the cell pressure probe to compare transport data from the different levels. Recently, this approach has been extended in that the measurements have been performed in the same experiment ("double pressure probe"; Zhu and Steudle, 1990). In osmotic experiments with maize roots, it was found that during the dehydration of the cortex following an increase of the osmotic pressure of the medium, most of the water was leaving the tissue <u>via</u> the apoplast. The apoplasmic hydraulic conductance was quite high (Table 1). The apparent reflection coefficient of cells in different layers of the cortex decreased with increasing depth of the cell in the root as one would expect. The results were in accordance with a model in which water moved in parallel in the apoplast and from cell to cell (see section III). Furthermore, a careful examination of the cell Lp in the root showed that Lp increased considerably in the cortex towards deeper cell layers, i.e., the water permeability of the cortical membranes was not constant.

Especially in long root systems, the contribution of the axial resistance would be also important besides the radial one. Using the root pressure probe, Frensch and Steudle (1989) examined the longitudinal resistance of maize roots. They measured the axial hydraulic conductance of short root segments which were open at both ends and determined the changes of the longitudinal resistance in dependence on the position in the root (root development). The data were compared with calculated values according to Poiseuille's law, and the relative contribution of the radial <u>vs</u>. the longitudinal resistance was evaluated. The authors described the total water uptake of the root and the contribution of different root zones in terms of a mathematical model. For the end segments of maize roots it turned out that, with the exception of the tip region, the radial rather than the axial resistance dominated the overall resistance. The technique may be used to analyze the "hydraulic architecture" of complex root systems which is much less known than that of shoots.

For environmental studies the root pressure probe technique may be of interest. Changes of the hydraulic resistance of the root in dependence on environmental changes (reduced water potential, acidification of soils, effect of pollutants, etc.) could be measured. Since the root pressure probe is much easier to handle than the cell pressure probe, these measurements could be also performed in the field on roots sitting in the soil. The hydraulic resistance of the soil may be separated from that of the root, if additional experiments are employed. On the other hand, the root pressure can be taken as a measure for the solute (nutrient) uptake into the root xylem. These data of the capacity of nutrient uptake and of the hydraulic resistance are of considerable practical interest for the growth and development of plants.

V. MEASUREMENT OF NEGATIVE PRESSURE

In the xylem of plants, the water vapor deficit of the atmosphere usually creates tensions which, in turn, drive the uptake of water from the soil and the flow of solution within the xylem. Only under conditions of very low or vanishing transpiration, positive root or stem pressures are established osmotically. This phenomenon is used in the root pressure probe technique as discussed in the previous section. Since the time when Böhm (1893) first proposed his cohesion theory of the ascent of sap, the proof and measurement of negative pressures (i.e. of pressures smaller than vacuum) is a problem in botany. Indirect evidence for the existence of negative pressures have come from measurents of the xylem water potential and osmotic pressure using the Scholander bomb, psychrometry, and other techniques. Conceptually, negative pressures have caused some controversies because tensions in liquids represent a metastable state and, hence, it may be difficult to accept that the xylem is a rather vulnerable pipeline.

Negative pressures have been also postulated in cells. However, as compared with vessels which are stiffened to withstand compressive forces created by the tensions, the walls of cells are constructed to withstand extension and, thus, negative pressures will be only possible, if the walls are stiffened such as in sclerophyllous tissue (Kreeb, 1963; Örtli, 1987) or during freezing, where the ice formed extracellularly may stiffen the tissue. Tremendous tensions of up to 300 bar in the liquid cell sap of frozen tissue have been postulated to balance the negative water potentials created at minus temperatures (Burke et al., 1983).

In principle, the cohesive forces of water are sufficient to withstand these tensions. By various methods, the tensile strength of water has been calculated to range between some 100 and 20,000 bar (for references see Zhu et al., 1989). Experimental values are much lower and range between some 10 and 300 bar. The difference is mainly brought about by the fact that the weakest point in the system is not the tensile liquid but the boundary between liquid and the surface of the vessel (Zimmermann, 1983). At this boundary, gas seeds of sufficient size may be formed. Thus, cavitations in the xylem may be rather frequent (Zimmermann and Milburn, 1982), and plants require mechanisms to reverse cavitations which interrupt the flow in the vessels. These mechanisms are not completely understood.

In order to study states of tension or negative pressure, we have introduced an artificial osmotic cell in which these states can be established and measured directly. The technique has been also used to simulate the events of freezing dehydration. Recently, the experiences with the osmotic cell have been used to measure negative pressures in the root xylem with the aid of the root pressure probe.

A. Artificial Osmotic Cell

The artificial osmotic cell has been constructed using reverse osmosis membranes (Steudle and Heydt, 1988). The cell consisted of a thin film of an osmotic solution (thickness: 100 to 200 μm) containing a non-permeating solute. The film was bounded between the membrane and the front plate of a pressure transducer which continuously recorded the "cell turgor" built up in the presence of a hypotonic solution outside the cell (Figure 6).

The membrane was supported by metal grids to withstand compression. Tensions could be created by applying hypertonic solutions (Figure 7). At maximum, minus 7 bar of absolute pressure (vacuum = 0 bar) could be established within the film on short-term and tension of up to minus 3 bar on long-term (several hours) prior to cavitation. Figure 7 also demonstrates that cavitations could be "healed" by establishing positive pressures in the cell for some time. This could be one of the mechanisms also occuring in the intact plant (root pressure). Upon cavitations, pressures suddenly jumped towards a value close to zero absolute pressure. This is theoretically expected, because the vapor pressure of water at room temperature is only about 20 mbar.

During the experiments, the proper function of the transducer could be tested. These tests proved that the negative pressures measured in the liquid were real and that artefacts could be excluded.

Using permeating solutes, biphasic pressure relaxations such as those given in Figures 2 and 4 could be produced either in the positive or in the negative range of pressures. The experiments indicated that the transport coefficients (Lp, P_s, and σ_s) did not change upon transition from the positive to the negative range of pressures.

The system may be used to simulate osmotic processes in living cells and to study phenomena of negative pressure as they occur in the xy-

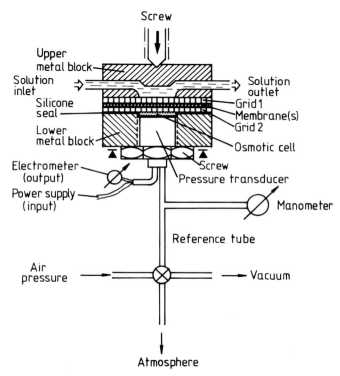

Figure 6. Artificial osmotic cell for measuring negative pressures and for simulating osmotic processes in plant cells. The cell is bounded by the metal membrane of a pressure transducer from the lower side and by a reverse osmosis membrane positioned between two metal grids from the upper side. The cell contains a solution of a non-permeating solute so that a "turgor" is etablished in the cell in the presence of hypotonic solutions which is continuously recorded. Changes of the osmotic concentration of the external solution result in changes of turgor which depend on the permeability of the solutes and the composition of the solution. The signals can be analyzed to yield Lp, P_s, and σ_s (as in Figures 2 and 4). With hypertonic solutions negative pressures (tensions) can be created in the cell (Figure 7).

lem and, perhaps, also in living cells. For example, Zhu et al. (1989) have studied the effect of extracellular freezing on the turgor pressure in order to simulate the situation in the living cell where the cell pressure probe cannot be used to measure tensions because of cavitation problems. The result was that tensions of up to 6 bar could be created

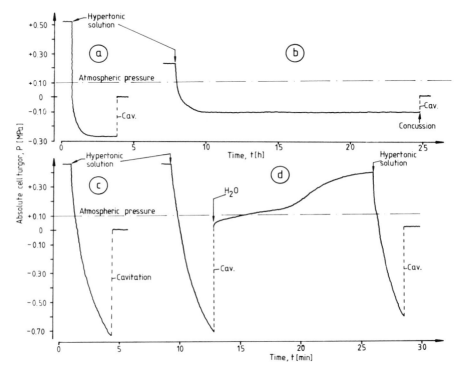

<u>Figure 7</u>. Negative pressures and cavitations in the artificial osmotic cell (Figure 6). Traces (a) and (b) demonstrate that low tensions (0 to -3 bar) can be kept for hours, whereas higher tensions result in cavitations in shorter terms (traces (c) and (d)). Fractures in the water film can be induced by concussion (trace (b)) and healed by pressurizing the cell for some time (trace (d)). Upon cavitation, the pressures in the cell instantaneously increase to values close to zero absolute pressure as theoretically expected.

by freezing dehydration prior to cavitation. Also, tensions within the cell could be maintained, while cooling the cell to minus temperatures. However, the establishment of tensions was only possible, if the membrane was supported from the inside to withstand a collapse. Thus, if negative pressures occur during freezing in living cells, this would require mechanisms for avoiding the removal of the plasmamembrane from the wall. One mechanism to achieve this could be that the tensions within the microcapillaries of the wall, upon a withdrawl of water, would always remain larger than those in the protoplasts so that the membrane remains attached to the wall (Oertli, 1987). However, this would only work, if the space between membrane and wall does not contain gas seeds.

Recently, the conventional cell pressure probe has been used to measure tensions in the xylem of <u>Plantago major</u> and <u>Nicotiana rustica</u> (Balling <u>et al.</u>, 1988; Zimmermann and Balling, 1989). The probe was introduced into the xylem of excised leaves, while the petiole was connected to an osmometer (dialysis membrane; solute: PEG 6000) to create tensions in the vessels. Only small tensions of up to -0.5 bar of absolute pressure could be created. With the osmotic cell itself, pressures of down to -2.5 bar (absolute) could be built up prior to cavitation. The cell pressure probe had to be filled completely with water in these experiments, since the boundary between oil and water (see Figure 1) was prone to cause cavitations. The pressures which have been measured in the xylem of <u>Nicotiana</u> were significantly different from those obtained from experiments using the Scholander bomb. Furthermore, changes in the rate of transpiration did not influence the absolute pressure in the xylem which has been interpreted by a variable supply of water from the root and/or an effect of the xylem parenchyma on the water potential of the xylem (Zimmermann and Balling, 1989).

B. Negative Pressures in the Root Xylem

Heydt and Steudle (unpublished results) have successfully created and measured tensions in the xylem of excised maize roots using the root pressure probe. Provided that gas was carefully removed from the tissue in contact with the probe, absolute values of down to -3 bar (absolute) could be measured over several hours. As for the artificial osmotic cell, transport properties of roots could be determined for both the positive and negative range of pressures (Lp_r, P_{sr}, and σ_{sr}). There was no change in the transport properties of the roots when the pressure changed to negative values. This is important when transport coefficients obtained in the positive range are used for calculations of flows in the intact transpiring plant.

When the root pressure probe was attached to roots of intact plants still having a shoot, positive root pressures were measured at zero transpiration. The onset of transpiration caused a rapid decrease of root pressure which was reversible. Thus, there was a good hydraulic contact between root and shoot as one would expect. These results are at variance with those given above for experiments in which the cell pressure probe was introduced into the xylem. However, in the experiments with the root pressure probe, it was not possible to establish tensions because of the presence of gas in the tissue and the formation of gas bubbles in the system as the pressure decreased towards negative values.

The results show that, at present, there are still considerable technical difficulties for the direct measurement of tensions in the xylem. Artefacts have to be carefully excluded. The difficulties are even larger with respect to plant cells, although the development of the artificial osmotic cell has been helpful to improve the methods. In the future, this will, perhaps, lead to techniques for measuring tensions of some ten bars as they occur in the vessels of transpiring plants in order to quantify the driving forces for the long distance transport of water in the xylem and the hydraulic interactions between root and shoot.

ACKNOWLEDGMENT

This work was supported by a grant from the Deutsche Forschungsgemeinschaft, Sonderforschungsbereich, 137.

VI. REFERENCES

Balling, A., Zimmermann, U., Büchner, K.-H., and Lange, O.L. (1988). Direct measurement of negative pressure in artificial-biological systems. Naturwissenschaften 75:409-411.

Böhm, J. (1893). Capillarität und Saftsteigen. Ber. Deutsch. Bot. Ges. 11:203-212.

Boyer, J.S. (1985). Water transport. Ann. Rev. Plant Physiol. 36:473-516.

Briggs, G.E. (1967). Movement of water in plants. Botanical monographs. Blackwell, Oxford.

Burke, M.F., Rajashekar, C., and George, M.F (1983). Freezing of plant tissues and evidence for large negative pressure potentials. Plant Physiol. 72S:44.

Cheung, Y.N.S., Tyree, M.T., and Dainty, J. (1976). Some possible sources of error in determining bulk elastic moduli and other parameters from pressure-volume curves of shoots and leaves. Can. J. Bot. 54:758-765.

Cosgrove, D.J. (1986). Biophysical control of plant cell growth. Ann. Rev. Plant Physiol. 37:377-405.

Cosgrove, D.J. (1987). Linkage of wall extension with water and solute uptake. In: Physiology of cell expansion during plant growth. D.J. Cosgrove and D.P. Knievel, eds. Proc. 2nd Symposium in plant physiology of the Amer. Soc. Plant Physiol., Rockville, MA, USA, pp. 88-100.

Cosgrove, D.J., and Steudle, E. (1981). Water relations of growing pea epicotyl segments. Planta 153:343-350.

Coster, H.G.L., and Zimmermann, U. (1976). Transduction of turgor pressure by cell membrane compression. Z. Naturforsch. 31c:461-463.

Dainty, J. (1976). Water relations of plant cells. In: Encyclopedia of plant physiology, Vol. 2, Part A. U. Lüttge and M.G. Pitman, eds. Springer-Verlag, Berlin, pp. 12-35.

Dainty, J. (1985). Water transport through the root. Acta Hort. 171:21-31.

Ferrier, J.M., and Dainty, J. (1977). A new method for measurement of hydraulic conductivity and elastic coefficients in higher plant cells using an external force. Can. J. Bot. 55:858-866

Ferrier, J.M., and Dainty, J. (1978). The external force method for measuring hydraulic conductivity and elastic coefficients in higher plant cells: application to multilayer tissue sections and further theoretical development. Can. J. Bot. 56:22-26.

Fiscus, E.L. (1977). Determination of hydraulic and osmotic properties of soybean root systems. Plant Physiol. 59:1013-1020.

Fiscus, E.L. (1986). Diurnal changes in volume and solute transport coefficients of Phaseolus roots. Plant Physiol. 80:752-759.

Frensch, J., and Steudle, E. (1989). Axial and radial hydraulic resistance in roots of maize (Zea mays L.). Plant Physiol. 91:719-726.

Glinka, Z., and Reinhold, L. (1972). Induced changes in permeability of plant cell membranes to water. Plant Physiol. 49:602-606.

House, C.R., and Jarvis, P. (1968). Effect of temperature on the radial exchange of labeled water in maize roots. J. Exp. Bot. 19:31-40.

Hüsken, D., Steudle, E., and Zimmermann, U. (1978). Pressure probe technique for measuring water relations of cells in higher plants. Plant Physiol. 61:158-163.

Jones H., Leigh, R.A., Wyn Jones, R.G., and Tomos, A.D. (1988). The integration of whole-root and cellular hydraulic conductivities in cereal roots. Planta 174:1-7.

Kamiya, N., Tazawa, M., and Takata, T. (1962). Water permeability of the cell wall in Nitella. Plant and Cell Physiol. 3:285-292.

Kreeb, K. (1963). Untersuchungen zum Wasserhaushalt der Pflanzen unter extrem ariden Bedingungen. Planta 59:442-458.

Läuchli, A. (1976). Apoplastic transport in tissues. In: Encyclopedia of plant physiology, Transport in plants II, Part B. Tissues and organs. U. Lüttge, M. G. Pitman, eds. Springer-Verlag, Berlin/Heidelberg/New York, pp. 3-34.

Lee-Stadelmann, O.Y., and Stadelmann, E.J. (1989). Plasmolysis and deplasmolysis. In: Methods of Enzymology, Vol. 174, Biomembranes, Part U: Cellular and subcellular transport: eukaryotic (nonepithelial) cells. S. Fleischer and B. Fleischer, eds. Academic Press, New York, pp. 225-246.

Matyssek, R., Maruyama, S., and Boyer, J.S. (1988). Rapid wall relaxation in elongating tissues. Plant Physiol. 86:1163-1167.

Mees, G.C., and Weatherley, P.E. (1957). The mechanism of water absorption by roots. I. Preliminary studies on the effects of hydrostatic pressure gradients. Proc. R. Soc. London, Ser. B, 147:367-381.

Miller, D.M. (1985). Studies on root function of Zea mays. III. Xylem sap composition at maximum root pressure provides evidence of active transport into the xylem and a measurement of the reflection coefficient of the root. Plant Physiol. 77:162-167.

Molz, F.J., and Boyer, J.S. (1978). Growth-induced water potentials in plant cells and tissues. Plant Physiol. 62:423-429.

Molz, F.J. and Ferrier, J.M. (1982). Mathematical treatment of water movement in plant cells and tissues: A review. Plant, Cell, Environment 5:191-206.

Molz, F.J., and Ikenberry, E. (1974). Water transport through plant cells and walls: Theoretical development. Soil Sci. Soc. Amer. Proc. 38:699-704.

Molz, F.J., and Klepper, B. (1972). Radial propagation of water potential in stems. Agron. J. 64:469-473.

Nonami, H., and Boyer, J.S. (1989). Turgor and growth at low water potentials. Plant Physiol. 89:798-804.

Nonami, H., Schulze, E.-D., and Ziegler, H. (1989). Mechanisms of stomatal movement in response to air humidity, light intensity and xylem water potential: role of the inner cuticle layer in the stomatal cavity. Planta, in press.

Oertli, J.J. (1987). Measurement of the resistance of cell walls to collapse during moisture stress. In: Proceedings of the international conference on measurement of soil and plant water status. Vol. 2, Plants. R.J. Hanks and R.W. Brown, eds., Utah State University, Logan, Utah, USA, pp. 193-198.

Passioura, J.B. (1988). Water transport in and to roots. Ann. Rev. Plant Physiol. Plant Mol. Biol. 39:245-265.

Peterson, C.A., Emanuel, M.E., and Humphreys, G.B. (1981). Pathway of movement of apoplastic fluorescent dye tracers through the endodermis at the site of secondary root formation in corn (Zea mays) and broad bean (Vicia faba). Can. J. Bot. 59:618-625.

Philip, J.R. (1958). Propagation of turgor and other properties through cell aggregations. Plant Physiol. 33:271-274.

Radin, J., and Matthews, M. (1989). Water transport properties of cortical cells in roots of nitrogen- and phosphorus-deficient cotton seedlings. Plant Physiol 89:264-268.

Schulze, E.-D., Steudle, E., Gollan, T., and Schurr, U. (1988). Response to Dr. P.J. Kramer's article, 'Changing concepts regarding plant water relations', Volume 11, number 7, pp. 565-568. Plant, Cell, Environment 11:573-576.

Steudle, E. (1980). Effect of cell diameter and length on the overall elasticity of cylindrical plant cells. Significance for membrane transport and growth. In: Plant membrane transport: Current conceptual issues. R.M. Spanswick, W.J. Lucas, and J. Dainty, eds. Elsevier/North Holland, pp. 483-484.

Steudle, E. (1985). Water transport as a limiting factor in extension growth. In: Control of leaf growth. SEB Seminar, Vol. 27, N.R.

Baker, W.D. Davies, C. Ong, eds., Cambridge University Press, pp. 35-55.

Steudle, E. (1989a). Water flow in plants and its coupling to other processes: an overview. In: Methods in Enzymology, Vol. 174, Biomembranes, Part U: Cellular and subcellular transport: eukaryotic (nonepithelial) cells. S. Fleischer and B. Fleischer, eds. Academic Press, New York, pp. 183-225.

Steudle, E. (1989b). Water transport in roots. In: Structural and functional aspects of transport in roots. B.C. Loughman, O. Gasparikova, and J. Kolek, eds. Kluwer Academic Publishers, pp. 139-145.

Steudle, E. (1989c). Water transport in roots. In: Plant water relations and growth under stress. M. Tazawa, M. Katsumi, Y. Masuda, and H. Okamoto, eds. Yamada Science Foundation Conference XXII. Myu K.K., Tokyo., pp. 253-260.

Steudle, E., and Boyer, J.S. (1985). Hydraulic resistance to radial water flow in growing hypocotyl of soybeans measured by a new pressure perfusion technique Planta 164:189-200.

Steudle, E., and Brinckmann, E. (1989). The osmometer model of the root: water and solute relations of roots of Phaseolus coccineus. Bot. Acta 102:85-95.

Steudle, E., Ferrier, J.M., and Dainty, J. (1982). Measurement of the volumetric and transverse elastic extensibilities of Chara corallina internodes by combining the external force and pressure probe technique. Can J. Bot. 60:1503-1511.

Steudle, E., and J. Frensch (1989). Osmotic responses of maize roots: Water and solute relations. Planta 177:281-295.

Steudle, E., and Heydt, H. (1988). An artificial osmotic cell: a model system for simulating osmotic processes and for studying phenomena of negative pressure in plants. Plant, Cell, Environment 11:629-637.

Steudle, E., and Jeschke, W.D. (1983). Water transport in barley roots. Planta 158:237-248.

Steudle, E., Oren, R., and Schulze, E.-D. (1987). Water transport in maize roots. Measurement of hydraulic conductivity, solute permeability, and of reflection coefficients of excised roots using the root pressure probe. Plant Physiol. 84:1220-1232.

Steudle, E., and Tyerman, S.D. (1983). Determination of permeability coefficients, reflection coefficients, and hydraulic conductivity of

Chara corallina using the pressure probe: Effects of solute concentrations. J. Membrane Biol. 75:85-96.

Steudle, E., Zimmermann, U., and Lüttge, U. (1977). Effect of turgor pressure and cell size on the wall elasticity of plant cells. Plant Physiol. 59:285-289.

Stout, D.G., Cotts, R.M., and Steponkus, P.L. (1977). The diffusional water permeability of Elodea leaf cells as measured by nuclear magnetic resonance. Can. J. Bot. 57:1623-1631.

Strugger, S. (1938). Die lumineszenzmikroskopi-sche Analyse des Transpirationsstromes in Parenchymen. Flora (Jena) 133:56-68.

Tanton, W., and Crowdy, S.H. (1972). Water pathways in higher plants. II. Water pathways in roots J. Exp. Bot. 23:600-618.

Tyerman, S.D., and Steudle, E. (1982). Comparison between osmotic and hydrostatic water flows in a higher plant cell: Determination of hydraulic conductivities and reflection coefficients in isolated epidermis of Tradescantia virginiana. Aust. J. Plant Physiol. 9:461-479.

Tyerman, S.D., and Steudle, E. (1984). Determination of solute permeability in Chara internodes by a turgor minimum method: Effects of external pH. Plant Physiol. 74:464-468.

Tyree, M.T. (1969). The thermodynamics of short-distance translocation in plants. J. Exp. Bot. 20:341-349.

Tyree, M.T., and Dainty, J. (1975). Phloem transport. Theoretical consideration. In: Encyclopedia of plant physiology. Vol. 1, Phloem transport. M.H. Zimmermann and J.A. Milburn, eds., Springer-Verlag, Berlin, pp. 367-392.

Tyree, M.T., and Richter, H. (1981). Alternative methods of analysing water potential isotherms. Some cautions and clarifications. I. The impact of nonideality and of some experimental errors. J. Exp. Bot. 32:643-653.

Tyree, M.T., Richter, H. (1982). Alternative methods of analysing water potential isotherms: Some cautions and clarifications. II. Curvilinearity in water potential isotherms. Can. J. Bot. 60:911-916.

Weatherley, P.E. (1982). Water uptake and flow in roots. In: Encyclopedia of Plant Physiology, Vol. 12 B, Physiological plant ecology II. O.L. Lange, P.S. Nobel, C.B. Osmond and H. Ziegler, eds. Springer-Verlag, Berlin, pp. 79-109.

Wendler, S., and Zimmermann, U. (1982). A new method for the determination of hydraulic conductivity and cell volume of plant cells by pressure clamp. Plant Physiol. 69:998-1003.

Westgate, M.E., and Steudle, E. (1985). Water transport in the midrib tissue of maize leaves: Direct measurement of the propagation of changes in cell turgor across a plant tissue. Plant Physiol. 78:183-191.

Zhu, G.-L., and Steudle, E. (1990). Water transport across maize roots: simultaneous measurement of flows at the cell and root level by double pressure probe technique. Plant Physiol., in press.

Zhu, J.J., Steudle, E., and Beck, E. (1989). Negative pressures produced in an artificial osmotic cell by extracellular freezing. Plant Physiol. 91:1454-1459.

Zimmermann, M.H. (1983). Xylem structure and the ascent of sap. Springer-Verlag, Berlin.

Zimmermann, M.H., and Milburn, J.A. (1982). Transport and storage of water. In: Encyclopedia of plant physiology, Vol. 12B, Physiological plant ecology II, O.L. Lange, P.S. Nobel, C.B. Osmond, and H. Ziegler, eds., Springer-Verlag, Berlin, pp. 135-151.

Zimmermann, U. (1989). Water relations of plant cells: Pressure probe technique. In: Methods of Enzymology, Vol. 174, Part U: Cellular and subcellular transport: eukaryotic (nonepithelial) cells. S. Fleischer and B. Fleischer, eds. Academic Press, New York, pp. 338-366.

Zimmermann, U., and Balling, A. (1989). Comparative measurements of the xylem pressure of Nicotiana plants by means of the pressure bomb and -probe. In: Plant membrane transport: the current position, J. Dainty, M.I. Demichelis, E. Marre, and F. Rasi-Caldogno, eds., Elsevier, Amsterdam, pp. 555-558.

Zimmermann, U., and Steudle, E. (1975). The hydraulic conductivity and volumetric elastic modulus of cells and isolated cell walls of Nitella and Chara spp. Pressure and volume effects. Aust. J. Plant. Physiol. 2:1-12.

Zimmermann, U., and Steudle, E. (1978). Physical aspects of water relations of plant cells. Adv. Bot. Res. 6:45-117.

MEASUREMENTS OF THE WATER ECONOMY
OF MANGROVE LEAVES

Susumu Kuraishi
Naoki Sakurai
Hiroshi Miyauchi

Department of Environmental Sciences
Faculty of Integrated Arts & Sciences
Hiroshima University
Hiroshima, Japan

Krisadej Supappibul

Mangrove Forest Management Unit
Lamngob, Namchew, Trat
Thailand

I. INTRODUCTION

Methods for the measurement of stomatal aperture and leaf transpiration have been available for many years (Hsiao and Fischer, 1975; Meidner, 1981; Slavik, 1974). Although many measurements have been made on crop plants and on temperate zone forests, few measurements have been made on leaves of trees in tropical forests. Even temperate zone forest trees suffer water stress and exhibit stomatal closure and high leaf temperatures during droughts (Nito, et al., 1979; Kuraishi and Nito, 1980). Among tropical trees mangroves are of particular interest because they grow in saline soils with low water potentials and are exposed to long periods of bright sunlight during the dry season. As their leaves appear to have no morphological characteristics that would reduce transpiration (Chapman, 1976), they probably are subjected to severe water stress. Thus, although mangrove leaves deserve study this has been

Measurement Techniques in Plant Science

151

discouraged by the fact that they grow far from laboratories
and easily portable instrumentation is necessary. The
development of such instrumentation in recent years makes
such studies possible.

II. MATERIALS AND METHODS

The mangroves used in this study were growing at Nomchew
(12°15'N) in Trat Province, Thailand. The monthly average
temperatures at Trat vary only from 25 to 28°C and the dry
season extends from November to April. The measurements
reported here were made during the dry season of 1984 on five
important species of the basin forest. These are *Lumnitzera
racemosa, Rhizophora apiculata, Sonneratia alba, Thespesia
populnea,* and *Xylocarpus granatum.*

Depths of brackish water at both spring and neap tides
at the location of the trees used in this study are shown in
Figure 1. The land where *Sonneratia* grows was flooded only

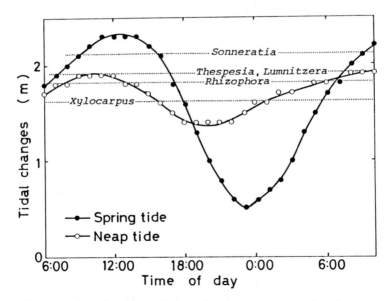

Figure 1. Depths of brackish water at the location of
trees of each species used in the present experiments in a
Namchew forest at both spring and neap tide. Depth of
brackish water is shown as a distance above the lowest low
tide level.

at the height of spring tides, but the land on which the
other species grow was also flooded at neap tide. Times of
high and low spring tides were 12:00 and 23:00 h. Salinity
was measured with a refraction salinometer and ranged from 19
to 32 parts per thousand regardless of tide conditions, which
is close to that of sea water.

Conductance of stomata was measured on fully expanded
mature leaves by a diffusion porometer with a LiCl sensor
(Lambda Inc., Lincoln, Neb., U.S.A.) (Impens, *et al.*, 1967;
Kanemasu, *et al.*, 1969; Maotoni and Machida, 1977).
Transpiration rate (E) was calculated by the method of von
Caemmerer and Farquhar (1981) as follows;

$$E = C \frac{e_i - e_a}{1 - (e_u + e_a)/2P}$$

where C is the stomatal conductance (cm/s), e_i and e_a are the
partial pressure of water vapor in the leaf internal air
space and in the ambient atmosphere, respectively. P is the
atmospheric pressure. e_u was assumed to be the saturated
water vapor pressure at the leaf temperature which was
measured with an infrared thermometer (Matsushita ER-2007
SD-1). Only sunlit leaves were used for the daytime
experiments. In each experiment, at least three leaves were
used. Measurements repeated three times gave almost
identical results.

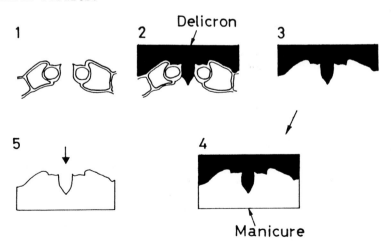

Figure 2. Schematic diagrams of preparation of replica
from a mangrove stoma.

Stomatal apertures were observed with a replica method.
Delicron (Bayer Japan Dentist Co., Osaka), a paste for dental
use, was smeared by a finger on the lower leaf surface, and
left for 3 min until the Delicron paste solidified. After
removing the solidified paste from the leaf, a replica of the
paste was made with transparent nail polish. The replica was
examined under the microscope to measure the stomatal
frequency and apertures shown in Figure 2. This is a
modification of the impression method first described by
Clements and Lang (1934). Shiraishi *et al.*, (1978) and
Pallardy and Kozlowski (1980) warned that cuticular ledges
over the guard cells can cause misinterpretation of replica
observations. However, inspection of Figure 2 suggests that
this is not a serious problem with mangroves.

III. RESULTS

Table 1 shows some morphological characteristics of the
five species studied. As *Xylocarpus* has compound leaves data
are given for a single leaflet. *Lumnitzera* has by far the
smallest leaves, and *Thespesia* the largest, but the latter
are much thinner than the former, as indicated by their lower
SLW in Table 1. *Rhizophora*, *Sonneratia*, and *Lumnitzera* all
have relatively thick leaves compared to those of *Thespesia*
and the leaflets of *Xylocarpus*. There are essentially no
hairs on the lower surface of these leaves and they
apparently have no xeromorphic characteristics. The
differences in water content were relatively small and
probably of no ecological or physiological significance.
Stomatal frequencies and changes in aperture in the
course of a day were observed by use of the Delicron replica
method and the results are summarized in Table 2. Stomatal
frequency was greatest in *Xylocarpus,* intermediate in
Rhizophora and *Thespesia*, and lowest in *Sonneratia* and
Lumnitzera. The frequencies observed on mangrove leaves are
within the range of frequencies observed in other kinds of
plants. The changes in stomatal aperture, the percentage of
stomata open and the pore area as a percentage of leaf area
during the course of a day are shown in Figure 3. The
stomata of *Thespesia* are unique in being already open at 6:00
A.M. and still largely open at 21:00 h when those of other
species are largely closed (Figure 3A) and many remain open
all night. Figure 3B shows that many stomata are never open,
that is never with apertures greater than 1 μm. The largest

Table 1. Fresh weight, dry weight and water content of leaves of five species of mangroves.

Species	Fr wt (g)	Dry wt (g)	Water content (%)	Leaf area (cm^2)	SLW in mg/cm^2
Rhizophora apiculate	3.02 ± 0.18	0.98 ± 0.04	67.4 ± 0.1	60.2 ± 2.1	16.3
Xylocarpus * *granatum*	1.22 ± 0.12	0.35 ± 0.04	72.0 ± 0.5	38.2 ± 3.5	9.2
Sonneratia ovata	2.90 ± 0.12	0.91 ± 0.04	68.2 ± 0.7	58.1 ± 2.1	15.6
Thespesia populnea	3.40 ± 0.14	1.03 ± 0.05	69.7 ± 0.4	131.0 ± 6.0	7.8
Lumnitzera racemosa	0.77 ± 0.07	0.19 ± 0.01	74.1 ± 0.8	12.3 ± 0.8	15.4

*: Data for a single leaflet of the compound leaf

Table 2. Characteristics of the stomata of five species of mangrove plants.

Species	Density (1/mm²)	Length (μm)	Width (μm)	Pore area* Stoma (μm²)	Pore area Leaf area (μm²/mm²)
Rhizophora apiculata	103 ± 3	36 ± 1	3.1 ± 0.2	88 ± 2	3180 ± 530
Xylocarpus granatum	276 ± 5	16 ± 1	2.5 ± 0.2	31 ± 1	3640 ± 530
Sonneratia ovata	97 ± 3	20 ± 1	3.9 ± 0.5	61 ± 3	2980 ± 380
Thespesia populnea	178 ± 12	21 ± 1	2.9 ± 0.1	48 ± 1	4800 ± 490
Lumnitzera racemosa	84 ± 2	15 ± 1	3.8 ± 0.2	45 ± 1	2220 ± 240

The density, length and width of mangrove stomata were measured on replicas of Delicron under a microscopic observation. *: Pore area was calculated by assuming the opening to be an ellipsoid.

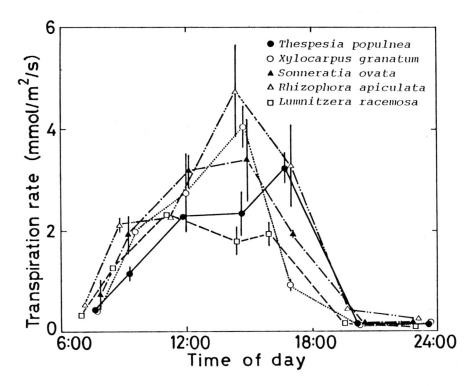

Figure 3. Diurnal changes in stomatal aperture of five
kinds of mangroves on a day of high tide. Stomatal apertures
were measured microscopically using a Delicron replica method.
A: stomatal width, B: ratios of open stomata to the total
number of stomata, C: stomatal pore area/unit leaf area.

percentage were open in *Thespesia,* the lowest percentage in
Rhizophora. Figure 3C shows the stomatal pore area of open
stomata per unit of leaf area. When the observations are
presented in this way *Thespesia* is still highest and
Sonneratia and *Rhizophora* are the lowest in number of open
stomatal pores per unit of leaf area.
     The daily course of transpiration of the five species is
shown in Figure 4. The curves on this figure are
particularly interesting when compared with those of Figure 3
showing stomatal aperture expressed in various ways.
Although the stomata are already half to fully open by 9:00
A.M. the transpiration rate is only half or less than half of
the maximum rate. This suggests that the rate of

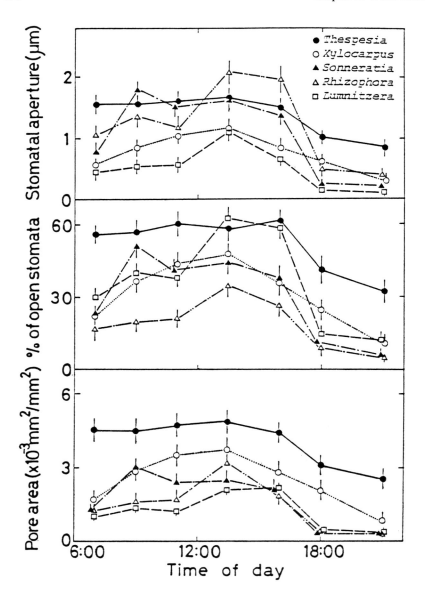

Figure 4.   The daily course of transpiration rate of leaves of mangroves on a day of the spring tide.   The transpiration rate was calculated by the method of von Caemmerer and Farquhar from measurements with a diffusion resistance porometer.   Time of sunrise and sunset were 6:15 and 17:45, respectively.

transpiration of these mangroves is limited in the morning by low radiation rather than by stomatal aperture. Overall, the rate of transpiration is not very well correlated with stomatal frequency, pore area of stomata, or pore area per unit of leaf area. *Xylocarpus*, *Rhizophora*, and *Sonneratia* all have higher maximum rates of transpiration than *Thespesia* which has a very large pore area relative to leaf area and stomata that stay open all day. Neither is the rate correlated with leaf thickness because *Lumnitzera* has a high specific leaf weight, but a low midday transpiration rate while *Rhizophora* which also has a high specific leaf weight has the highest rate of transpiration at midday.

In order to confirm the relationship between the transpiration rates measured by a diffusion porometer and those calculated from the stomatal parameters observed by the replica method, we calculated the conductance (C) to H_2O from the observed stomatal parameters by the method of Kanemasu, *et al.*, (1969), as follows:

$$C = \frac{A \ p \quad a \ b}{4(d + \ a \ b/8)} \tag{2}$$

where A, vapor diffusion constant (0.22 cm^2/s); p, stomatal density; a and b, short and long axis of an elipsoidal stomata; d, thickness of one layer of epidermis. The thickness of epidermis was microscopically determined with a cross section of mangrove leaf for five species (*Thespesia*, 33.3 μ m; *Xylocarpus*, 24.5 μ m; *Lumnitzera*, 46.6 μ m; *Rhizophora*, 18.1 μ m; *Sonneratia*, 20.6 μ m). After substituting the conductance into equation (1), diurnal changes in the transpiration rates from stomatal parameters are calculated and shown in Figure 5. The diurnal changes are quite similar to those determined by a diffusion porometer (Figure 4). The correlation between the transpiration rates determined by a diffusion porometer and calculated by stomatal parameters is shown in Figure 6. The correlation coefficient was 0.82 and the slope of the regression line was nearly one.

The relatively high transpiration rate of mangroves measured in this study suggests that mangroves growing in tropical areas generally have comparatively high rates of water loss. The rates observed here were higher (2 - 5 mmol/m^2/s in the daytime) than those observed in the past (1.7 mmol/m^2/s by Attiwill and Clough (1980). They used mangroves grown at 38'21'S which is far from the equator. Moore et al., used mangroves grown in an area where seasonal temperature variation exists. Ball and Farquhar (1984a,

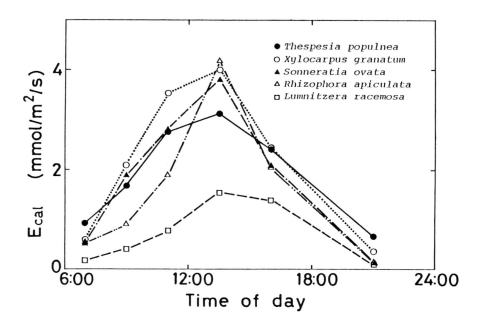

Figure 5. Diurnal changes in transpiration rates (Ecal) calculated with stomatal parameters observed microscopically (Fig. 3).

1984b) used young mangrove seedlings grown in laboratory conditions. Mangroves in Trat, Thailand, grow in a tropical area with almost no seasonal changes in temperature. Furthermore, mature intact leaves of mangroves were used in the present experiments. These differences can explain why the transpiration rates of mangroves obtained in this experiment differ from those obtained in other experiments.

Evidence for high transpiration rate of mangrove leaves in this experiment was obtained by two totally different measuring systems, a replica method and a diffusion porometer. Values of the transpiration rates calculated with stomatal parameters from a replica method were in good agreement with those obtained by a diffusion porometer, because the effective depth of stomata is the thickness of only one layer of epidermis. The stomata in all five mangrove species are flush with the epidermis rather than sunken in pits. Thus, estimation of the effective stomata depth as the thickness of one layer of epidermis seems to be proper.

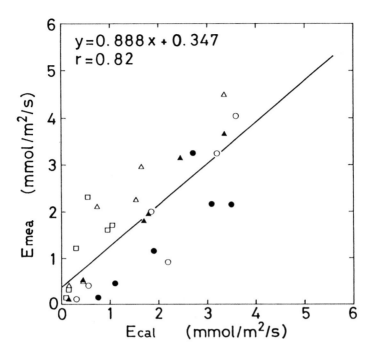

Figure 6.   Correlation between the transpiration rate
measured by a diffusion porometer and calculated with
stomatal parameters observed microscopically.   Ecal, the
transpiration rate calculated with stomatal parameters from a
microscopic observation (see Fig. 4 and Table 2).   Emea, the
transpiration rate measured with a diffusion resistance
porometer.   ●, *Thespesia populnea*; ○, *Xylocarpus granatum*;
▲, *Sonneratia ovata*; △, *Rhizophora apiculata*;   □,
Lumnitzera racemosa.

REFERENCES

Attiwill, P.M. and Clough, F.B., (1980).   Carbon dioxide and
    water vapour exchange in the white mangrove.
    Photosynthetica 14:40-47.
Ball, M.C., and Farquhar, G.D., (1984a).   Photosynthetic
    and stomatal responses of two mangrove species, *Aegiceras
    comiculatum* and *Avicennia marina,* to long term salinity
    and humidity conditions. *Plant Physiol. 74:1-6.*

Ball, M.C., and Farquhar, G.D., (1984b). Photosynthetic and
    stomatal responses of the grey mangrove, *Avicennia marina*,
    to transient salinity conditions. *Plant Physiol.* 74:7-11.
Chapman, V.J., Mangrove Vegetation. *J. Cramer, Vaduz,*
    (1976).
Clements, F.E. and Long, F.L. (1934). The method of
    collocion films for stomata. *Ann. J. Bot.* 21:7-17.
Hsiao, T.C., and Fischer, R.A., Measurement of Stomatal
    Aperture and Diffusive Resistance. P. 5. Bull. 809,
    Coll. Agric. Res. Ctr., Washington State Univ. 1975.
Impens, I.I., Stewart, D.W., Allen, Jr. L.H., and Lemon, E.R.,
    (1967). Diffusive resistances at, and transpiration
    rates from leaves *in situ* within the vegetative canopy of
    a corn crop. *Plant Physiol.* 42:99-104.
Kanemasu, E.T., Thurtell, G.W., and Tanner, C.B. (1969).
    Design, calibration and field use of a stomatal diffusion
    porometer. *Plant Physiol.* 44:881-885.
Kuraishi, S., and Nito, N. (1980). *Bot. Mag. Tokyo*
    *93:*209.
Maotani, T., and Machida, Y. (1977). *J. Japan. Soc.*
    Hort.Sci. *46:*1.
Meidner, H., Stomatal Physiology (P.G. Jarvis and T.A.
    Mansfield, ed.), p. 26. Cambridge Univ. Press, Cambridge,
    1981.
Moore, R.T., Miller, P.C., Albright, D., and Tieszen, L.L.
    (1972). Comparative gas exchange characteristics of
    three mangrove species during the winter.
    Photosynthetica 6:387-393.
Moore, R.T., Miller, P.C., Ehleringer, J., and Lawrence,
    W.(1973). Seasonal trends in gas exchange
    characteristics of three mangrove species.
    Photosynthetica 7:387-394.
Nito, N., Kuraishi, S., and Sumino, T., (1979). Daily
    changes in the highest leaf surface temperature of
    plants growing at Heiwa Avenue, Hiroshima. *Environ.*
    Control in Biol. 17:59-66.
Pallardy, S.G. and Kozlowski, T.T. (1980). Cuticle
    development in the stomatal region of *Populus* clones.
    New Phytol. 85:363-368.
Shiraishi, M., Hashimoto, Y., and Kuraishi, S. (1978).
    Cyclic variations of stomatal aperture observed under the
    scanning electron microscope. *Plant and Cell Physiol.* 19:
    637-645.
Slavik, B. (1974). Methods of Studying Plant Water Relations
    p. 314. Springer-Verlag, New York.

Von Caemmerer, S., and Farquhar, G.D. (1981). Some relationships between the biochemistry of photosynthesis and the gas exchange of leaves. *Planta 153*:376-387.

MEASUREMENT OF SOIL MOISTURE
AND WATER POTENTIAL

Yoshio Kano

Department of Electrical Engineering
Faculty of Technology
Tokyo University of Agriculture and Technology
Koganei, Tokyo

I. INTRODUCTION

Measurement of soil moisture is very important for
studies of the physiological ecology of plants. Techniques
vary from gravimetric measurement of water content on a mass
or volume basis to determination of the free energy of water
by psychrometric measurement of water vapor pressure in soil
air spaces. This paper will briefly review some of the
common methods of measuring soil moisture and will present
some new methods being developed and tested.

II. REVIEW OF SOME CURRENT METHODS
OF MEASURING SOIL WATER CONTENT

This is a very brief review of the major techniques
being used for the measurement of soil water status. No
attempt is made to identify particular instruments.

A. Electronic Methods

The electrical properties of soil are strongly dependent on the moisture content of the soil. Various methods
have been devised to minimize electrical effects of other

Measurement Techniques in Plant Science

physical properties (e.g., bulk density, soil particulate
structure, temperature).

   1. *Resistance Type.*   There are two primary ways of
measuring the electrical resistance or capacitance of soil.
One way utilizes two electrodes; the other four electrodes.
when resistance is measured, it is important to control or
measure electrode impedance and the electromotive force (emf)
on the surface of the electrodes. All electronic resistance
probes are influenced by temperature and the solute
concentration of the soil solution. Utilizing high A.C.
current and voltage across the two electrodes of that type
will overcome sensor characteristics and allow electrical
resistance due to soil .water content to be accurately
determined. In the 4 electrode method, a null-type impedance
meter is used to zero output voltage of the circuit, allowing
soil electrical impedance due to soil water content to be
accurately determined (Baba *et al.*, 1970; Grahame, 1952, 1964;
Mayell and Langer, 1964; Mine *et al.*, 1969; Morris 1964;
Schwan, 1968; Toshima and Uchida, 1970; Yamamoto *et al.*,
1970).

   2. *Capacitance Type (L-C resonance type).* Because the
dielectric constant of water is very high and that of soil is
almost strictly controlled by soil water content, soil
moisture is detected by a capacitance between the electrodes
in the soil. The frequency of L-C resonance oscillator
changes in proportion to the soil water content. (Ambrus *et
al.*, 1981; Hamid, 1973; Kuraz *et al.*, 1976; Kuraz and
Matousek, 1977; Pfeifer and Fisher, 1978; Topp, 1980; Vlaby,
1974).

                          B. Light Method

   Water has unique reflectance properties in the infrared
region of the spectrum. Because of difficulties in detecting
the longer infrared wavelengths, 1.93 μ m usually is used.
Since the reflectance of soil is affected by the condition of
the surface as well as its water content, two wavelengths are
used. The wavelength of 1.93 μ m responds to water and some
other wavelength is used that will not be responsive to water.
If the absorption ratio of these two wavelengths is recorded,
the water content can be distinguished from other properties
of the soil sample (Ben-gera and Norris, 1968; Beutler, 1965;
Horbert *et al.*, 1974).

C.   Radioisotope (Neutron) Method

Neutrons passing through soil are scattered by hydrogen atoms contained in the soil. Gamma radiation passing through soil is absorbed by soil particles and hence is inversely proportional to the density of the soil. The moisture content of a soil, therefore, can be determined independently from its density, by determining the ratio of the scattered neutron radiation to the absorbed gamma radiation (Nagy and Razga, 1966).

III.   REVIEW OF CURRENT METHODS OF
MEASURING SOIL WATER POTENTIAL

Water potential is a measure of the free energy of water in a mixture. The free energy of pure water is maximal. Any solute or matric surface (e.g. cellulose fibers or clay particles) added to pure water will decrease the free energy of that water. The amount of decrease may be measured in pressure terms (pascals) and expressed as water potential. By definition the water potential of pure water is zero.   As a solution is created by adding solute to the pure water, the free energy of water and therefore, its water potential declines (Richards, 1928; Richards and Wadlieigh, 1952).

A.   Pressure Measurement of
Soil Moisture Tension

1. *Mercury tensiometer.* The classical tensiometer uses a porous cup and a mercury manometer to measure the tension on the soil water column created by matric potential of solid surfaces, the osmotic potential of the soil solution, and vapor pressure deficit of the soil atmosphere. The most common device for the determination of the tension on the column is a Bourdon vacuum gauge.

2. *Electronic tensiometer.* There are various electronic tensiometers which are combinations of a porous cup and a bellows or a diapharm and various electronic methods of measuring distention of the systems by tension from the water column. (Bianchi and Tovey, 1968; Klute and Peters, 1962; Whitney and Porterfield, 1968).

B.   Vapor Pressure or Dew Point
Methods for the Measurement of Soil Moisture

Air in vapor pressure equilibrium with soil may be
measured psychrometrically or hygrometrically. The resultant
vapor pressure or dew point will indicate the water potential
of this surrounding soil if the system is calibrated to a
known water potential, (Hoffman and Herkelrath, 1968; Hoffman
and Splinter, 1968; Riggle and Slack, 1980).

1. *Psychrometers.* Thermocouples are used to obtain dry-
and wet-bulb temperature readings of porous cups inserted
into the soil to be measured. Water vapor in the cup comes
to equilibrium with that of the surrounding soil and the
relative humidity of the air is measured psychrometrically.

2. *Hygrometers.* Dew point of the air inside of a porous
cup in equilibrium with surrounding air may be determined
with a thermocouple system arranged as shown in Figure 1.
Peltier cooling is applied in pulses until the dew point is
reached and the system stabilizes at dew point. When dew
point is calibrated against solutions of known water
potential it may be used to measure unknown water potential
of soil samples (Hdlbo, 1980).

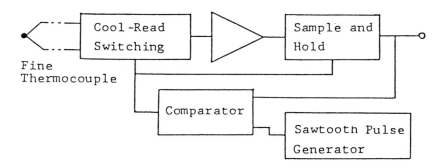

Figure 1.   Schematic diagram of dew point hygrometer.

IV.   NEW METHODS OF MEASURING SOIL WATER

A.   The Four Electrodes Method

As reviewed earlier in this paper, the electric
conductance of soil may be used to measure its moisture
content. The circuit used to measure electrical conductance
must be carefully selected, however, because electrode
impedance and emf at the surface of the electrodes are
critical sources of error. I recommend the four electrode
method (Figure 2) for measuring electrical conductance of
soil. Of the four electrodes, the first and fourth are
called the current pair; the second and third are called the
voltage pair. The circuit diagram for the sensor (Figure 3)
shows the electrode impedance (Zc1, Zc2, Zv1, and Zv2).

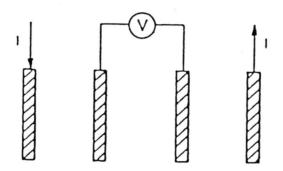

Figure 2.   Four electrodes method.

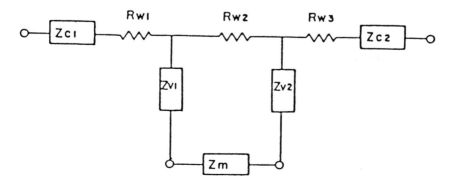

Figure 3.   Circuit diagram of electrodes.

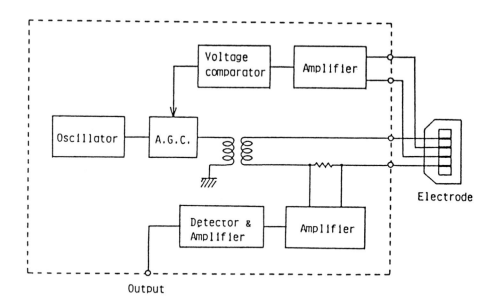

Figure 4.   Schematic diagram of the measurement circuit.

The input impedance of the voltage detector is Zm.   Rw1, Rw2
and Rw3 are the resistances in the soil.   The circuit diagram
for the conductivity meter is shown in Figure 4.   As the
voltage detector (Zm in Figure 3) is set at a high input
impedance, no current will flow through Zv1, Zv2 and Zm,
consequently, no voltage   drop will occur on the impedance at
Zv1 and Zv2. Even if an impedance change does occur at Zc1
and Zc2, the voltage between the second and third electrodes
is adjusted by the Automatic Gain Control (A.G.C.) circuit.
The conductivity of the soil (G) may then be calculated from:

$$G = k \frac{I}{V} = k'I$$

where I is current measured and k and k' are constants.   The
system is stable in different soil types and over a large
range of water contents as shown in Figure 5.   Additional
information on the four electrode sensor may be found in
Valder (1954) and Lògan (1961).

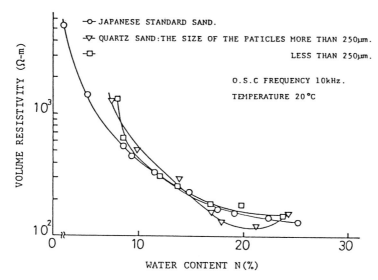

-o- JAPANESE STANDARD SAND.

-▽- QUARTZ SAND: THE SIZE OF THE PATICLES MORE THAN 250μm.

-□-   LESS THAN 250μm.

O.S.C FREQUENCY 10kHz.

TEMPERATURE 20°C

Figure 5.   Characteristic of volume resistivity vs water content.

B.   Phase Lock Loop Method

Since the capacitance of soil is dependent upon water content as reviewed earlier in this paper, the Phase Lock Loop (PLL) method is appropriate. With the availability of excellent and inexpensive integrated circuits (I.C.), PLL now provides an accurate and easily used method. Most electric circuits for PLL are packaged in one I.C. chip. A new meter that is being tested for soil moisture determination is shown in Figure 6. The electric constant Cx between the electrodes is dependent on the water content of the soil. As the electric constants Cx, Co and L generate a resonance circuit of frequency fo, PLL circuits can be used to detect fo, where Co is the stray capacitance parallel to Cx and L is a known stable inductance. Thus Cx can be used to measure soil water content because:

$$fo = \frac{1}{2\pi\sqrt{(Co + Cx)\ L}} \quad (Hz)$$

Frequency (fo) is not influenced by Rx (i.e. the solutes in the soil water). Characteristics of this meter are shown in Figure 7. Cx is the known capacitance and N is the soil water content of a specific soil sample. The curve fx vs. Cx

is the resonance curve of fo. The curve fx vs. N is the specific relationship obtained for the sample used (Kano *et al.*, 1974) and must be specifically calibrated for different soil type.

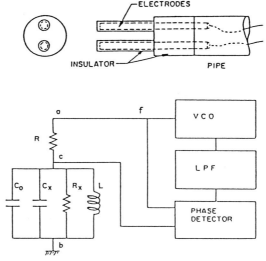

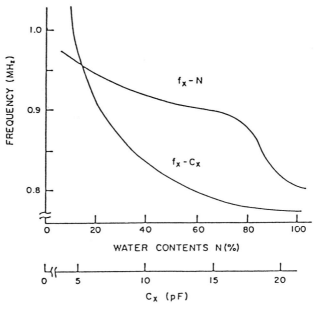

Figure 6. soil moisture meter using PLL and its probe.

Figure 7. Characteristics of the soil moisture meter.

C. Phase Sensitive Detector (PSD) method

Another method for measuring electrical capacitance of the soil is Phase Sensitive Detector (PSD)utilized with a porous sensor. When the electrodes are in direct contact with the soil (as in the PLL described above) the relationship between capacitance and water is affected by density and chemistry of the soil.

By placing the electrodes on a porous cup as shown in Figure 8, the cup capacitance comes to equilibrium with the water potential of the surrounding soil independent of soil density and/or chemistry. The output voltage (V_{o1}) of the circuit is:

$$V_{o1} = -(GxR + jCxR)e_i$$

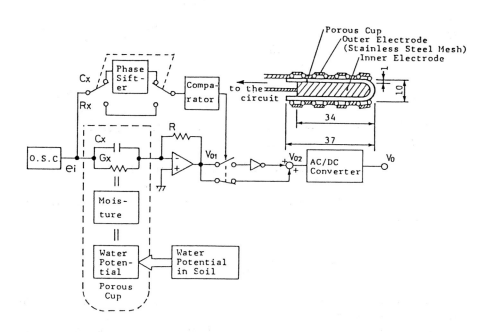

Figure 8. The probe and the block diagram of electric circuit.

where Gx is conductivity, Cx is capacitance and R is feedback resistance of the amplifier. the imaginary part of Vo_1 is Vo_2 and they are separated by the PSD. The voltage Vo_2 measures the Cx and can be calibrated to represent the soil water potential. Characteristics of this system are shown in Figure.9 for the water potential range of 5 MPa (i.e., pF 4.5) to 1×10^3 MPa (i.e., pF 7.0) comparing the output voltage of GxRe.

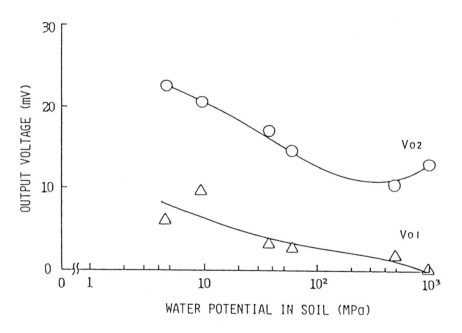

Figure 9. Characteristics of the water potential meter by the PSD method.

D. Near Infra-Red, Light Emitting Diode
(NIR LED) Method

A new system using a near infra-red LED, a PIN photodiode, an integrating sphere and an appropriate electric circuit is shown in Figure 10. The output voltage of the instrument is linear and inversely proportional to its water content of soil in the range 20% to 50%. As the output voltages are influenced by the color of soil, outputs are normalized (Figure 11).

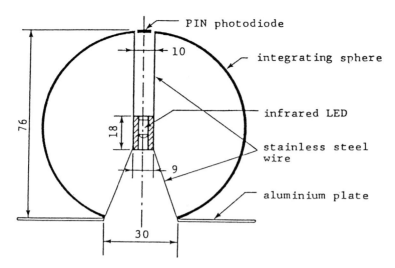

Figure 10   Moisture meter using NIR LED.

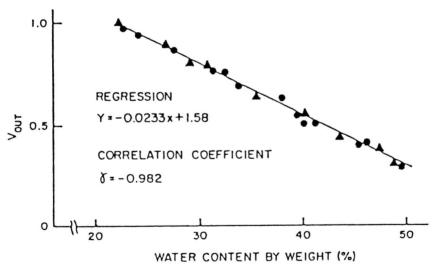

Figure 11.   Normalized output voltage.

Inserting the NIR LED system in a porous cup (Figure 12) produces a sensor that can be calibrated to soil water potential (Figure 13).   the system can be calibrated over a water potential range from field capacity (i.e., 6.4×10^{-3} MPa or pF 1.8) to the permanent wilting point (i.e., 1.6 MPa or pF 4.2) and beyond to completely dry soil with 1×10^3 MPa (i.e., pF 7.0).

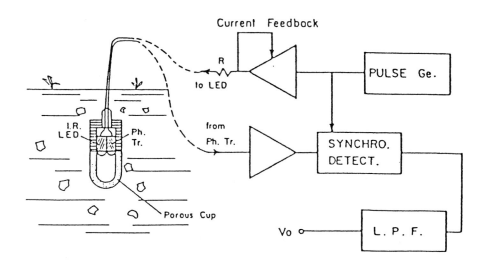

Figure 12.   Schematic diagram of water potential meter.
(sensor and meter)

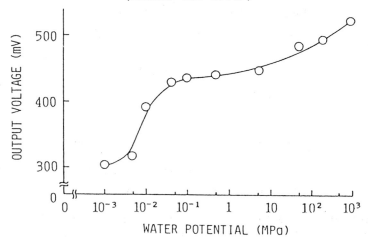

Figure 13.   Output voltage characteristic of the water
potential meter used with porous cup.

V. APPLICATIONS

A. Electronic Measurement

Measurement of root growth is as important as measurement of shoot growth, but more difficult because roots are hidden in the soil. However, roots remove water, reducing the electrical conductance of the soil in their vicinity as compared with that of the soil mass. Thus periodic measurement of soil conductance should indicate when roots invade a soil mass containing the equipment for measuring conductance (Chloupek, 1972, 1977).

This approach was tested with the improved conductivity meter described earlier in this paper, using the comb arrangement of four electrodes. One set of electrodes were buries in a similar container filled with gelatin or sand on which seeds were planted while the reference electrodes were buried in a similar container in which no seeds were planted (Figure 14). The decrease in voltage at point P, indicating a decrease in conductance, as water was removes from the soil is indicated in Figure 15.

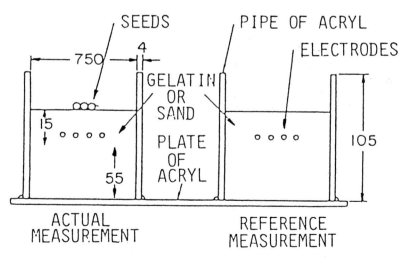

Figure 14.   Arrangement of measuring the growth of root.

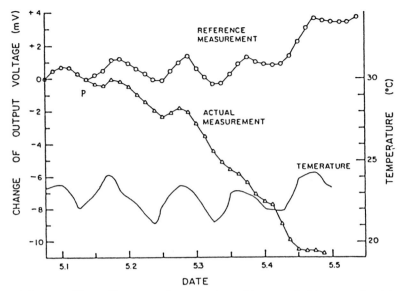

Figure 15. Change of output voltage of the conductivity meter measuring the growth of a root.

B.   Stem   Water Potential

the soil moisture meter using NIR-LED was adapted to measurements of stem water potential. The porous cup used for soil moisture measurements was replaced by a ceramic plate and the circuitry is shown in Figure 16.

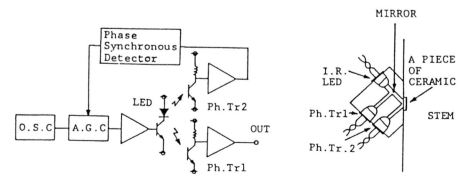

Figure 16 Diagram of circuit and anew probe using a piece of ceramic plate and NIR LED.

As the response is influenced by temperature this is
compensated by photo-transistor1 (Ph.Tr1) that receives light
directly from the LED while the signal from the ceramic plate
is received by photo-transistor2 (Ph.Tr2).

REFERENCES

Ambrus, L., Antal,E., and Karsal, H.A. (1981). New
    electronic evaporation and rain measuring equipment.
    Agri. Meteorology 25:35-43
Baba, Y., Mine, T,. Yamamoto, T., and Yamamoto, y. (1970).
    Correction Factors for the resistances measurement with
    four point probes. *J. of Elect. Eng. of Japan*
    90:1775-1777.
Ben-gera, I., and Norris, K.H. (1968). Direct
    spectrophotometric determination of fat and moisture in
    meat products. *J gage of Food Science 33*:64-67.
Beuter, A.J. (1965). An infrared backscatter moisture gage.
    Tappi. 48:490- 493.
Bianchi, W.C., and Tovey, R. (1968). Continuous monitoring of
    soil moisture tension profiles. *Trans. of ASAE.*
    11:441-443 and 447.
Chloupek, O. (1972). The relationship between electric
    capacitance and some other parameters of plant roots.
    Biologia Plantarum 14:229-230.
Chloupek, O. (1977). Evaluation of the size of a plant's
    root system using its electrical capacitance. *Plant and
    Soil 48*:525-532.
Grahame, D.C. (1952). Mathematical theory of the faradic
    admittance. *J. of the Electrochemical Society*
    99:370C-385C.
Graja, E.D.C. (1964). Properties of the electrical double
    layer of a mercury surface. The effect of frequency on
    the capacity and resistance of ideal polarized electrodes.
    J. Am. Chemn. Soc. 68:301-310.
Hamid, M.A.K. (1973). Sarvey of radio frequency techniques
    for teledetection of soil moisture. *J. of Microwave
    Power 8*:217-225.
Hdbo, H.R. (1981) A Dew-point hygrometer for field use.
    Agric. Meteorology 24:117-225.
Harbert, F.C., Sjohorg, Bal, J (1974). Measuring moisture
    content on-line by infra-red method. *Control &*

Instrumentation. pp, 36-37.

Hoffman, G.J., and Splinter, W.E (1968). Instrumentation for measuring water potential of an intact plant-soil system. *Tranc. of ASAE.* *11*:38- 40 and 42.

Hoffman, G.J., and Herkelrath, W.N, (1968). Design features of intact leaf thermocouple psychrometers for measuring water potential. *Trans. pf ASAE.* *11*:631-634.

Kano, Y., Hasebe, S., Kariya , M., and Kobayashi, K. (1974). A dielectric soil moisture meter using PLL. of Inst. *Elect. Eng. of Japan.* *100-A:*60.

Klute, A., and Peters, D.B. (1962). Arecordeng tensiometer with a short response time. *Soil Sc. Am. Proc.* *26:*87-88.

Karaz, V., Kutilek, N.M., and Kaspar, I., (1976) Resonancecapacitance soil moisture meter. *Soil Science* *110:*278-279.

Kuraz, V., and Matousek, J. (1977). New dielectric soil moisture meter for field measurement of soil moisture. *ICID Bullentin* *26*:76--79.

Logan, M.A. (1961). An AC bridge for semiconductor resistivity measurements using a four-point probe. *Bell Syt. Tech. J.* *40*:885-919.

Mayell, J.S., and Langer, S.H. (1964). Effect of dilute chloride ion on platinum electrodes. *J. of Electroanalytical Chemistry* *7*:288-296.

Mine, T., Yamamoto, T., Yamamoto, Y., and Baba, Y. (1969). A method of impedance measurement of electrolytic conductors. *J. of Inst. of Elect. Eng. of Japan* *89*:1961-1969.

Morris, M.D. (1964). The rate of electrode oxide formation during chronopotentiogrames at a platinum anode. *J. of Electroanalytical Chemistry* *8*:350-358.

Nagy, A.Z. and Razga, T. (1966). Radioisotopic combined moisture-density meter. *J. of Sci. Instrum.* *43*:383-387.

Pfeifer, G., and Fischer, H. (1978). Verfahreu zur elektrischen. *Praezisionsmessung Kleiner Wege* *27*:500-503.

Richards, L.A. (1928). The usefulness of capillary potential to soil-moisture and plant investigation. *J. Agri. Research.* 37:719-762.

Richards, L.A., and Wadlieigh, G.H. (1952). Soil water and plant growth. pp. 73-251. Academic Press (New York).

Riggle, F.R., and Slack, D.C (1980). Rapid determination of soil water characteristic by thermocouple psychrometry. *Trans. of ASAE.* 23:99-103.

Schwan, H.P., (1968). For-electrode null techniques for impedance measurement with high resolution. *The Review of Scientific Instruments.* *39*:481-485.

Topp, G.C. (1980). Electromagnetic determination of soil

water content: measurements in coaxial transmission Lines. *Water Resources Research 16*:574-582.

Toshima, S., and Uchida, I. (1970). Specific absorption halide ions at the germanium electrolyte-solution interface; frequency dispersion of interfacial impedance. *Electrochimica Acta 15*:1717-1732.

Ulaby, F.T. (1974). Radar measurement of soil moisture content. *IEEE Tran. Antennas and Propagation. AP-22*:257-265.

Valdes, L.B. (1954). Resistivity measurements of germanium for transistors, *Proc. of the IRE. 42*:420-427.

Whitney, J.D., and Porterfield, J.G. (1968). Moisture movement in a porous, hygroscopic solid. *Trans. of ASAE. 11*:716-719 and 723.

Yamamoto, Y., Mine, T., Yamamoto, T., and Baba, Y. (1970). conductors. *J. of Inst. of Elect. Eng. of Japan. 90*:2569-2576.

Chapter 3
Photosynthesis

MEASURING PHOTOSYNTHESIS
UNDER FIELD CONDITIONS:
PAST AND PRESENT APPROACHES

C. B. Field

Department of Plant Biology
Carnegie Institution of Washington
Stanford, CA 94305

H. A. Mooney

Department of Biological Sciences
Stanford University
Stanford, CA 94305

I. INTRODUCTION

For many years, scientific questions concerning the CO_2 and water vapor exchange of naturally growing plants have outpaced technological solutions for satisfactory measurements. Recent advances, especially in electronics and materials science, have narrowed but not closed the gap between ideal and realistic objectives. As was the case in the beginning of the era of field gas exchange, nearly 40 years ago, the primary challenge in instrument design is reaching a balance among characteristics that are often incompatible -- including accuracy, level of environmental control, range of parameters measured, portability, and measurement speed. Modern instruments, designed for optimum performance with regard to some characteristics, may be excellent for some research tasks but poorly suited for others. Here, we use an historical perspective to introduce conceptual and technical issues in field gas exchange and then consider modern systems with respect to these concepts.

Many papers review techniques for measuring photosynthesis under field conditions. Sestak, Catsky, and Jarvis (1971) edited a definitive treatise on techniques for measuring plant productivity. Much of the information in this classic treatment is still useful. Chapters in Coombs et al. (1985), Marshall and Woodward (1985) and Pearcy et al. (1989) cover many topics in gas-exchange research in more detail than is practical here. Many of the recent advances in field gas-exchange research have involved meteorological, rather than chamber-based techniques. Baldocchi et al. (1988), Matson and Harriss (1988), and Gosz et al. (1988) discuss the potential and limitations of large scale, chamberless gas-exchange technologies.

II. BRIEF HISTORY

Modern studies of plant gas exchange began with the development of the infrared gas analyzer (IRGA). Earlier techniques utilized CO_2's strong absorption in the infrared as the basis for measuring concentration, but it was not until the mid-1940's that instruments utilizing broad-band (non-dispersed) sources, positive filtration, and Luft's (1943) combination of a gas-filled detector with interrupted or "chopped" radiation became rugged, reliable, and practical enough for field studies. Before the application of IRGAs to field studies, photosynthesis was only measurable as a time-averaged parameter, over times ranging from minutes for spectrophotometric titration (Bowman, 1968) to weeks with growth analysis. IRGAs were first utilized for studies of photosynthesis by Egle and Ernst (1949) and were employed for field studies as early as 1953 by Bosian in Germany, 1957 by Tranquillini in Austria, and 1958 by Went in the United States. Bosian's (1955) "Fliegende Laboratorium" or flying laboratory was one of the first mobile laboratories for gas-exchange experiments. His pioneering studies (Bosian, 1960) demonstrated the importance of climate-controlled chambers for the accurate measurement of photosynthesis and led to an elaborate system for controlling chamber temperature, humidity, and wind speed (Bosian, 1965). Concurrent with the development of large mobile laboratories, Billings and his students began utilizing more portable IRGA-based systems for field studies in alpine regions of the Rocky Mountains and elsewhere (Mooney et al., 1964; Billings et al., 1966).

Through the 1960's and 1970's, the dominant trend in photosynthesis systems for field research was the development of increasingly large and powerful mobile laboratories, often designed for use in extreme environments. Eckardt's (1966) mobile laboratory, developed in France, utilized an array of chamber types, including a whole-plant chamber. This system measured CO_2 and H_2O fluxes with a compensating or null-

balance approach. Lange et al. (1969), for their studies in
the Negev desert, constructed a mobile laboratory capable of
simultaneously operating several climate-controlled chambers.
Their system relied on thermoelectric or peltier-effect
modules for temperature control. The system of Lange et al.
(1969) became one of the first commercially-available gas
exchange units (Koch et al., 1968). In the United States,
Strain (1965) and Mooney et al. (1971) constructed
increasingly complex mobile laboratories. The unit described
by Mooney et al. (1971), with improvements described by
Björkman et al. (1973), was eventually capable of controlling
and measuring light, temperature, humidity, and CO_2.
Employing computerized data acquisition and automatic
computation, this system was competent to perform, in the
field, most of the experiments expected of a modern
laboratory system, including analytic studies based on the
interpretation of CO_2 response curves.

The increasing complexity of these systems brought
increases in size and power requirements. The Björkman et al.
(1973) system, fully equipped for desert conditions, consumed
nearly 10 kW of input power.

The 1980's have seen a reversal in the size escalation
of measurement systems and the development of small, truly
portable units that can be operated by a single individual.
Many technical developments have contributed to this
miniaturization. Integrated circuits, combining high
accuracy, low power consumption, and excellent temperature
stability have been among the most important. Other keys to
miniaturization have been the broader application of peltier
systems for temperature control (replacing large
compressors), mass flow controllers (in place of rotameters
and mixing pumps), and miniature plumbing developed for
industrial fluidics applications. Low cost, yet
sophisticated, data acquisition systems have replaced
minicomputers. New IRGAs are light, yet relatively
insensitive to vibration and frequency variation. Coincident
with these developments has come increased commercialization
of field systems. We concentrate our remarks here on new
developments in IRGAs and commercial field systems, focusing
more on operational than technical aspects.

III. RECENT IMPROVEMENTS IN INFRARED GAS ANALYZERS

The earliest IRGAs were a big advance over alternative
techniques, but their limited sensitivity restricted their
application to photosynthesis experiments that did not
require measuring CO_2 concentration to more precision than
plus or minus several ppm. The major improvement increasing
sensitivity to 1 ppm of CO_2 or better was the development of
differential analyzers in which the sample gas is compared to

a slightly different reference gas, rather than a CO_2-free reference. Incorporating this advance, commercial IRGAs sold during the 1960's were potentially as accurate as today's instruments, but had several features that presented problems, especially under field conditions.

Many of the characteristics that made the IRGAs of a decade ago less than ideal for field research were intimately connected with the same design features that conferred high sensitivity (Table 1). Overcoming these problems, for example the vibration sensitivity inherent in the Luft detector, has required fundamental changes in instrument design. Other problems, for example the frequency sensitivity and the large power requirement, have been solved more-or-less automatically with general advances in electronics. Sadly, none of the solutions is ideal, and some cause other problems. One example of this is that many of the solid-state detectors are less sensitive than the Luft detector. Increasing the cell length counteracts the loss of sensitivity, but at the cost of a larger instrument. Designing an excellent field instrument, of which there are several, is now as much an exercise in judicious choice among specifications to emphasize and limitations to accept as it is a challenge to produce a fundamentally new technology. However, new technologies on the horizon, including tuned-laser sources (Sheehy 1985) and folded optical paths (Bingham et al., 1978), may dramatically change these trade-offs in the next few years.

The principal IRGAs utilized today in field systems are manufactured by The Analytical Development Co., Limited (Hoddesdon, England), Leybold-Heraeus GMBH (Hanau, Federal Republic of Germany), and LI-COR Inc. (Lincoln, Nebraska, USA).

1. The Leybold-Heraeus BINOS 100. Leybold-Heraeus manufactures a large range of industrial process analyzers, available with detectors for many gases. The portable BINOS analyzers suitable for field gas exchange can be equipped with either one or two analysis channels, including differential CO_2, absolute CO_2, differential water vapor, absolute water vapor, or some combination. Depending on the configuration, the BINOS weighs as little as 5 kg and consumes as little as 15 W of input power, making it suitable for operation from a battery, a generator, or power mains. The BINOS detector is in essence a mass-flow meter connecting two chambers alternately exposed to the IR source. Because the detector operates at a high chopping frequency, it is relatively vibration insensitive. Control of the chopper motor by an internal microprocessor renders the BINOS insensitive to AC frequency variation. The combination of a single IR source, reference filters, and a microprocessor-based internal stabilization system that corrects for changes in source output or detector sensitivity (Fig. 1) keep drift

Table 1. Traditional problems associated with IRGA function
in the field, and a summary of solutions available in many,
but not all, instruments.

Problem	Cause	Solution
Vibration sensitivity	Instrument motion exaggerates oscillation of Luft detector	Solid-state detectors, including mass flow and photovoltaic
Frequency sensitivity	Low frequency radiation choppers tied to line frequency	Higher chopping speeds and internal speed regulation
Sensitivity to ambient CO_2	Open optical path allows ambient CO_2 to appear as noise	Optical paths either sealed or purged with CO_2-free air
Zero and gain drift	Use of separate IR sources for sample and reference, and lack of feedback source regulation	Designs using one IR source and one detector plus feedback regulation of source output
Temperature sensitivity	Lack of correction for ideal gas behavior plus temperature sensitive electronics	Temperature corrected calculations or temperature controlled cells and temperature stable electronics
Large power requirements	Thermostating, chopper motors, and IR sources	IC electronics, low power sources, and temperature correction rather than control

to very low levels. The BINOS can be operated after a 2
minute warm-up.

    2. LI-COR. LI-COR manufactures three IRGA models
specifically for field gas-exchange measurements. All use
the same basic IRGA, a single source design with an
internally regulated, high frequency chopper and a solid
state detector, making them very insensitive to vibration and
AC frequency variation (Fig. 1). The basic model (LI-6251)
provides non-linear outputs for CO_2 and temperature while the
LI-6252 and LI-6262 are digital instruments that provide
corrected and linearized outputs. The LI-6262 has detectors

for both CO_2 and water vapor. The LI-COR IRGAs have several
features that allow them to achieve very low noise (0.2 ppm
peak to peak at 350 ppm) and very high repeatability (0.2 ppm
at 350 ppm). These include thermoelectric cooling to depress
the noise from the solid state detector, focusing optics at
both ends of the cells, and feedback regulation of the source
output to maintain a strong signal at the detector,
regardless of the IR absorption. Feedback regulation of the
source output allows the LI-COR IRGAs to function effectively
in either absolute (up to 1000 ppm) or differential mode. At
approximately 5 kg, the LI-COR IRGAs are comparable in mass
to the BINOS 100, but the capacity to operate a one-channel
instrument in either absolute or differential mode confers
high flexibility at a modest price. Though not based on
radically new technology, the LI-COR IRGAs are well optimized
for field photosynthesis studies.

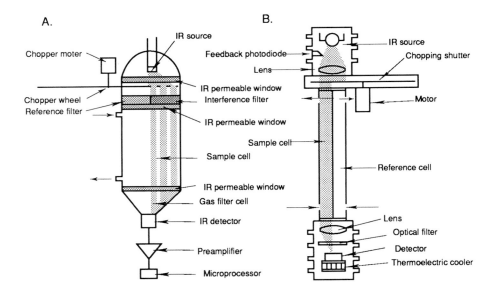

Fig. 1. IRGA designs. A. BINOS 100 in absolute
configuration. Light shading indicates IR path when the
chopper exposes the interference filter. B. LI-COR LI-6251
with focusing optics at source and detector. Shading
indicates the IR path when the chopper exposes the sample
cell.

3. The Analytical Developments Company's LCA-2. This
instrument is a small, light (2.8kg), battery-powered IRGA
with a solid-state detector. The LCA-2 has a single cell, but
utilizes a unique gas switching system to provide both
chopping and differential measurements. The IR source in the
ADC-LA is continuously energized and continuously exposed to
the sample cell, but the sample cell is alternately flushed
with sample and reference gases. If the reference gas is CO_2
free, the analyzer functions in absolute mode. If the
reference gas contains CO_2, the analyzer output is the
difference between the reference and sample concentrations.
The cycling time for sampling both gases is 6 seconds, making
the ADC-LA unacceptably slow for some applications. The ADC-
LA functions over a range of 0-1000 ppm CO_2, yielding
precision of 0.5% in the 0-500 ppm range.

IV. MEASUREMENT APPROACHES

       Nearly all of the photosynthesis systems so far
developed operate in a closed, open, or isotope
configuration. Here, we discuss only IRGA-based, or closed
and open systems (Fig. 2). For a discussion of isotope
techniques, see Hällgren (1982). Closed and open systems
have been designed for measuring CO_2 exchange, H_2O exchange,
or both. For simplicity, we summarize the concepts for
measuring CO_2 exchange on a single leaf, but the principles
are the same for water vapor and for measurements at the
level of the branch, the whole plant, or even the community.
The equations for calculating photosynthesis and related
parameters are presented here in simplified form. Von
Caemmerer and Farquhar (1981), Ball (1987), and Field et al.
(1989) present detailed calculations, with corrections.
       In the closed system, an air stream is maintained in a
closed loop, including the chamber, for a period during which
CO_2 is depleted by photosynthesis. The rate of
photosynthesis is calculated from the CO_2 depletion rate(Fig.
2). Chamber conditions change continuously during a closed
system measurement, so the measurements never represent truly
steady state responses. However, if the leaf is in the
chamber for a short enough period that the CO_2
depletion by photosynthesis is small, and if the chamber
approximates ambient conditions, closed systems can provide
accurate measurements of photosynthesis under ambient
conditions. The major problem with closed-system measurements
is that it is very difficult to know whether the observations
reflect steady-state responses to current conditions, past
conditions, or transient responses to changing conditions.
On the other hand, closed systems provide very attractive

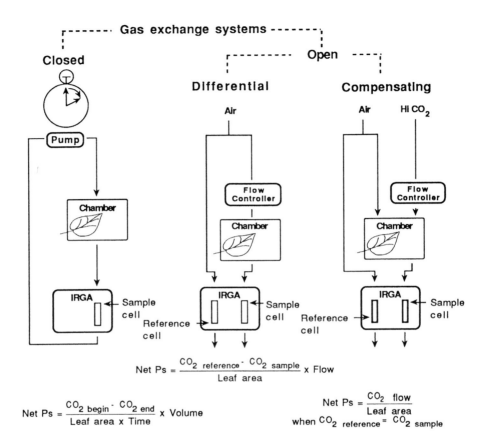

Fig. 2. The principal measurement philosophies used in IRGA
photosynthesis measurements. The equations are intended to
convey the fundamental principles of the measurement
philosophy but do not include important corrections and
compensations.

technologies for surveying photosynthesis on large samples of
leaves.

In an open system, air passes through the chamber only
once, and photosynthesis is equal to the difference between
the amount of CO_2 leaving and entering the chamber (Fig. 2).
If the composition of the air entering the chamber is
constant and if the leaf in the chamber equilibrates, the
chamber attains a steady-state. If the air flow rate through
the chamber is high, then photosynthesis causes only a

moderate depletion in CO_2, and an open system can be run in differential mode (Fig. 2). As the rate of air flow decreases, the CO_2 depletion due to photosynthesis increases. To offset this depletion, compensating systems inject CO_2 into the chamber. In compensating systems, photosynthesis is equal to the amount of CO_2 added, as long as the CO_2 concentrations in the air entering and leaving the chamber are the same (Fig. 2). Both types of open systems provide steady-state conditions and are suitable for extended measurements (e.g., diurnal courses or response curves) on single leaves. A compensating system is often more economical with gas than a differential system, a frequent advantage in the field, but differential systems are typically superior for tracking rapid changes. Several modern instruments blur the distinction between differential and compensating approaches by incorporating features of each.

Each of these sampling designs can be incorporated in a system with or without environmental control. However, intrinsic differences among these system philosophies shape the reasonable expectations for systems with different levels of environmental control. For example, since closed systems do not make steady-state measurements, there is little justification for designing a closed system to maintain constant temperatures over prolonged periods. Exceptions to this include semi-closed systems like those of Musgrave and Moss (1961) or Bazzaz and Boyer (1972) and the use of the CO_2 depletion by photosynthesis in a closed system to drive a CO_2-response curve (Davis et al., 1987). In compensating systems requiring several minutes for equilibration, we expect limited utility from a device for quick sampling under ambient conditions (but see Griffiths and Jarvis (1982) for an application which reduces equilibration time).

The appropriate level of environmental control depends on the goals of the measurement (Fig. 3). Systems designed for rapid sampling should have sufficient environmental control to maintain chamber conditions, especially temperature, near ambient. Effective strategies with minimum power requirements include utilizing forced-air cooling, as in the LI-1600 porometer (LI-COR, Inc., Lincoln , Nebraska, USA) or IR-transmitting plastics for chamber construction. The earliest field systems designed for response curves controlled temperature and light, but the discovery that humidity control is essential for reliable measurements of photosynthesis and conductance led to instruments of substantially increased complexity. Problems with humidity control stem from both the complexity of the control technology and the tendency for many commonly used construction materials to adsorb and desorb water vapor (Bloom et al., 1980). As the emphasis in field studies has expanded to include increasingly mechanistic approaches, it

has become necessary to add CO_2 concentration to the list of
controlled parameters.

    The consequences of increased environmental control
usually include decreased portability, decreased sample
handling capacity, and increased set-up time (Fig. 3).
Whether or not these drawbacks are offset by the advantages
of a controlled-environment system depends on the goals of
the research program.

V. OBJECTIVES OF FIELD GAS-EXCHANGE MEASUREMENTS

    Most of the photosynthesis measurements practical under
laboratory conditions are now practical in the field.
In fact, field systems are increasingly finding applications
in laboratory studies. The commercial systems, however,
differ in their capabilities and in their suitability for
various sampling programs. Here, we outline some of the
kinds of data that can be obtained with field measurements,
discuss how these data can be used, and describe some
commercial systems.

A. Descriptive Characterizations

    The earliest field photosynthesis measurements were
intended to determine rates under ambient conditions. These
were frequently used to characterize differences between
species or genotypes, or responses to growth under
contrasting environments. With the development of climate-
controlled chambers, it became possible to determine
photosynthetic responses to temperature, light, and humidity.
Over many years, comparative analysis of response curves has
yielded many insights into photosynthetic adaptations.
Recently, however, the emphasis has shifted to the use of
models for a more quantitative analysis of photosynthetic
responses.

B. Mechanistic Studies and Model Building

    The addition of CO_2 control to field photosynthesis
systems added a new dimension of analytic capacity. It
provides a means to distinguish between stomatal and non-
stomatal limitations and allows parameterization of
mechanistic models based on the biochemistry of
photosynthesis (cf. Farquhar et al., 1980). Models can be
used both to identify the probable biochemical basis of
differences in gas-exchange responses and to predict
photosynthetic performance over any period for which climatic
data are available (cf. Caldwell et al., 1986). Models also

	Ehleringer & Cook 1980	LI-COR LI-6200	Schulze et al. 1982	Field et al. 1982	Björkman et al. 1973
	Closed system with syringe sampling	Closed or mixed system with integral IRGA	Differential system	Compensating system	Differential system Mobile lab
Single Leaf Responses					
Ps 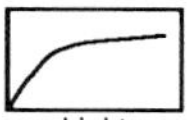 — Light	+	+	+	+++	+++
Ps 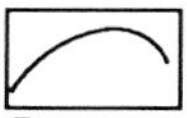 — Temperature	+	+	+	+++	+++
Ps 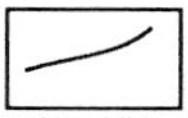 — Humidity		+	+	+++	++
Ps 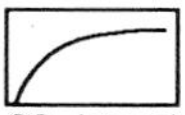 — CO_2 internal		++	+	++	+++
Ps 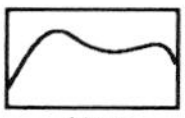 — Hour	++	+++	+++	++	+++
Ps 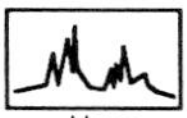 — Hour		++	++	+	++
Ps 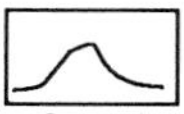 — Seconds			++	+	+++
Leaf Population Studies					
Maximum Photosynthesis	+	+++	++	++	+
Diurnal Course (Many Leaves)	++	+++	+++		
Ps 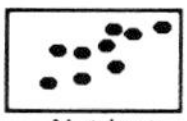 — Nutrient	+	+++	+++	++	+
Ps 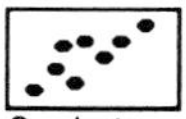 — Conductance		+++	+++	++	+
Performance Characteristics					
Leaves per Day	200	200	200	20	2
Set Up (hours)	0.5	0.1	0.5	1	5
System Mass (kg)	50	10	20	100	4000
Power Requirement (W)	100	10	50	400	5000
Initial Investment ($)	10,000	15,000	20,000	20,000	50,000

Fig. 3.   Comparative uses and attributes of selected photosynthesis systems.   A larger number of pluses indicates that the system is better suited for a particular objective. The numerical values are approximate and change frequently with technological advances.   Redrawn from Field et al. (1989).

play a critical role in integrating leaf-level responses to
the the canopy, ecosystem, or biosphere (cf. Norman, 1989;
Collatz et al., 1990).

C. Time Course Measurements

Laboratory gas-exchange studies have traditionally
emphasized steady-state responses. Often, the measurements
shown in a response curve are obtained after an hour or more
of equilibration under constant conditions. Advances in gas-
exchange technology have made it possible to examine dynamic
as well as steady-state components of leaf responses (Pearcy,
1990). The time-constants for dynamic gas-exchange responses
range from seconds for responses to natural or artificial
sunflecks to minutes or hours for responses to diurnal
changes in light, temperature or water stress. Gas exchange
systems differ in their ability to measure dynamic responses
to fast or slow changes in the environment. Differential
systems, configured for high flow rates, minimum volume, and
fast data logging, can deal effectively with responses on the
order of one second (Pearcy, 1988). Mathematical
compensation can correct for many unavoidable instrument and
gas-flow time lags (Bartholomew et al., 1981), but absorption
and desorption properties of materials in the system
potentially add variability to the lags (see Parkinson, this
volume).

Closed and compensating systems may be appropriate for
following diurnal courses but are generally unsuitable for
resolving changes faster than one to several minutes (cf.
Field et al., 1982). The utility of these systems for
diurnal measurements depends on a variety of factors,
including the dynamics of cloudiness, the architecture of
canopy, and the position of the target leaves in the canopy.
Artificially illuminating or shading leaves prior to a
measurement can remove some of the uncertainty associated
with dynamic responses, but it makes the measurements less
relevant to ambient conditions. One important exception to
the general unsuitability of closed systems for dynamic
responses is the relatively good performance of the LI-6200
(LI-COR, Inc., Lincoln, Nebraska, USA) for determining rapid
CO_2 response curves without stomatal equilibration (Davis et
al., 1987).

D. Population Studies

It is now possible to measure photosynthesis under
natural conditions in periods of one or a few minutes. This
means that one can probe populations for variation in
photosynthesis or simultaneously follow the daily course of
photosynthesis and conductance on several leaves. Perhaps
the major obstacle in population studies is distinguishing

between variation due to genotype and to current or past environment. Nonetheless, instruments for rapid sampling open a whole new realm of analysis, especially when they are supported by good experimental design and are used in combination with other techniques.

VI. MODERN FIELD SYSTEMS

Currently available gas exchange systems include a range of commercial and user-assembled instruments. All have the potential to make highly accurate photosynthesis measurements under some conditions. The factors that go into choosing an instrument should, therefore, focus on the suitability of the instrument for its intended application rather than on technical specifications in isolation.

A. LI-6200, Portable Photosynthesis System (LI-COR, Inc., Lincoln, Nebraska, USA).

The LI-6200 is a closed or mixed closed and compensating photosynthesis system with the potential to make limited open-system measurements. It is usually used to make quick measurements under ambient conditions or to record the CO_2 response of photosynthesis, over a limited range of CO_2 concentrations (Table 2). Depending on the choice of batteries and accessories, the total system weighs 9 - 16 kg, and can be operated by one person. The system consists of the innovative LI-6250 IRGA, discussed above, a sophisticated data acquisition unit, and a range of chambers.

Prior to the introduction of the LI-6200's predecessor, the LI-6000 in 1982, closed systems were relatively unpopular. Uncertainties concerning the chamber environment and the need for large CO_2 depletions made closed systems only rarely the best choice for careful studies. With the LI-COR systems, however, the data acquisition and control unit provides nearly instant updates on chamber conditions, rapidly informing the operator of problems that need to be corrected. The combination of sampling as rapidly as necessary with the very low-noise IRGA allows photosynthesis measurements with minimal CO_2 depletions. In addition, the LI-6200 has the potential to operate in essentially compensating mode for transpiration measurements, eliminating all or most of the humidity increase characteristic of earlier closed-system measurements.

In a closed system, total system volume is a critical determinant of system performance, because the rate of CO_2 depletion is inversely proportional to the volume. LI-COR offers two approaches to matching the chamber volume to the rate of photosynthesis. First, they market 3 chambers spanning a 16-fold volume range. Second, the one-liter

Table 2.   Environmental measurement and control capabilities of selected photosynthesis systems. M = measure, C = control, ±C = limited control.

	Temp.	Hum.	CO_2	Light
Closed Systems				
Ehleringer and Cook (1980)	--	--	M	M
LI-COR LI-6200	M	M ±C	M ±C	M
Differential Systems				
Mooney et al. (1971)	M C	M C	M C	M C
Schulze et al. (1982)	M	M	--	M
Compensating Systems				
Field et al. (1982)	M C	M C	M C	M C

chamber has inserts that allow sampling an adjustable leaf area. The constant-area inserts represent a significant time saver because measuring leaf area can represent a large proportion of the total sample time.

To minimize the costs of changing chambers, all three utilize the same instrument package, which contains sensors for temperature, humidity, and photon flux density. The chambers are constructed from polycarbonate for good transmission in the infrared and are lined with teflon to minimize CO_2 and H_2O exchange.

B. The ADC LCA-2 and LCA-3 Portable Photosynthesis Systems
   (Analytical Development Co., Limited, Hoddesdon, England)

The LCA photosynthesis systems are open differential systems. Originally developed for rapid measurements under ambient or nearly ambient conditions, they have now been upgraded to supply the chamber with gas of controlled humidity and CO_2 concentration. The primary difference between the two systems is that the newer LCA-3 accomplishes with a single integrated instrument tasks that require a range of accessories with the LCA-2. Data acquisition and storage, as well as IRGA precision, are also upgraded in the LCA-3.

Both instruments consist of 4 basic units -- an IRGA, a leaf chamber, an air supply, and a data logger. The IRGA in both is the LCA-2, described earlier. Four Parkinson Leaf Chambers are available, with versions optimized for narrow leaves, broad leaves, cereal heads, and conifer needles. The narrow-leaf and broad-leaf chambers can sample a fixed leaf area for some leaves, eliminating the need for area measurements. Each chamber incorporates sensors for temperature, photon flux density, and ingoing and outgoing humidity. Because the system is open, the requirement for a tight seal is much less critical than in the LI-COR system.

Chamber temperature is kept close to ambient by a heat exchanger at the base of the chamber and an infrared-absorbing acrylic window on top of the chamber. The air supply unit provides a measured flow of air to the system and can condition the air by removal of water vapor, CO_2, or both. Under normal operation, air passes through the chamber only once and is drawn through a mast from 4 m above the ground to minimize supply variations in CO_2 concentration. The data logger for the LCA-2 stores up to 232 readings while that for the LCA-3 has infinite storage capacity with removable RAM cards. The LCA-3 also provides automatic diagnostics and error checking.

The LCA-3 contrasts with the LI-COR LI-6200 in operating philosophy (open versus closed) but it is quite similar in terms of capabilities. Both systems are powerful units for sampling photosynthesis under ambient conditions and both can potentially be modified to improve their performance for response curves, but neither is ideally suited for this application. Recent and in-progress developments from both manufacturers make measuring response curves increasingly feasible.

C. The Schulze et al. (1982) Photosynthesis System.

Two systems based on the one described in this paper are available commercially from H. Walz, Mess- und Regeltechnik, D-8521 Effeltrich, Federal Republic of Germany. Both are open, differential system with regard to both CO_2 and water-vapor exchange (Table 2). The CO_2/H_2O Porometer, which is essentially similar to the instrument described by Schulze et al. (1982) was designed primarily for rapid sampling under ambient conditions (Fig. 3). The Compact Minicuvette System, which adds control of humidity, light, temperature, and CO_2 concentration, is suitable for studying functional responses. Both systems are built around a BINOS 100 IRGA with differential channels for CO_2 and water vapor. In the simplest operating mode, both systems follow the pure differential model. A pump draws ambient air and partitions it between 2 pathways. One path leads through a mass flow meter, a leaf chamber, and finally through the sample cells of both IRGA channels. The other passes first through a buffer volume to insure that gas in the sample and reference cells entered the system at the same time and then through the IRGA reference cells. The Compact Minicuvette System allows the option of regulating chamber humidity with a feedback regulated water scrubber, similar to that in the mobile laboratory of Lange et al. (1969).

The system of Schulze et al. (1982) is portable and battery powered but is typically operated with the chamber hand-held but the IRGA set at one or a few locations during a day. The most recent version of the CO_2/H_2O Porometer, with integrated computer and data logger weighs 11.5 kg and is

approximately as portable as the LI-6200 or the LCA-3. The
Compact Minicuvette System, which weighs 40 kg, is comparable
in capabilities and portability to the following systems.

D. The Bingham et al. (1980)
Field et al. (1982)
Photosynthesis Systems.

     Commercial controlled-environment compensating systems
for the simultaneous measurement of CO_2 and water-vapor
exchange are now marketed by Armstrong Enterprises (Palo
Alto, CA), Data Design Group (La Jolla, CA) and Opto-
Diagnostics, Inc. (Logan, UT). These instruments are only
marginally portable (often called transportable), but they
provide the most extensive and flexible environmental control
(Table 2, Fig. 3). Total systems weigh approximately 40-100
kg and require 100-300 W of input power without artificial
lighting but approximately 400 W with artificial lighting.
Variability in natural light environments is probably the
factor that most compromises the utility of field gas-
exchange data, but artificial lighting is clearly the largest
power consumer. In terms of portability, a commitment to
artificial lighting moves one most of the way from the most
portable to the least portable systems. We consider
controlled illumination to be a necessity for response curves
in all but the clearest conditions, reducing the priority for
modifying very portable systems for response curves.
     Systems requiring several hundred watts of input power
can be battery operated, but portable generators provide a
more satisfactory power source. To date, all of the leaf
chambers used with controlled-environment systems have been
tripod mounted and cannot be moved between leaves in less
than about five minutes. It is currently possible to build
hand-held controlled-environment chambers, but plant
responses to changes in the environment, especially of
stomatal conductance, may be so slow that a hand-held chamber
is totally impractical. Sample handling capacity in these
systems depends on the type of measurements. Typical outputs
are replicated measurements at a single set of conditions for
10-20 leaves per day or 2-10 response curves per day,
depending on the time allowed for equilibration of the plant
material.
     Bingham et al. (1980) and Field et al. (1982) describe
systems in which transpiration is the only source of water
vapor for humidifying the chamber and in which CO_2 depletion
by photosynthesis is partly or fully compensated by injecting
CO_2 into the leaf chamber. Their approach provides a minimum
hardware solution for full control of chamber conditions but
is somewhat limited in terms of flexibility and speed.
Recent improvements in the technology for flow and humidity

control have made it possible to upgrade these systems to provide incoming air containing any amount of CO_2, H_2O, and even pollutant gases.

VII.   TO BUY OR TO BUILD?

In general, the trade-off between buying and building portable systems for gas-exchange research has shifted strongly toward buying in the last few years. Commercially available units are carefully engineered and optimized for the most frequent tasks in gas-exchange research. Building a system typically offers no more than modest cost savings but presents many challenges, from the level of locating suppliers to the level of testing chamber aerodynamics without many iterations of a chamber design. In the most portable systems, the engineering invested in compactness is very difficult to duplicate without a major effort.

For controlled environment systems, the balance between buying and building is still reasonably close. The larger mass of these systems relaxes the constraints on miniaturization, and the broad range of questions for which these systems may find application increases the potential range of options and user modifications.

This chapter was originally prepared for a meeting in 1985. Since then, improvements in commercial systems for gas-exchange research have been dramatic. Especially impressive is the extent to which increased computerization has enhanced the function of instruments incorporating measurement philosophies that have been around for decades.

ACKNOWLEDGEMENTS

We gratefully acknowledge support from the National Science Foundation (grant BSR 8717422 to CBF and 83-15675 to HAM) and to the Electric Power Research Institute (contract EPRI RP-1313 to HAM), which assisted in developing our knowledge in this area. This is CIWDPB publication number 1073.

REFERENCES

Baldocchi, D.D., B.B. Hicks and T.P. Meyers (1988). Measuring biosphere-atmosphere exchanges of biologically related gases with micrometeorological methods. Ecology 69: 1331-1340.

Ball, J.T. (1987). Calculations related to gas exchange. pp.
    445-476 In: Stomatal Function (E. Zeiger, G. D.
    Farquhar, and I. R. Cowan (eds)) Stanford University
    Press, Stanford, California.
Bartholomew, G.A., D. Vleck, and C.M. Vleck (1981).
    Instantaneous measurements of oxygen consumption during
    pre-flight cooling in spingid and saturniid moths. J.
    of Expt. Biol. 90:17-32.
Bazzaz, F.A. and J.S. Boyer (1972). A compensating method for
    measuring carbon dioxide exchange, transpiration, and
    diffusive resistances of plants under controlled
    environmental conditions. Ecology 53:343-349.
Billings, W.D., E.C.C.C. Clebsch, and H.A. Mooney (1966).
    Photosynthesis and respiration rates of Rocky Mountain
    alpine plants under field conditions. 1966. Amer. Mid
    Nat. 75:34-44.
Bingham, G.E., P.I. Coyne, R.B. Kennedy, and W.L. Jackson
    (1980). Design and fabrication of a portable
    minicuvette system for measuring leaf photosynthesis
    and stomatal conductance under controlled conditions.
    Lawrence Livermore National Laboratory, Livermore,
    California, UCRL-52895.
Bingham, G.E., C.H. Gillespie, J.H. McQuaid, and D.F. Dooley
    (1978). A miniature, battery powered, pyroelectric
    detector-based differential infra-red absorption sensor
    for ambient concentrations of carbon dioxide.
    Ferroelectrics 34:15-19
Björkman, O., M.Nobs, J. Berry, H.A. Mooney, F.Nicholson, and
    B. Catanzaro (1973). Physiological adaptation to
    diverse environments: approaches and facilities to
    study plant response in contrasting thermal and water
    regimes. Carnegie Inst. Washington Ybk. 72:393-403.
Bloom, A.J., H.A. Mooney, O. Björkman, and J. Berry (1980).
    Materials and methods for carbon dioxide and water
    exchange analysis. Plant, Cell Env. 3:371-376.
Bosian, G. (1953). Über die Vollautomatisierung der CO_2 --
    Assimilationsbestimmungen. Ber. Dtsch. Bot. Ges., Gen.
    Vers. 66:35.
Bosian, G. (1955). Über die Vollautomatisierung der CO_2-
    Assimilations-bestimmung und zur methodik des
    küvettenklemas. Planta 45:470-492.
Bosian, G. (1960). Zum Kuvettenklimaproblem: Beweisführung
    für die Nichtexistenz 2-gipfeliger Assimilationskurven
    bei Verwendung von klimatisierten Küvetten. Flora
    149:167-188.

Bosian, G. (1965). Control of conditions in the plant
    chamber: fully automatic regulation of wind velocity,
    temperature and relative humidity to conform to
    microclimatic field conditions. pp 233-238. In:
    Methodology of Plant Eco-Physiology (F.E. Eckardt
    (ed.)) UNESCO, Paris.
Bowman, G.E. (1968). The measurement of carbon dioxide
    concentration in the atmosphere. pp 131-139 In: The
    Measurement of Environmental Factors in Terrestrial
    Biology (R. M. Wadsworth et al. (eds)) Blackwell
    Scientific Publications, Oxford.
Caldwell, M. M., H.-P. Meister, J. D. Tenhunen, and O. L.
    Lange (1986). Canopy structure, light microclimate and
    leaf gas exchange of Quercus coccifera L. in a
    Portuguese macchia: Measurements in different canopy
    layers and simulations with a canopy model. Trees 1:25-
    41.
Collatz, G.J., J.T. Ball, C. Grivet, and J. A. Berry (1990).
    Regulation of stomatal conductance and transpiration: A
    physiological model of canopy processes. Agric. For.
    Meteorol. (in press).
Coombs, J., D.O. Hall, S.P. Long, and J.M.O. Scurlock (eds).
    (1985). Techniques in Bioproductivity and
    Photosynthesis, 2nd edition. Pergamon Press,Oxford.
Davis, J.E., T.J. Arkebauer, J.M.Norman, and J.R. Brandle
    (1987). Rapid measurement of the assimilation rate
    versus internal CO_2 concentration relationship in green
    ash (Fraxinus pennsylvanica Marsh.): the influence of
    light intensity. Tree Physiol. 3:387-392.
Ehleringer, J. and C.S. Cook (1980). Measurements of
    photosynthesis in the field: utility of the CO_2
    depletion technique. Plant, Cell Env. 3:479-482.
Eckardt, F.E. (1966). Le principe de la soufflerie
    aerodynamique climatisee applique' a l'etude des
    echanges gazeux de la couverture vegetale. Oecol. Plant
    1:369-399.
Egle, K. and A. Ernst (1949). Die Verwendung des
    Ultrarotabsorptions-schreibers für die
    vollatutomatische und fortlaufende CO_2-Analyse bei
    Assimilations-und Atmungmessungen an Pflanzen.
    Zeitschr. f. Naturforschg. 46:351-360.
Farquhar, G.D., S. von Caemmerer, and J.A. Berry (1980). A
    biochemical model for photosynthetic CO_2 assimilation
    in leaves of C_3 species. Planta 149:78-90.
Field, C.B., J.T. Ball, and J.A. Berry (1989).
    Photosynthesis: Principles and field techniques. pp.
    209-253 In: Plant Physiological Ecology: Field Methods
    and Instrumentation (R.W. Pearcy, H.A. Mooney, J.R.
    Ehleringer, and P.W. Rundel (eds.)) Chapman and Hall,
    London.

Field, C., J.A. Berry, and H.A. Mooney (1982). A portable
    system for measuring carbon dioxide and water vapour
    exchange of leaves. Plant, Cell Env. 5:179-186.
Griffiths, J.H., and P.G. Jarvis (1982). A null balance
    carbon dioxide and water vapour porometer. J. Expt.
    Bot. 32:1157-1168.
Gosz, J. R., C.N. Dahm, and P.G. Risser (1988). Long-path
    FTIR measurement of atmospheric trace gas
    concentrations. Ecology 69:1326-1330.
Hällgren, J.-E. (1982). Field photosynthesis; monitoring with
    $^{14}CO_2$. pp 36-43 In: Techniques in Bioproductivity and
    Photosynthesis (J. Coombs and D.O. Hall (eds.))
    Pergamon Press, Oxford.
Koch, W., E. Klein, and H. Walz. (1968). Neuartige
    Gaswedisel-Messanlage für Pflanzen in Laboratorium und
    Freiland. Siemens-Z. 42:392-404.
Lange, O.L., W. Koch, and E.-D. Schulze (1969). CO_2-Gaswechsel
    und Wasserhaushalt von Pflanzen in der Negev-Wuste am
    Ende der Trockenzeit. Ber. Dtsch. Bot. Ges. 82:39-61.
Luft, K.F. (1943) Über eine neues Methode der registrierenden
    Gasanalyse mit Hilfe der Absorption ultrarot Strahlen
    ohne spectrale Zerlegung. Zeitschrift fr tech. Phys.
    24:97-104.
Marshall, B., and F.I. Woodward (eds) (1985).
    Instrumentation for Environmental Physiology.
    Cambridge University Press, Cambridge.
Matson, P. A. and R.C. Harriss (1988). Prospects for
    aircraft-based gas exchange measurements in ecosystem
    studies. Ecology 69:1318-1325.
Mooney, H.A., R.D. Wright, and B.R. Strain (1964). The gas
    exchange capacity of plants in relation to vegetation
    zonation in the White Mountains of California. Amer.
    Mid. Nat. 72:281-297.
Mooney, H.A., E.L. Dunn, A.T. Harrison, P.A. Morrow, B.
    Bartholomew, and R. Hays (1971). A mobile laboratory
    for gas exchange measurements. Photosynthetica 5:128-
    132.
Musgrave, R.B. and D.N. Moss. (1961). Photosynthesis under
    field conditions. I. A portable, closed system for
    determining net assimilation and respiration of corn.
    Crop Sci. 1:37-41.
Norman, J. M. (1989). Synthesis of canopy processes. pp. 161-
    175 In Plant Canopies: Their Growth, Form and Function
    (G. Rusell, B. Marshall, and P. G. Jarvis (eds.))
    Cambridge University Press, Cambridge.
Pearcy, R. W. (1988). Photosynthetic utilisation of
    lightflecks by understory plants. Aust. J. Plant
    Physiol. 15:223-238.
Pearcy, R. W. (1990). Sunflecks and photosynthesis in plant
    canopies. Ann. Rev. Plant Physiol. 41:421-453.

Pearcy, R.W., J. Ehleringer, H.A. Mooney, and P.W. Rundel
    (eds) (1989). Plant Physiological Ecology: Field
    Methods and Instrumentation. Chapman and Hall, London.
Schulze, E.-D., A.E. Hall, O.L. Lange, and H. Walz (1982). A
    portable steady-state porometer for measuring the
    carbon dioxide and water vapour exchanges of leaves
    under natural conditions. Oecologia 53:141-145.
Sestak, Z, J. Catsky, and P. Jarvis (1971). Plant
    Photosynthetic Production. Manual of Methods. Dr. W.
    Junk Publ., The Hague.
Sheehy, J. E. (1985). Radiation. pp. 5-28. In:
    Instrumentation for Environmental Physiology (B.
    Marshall and F. I. Woodward (eds)) Cambridge University
    Press, Cambridge.
Strain, B.R. (1965). Another mobile laboratory. Bull. Ecol.
    Soc. Amer. 46:190.
Tranquillini, W. (1957). Standortsklima, Wasserbilanz und
    CO_2-Gaswechsel junger Zirben (Pinus cembra L.) an der
    alpinen Waldgrenze. Planta 49:612-661.
von Caemmerer, S. and G.D. Farquhar (1981). Some
    relationships between the biochemistry of
    photosynthesis and the gas exchange of leaves. Planta
    153:376-387.
Went, F.W. (1958). A mobile desert laboratory. Plant Science
    Bulletin 4:1-3.

DESIGN AND TESTING OF LEAF CUVETTES FOR USE IN MEASURING PHOTOSYNTHESIS AND TRANSPIRATION

K.J. Parkinson[1]
W. Day[2]

Rothamsted Experimental Station
Harpenden, Herts, U.K.

I.   INTRODUCTION

It is a well known physical principle that the presence of a measuring instrument always influences the quantity being measured to a greater or lesser extent. Photosynthesis of leaves in the field can be particularly perturbed by the measuring instrument.

Micro-meteorological methods cause the least disturbance but they are extravagant in instrumentation and land area and cannot be easily used to examine the effects of agronomic treatments, nor at all for measurement of photosynthesis of individual plant parts. To do this we enclose leaves in cuvettes, which immediately and drastically modify the leaf micro- climate.

Solar radiation on the leaf declines by an amount dependent upon the absorption characteristics of the cuvette windows. The PAR component declines by at least the 10% that is commonly reflected from the window surface. Back radiation from the leaf in the far infra-red wavelengths is affected by the change in the leaf surroundings and by the long-wave transmission properties of the windows. In the

---

[1]Present address: PP systems, 24-26 Brook ST, Stotfold, Hitchin, Herts. U.K. SG54LA
[2]Present address: Institute of Engineering Research, Wrest Park, Silsoe Bedford U.K. MK45 4HS

Measurement Techniques in Plant Science
Copyright © 1990 by Academic Press, Inc.
All rights of reproduction in any form reserved.

field, the leaf is surrounded by eddies of air, usually turbulent and variable in size and velocity. In a cuvette air movement is generally induced by a fan and is uniform in velocity and direction. The ventilation rate controls the boundary layer resistance and influences both the gas and the energy exchanges of the leaf.

Modification of the energy exchange, by changes in both the gain and loss of energy, leads to a leaf temperature different from that before enclosure. Shortly after enclosure the leaf begins to respond to its new surroundings and not only are we now measuring under different environmental conditions but as a consequence, with a leaf of differing physiological characteristics.

We cannot hope to simulate within a cuvette the natural micro-climate of the leaf. The best we can do is make our measurements over a range of carefully measured conditions that encompass the field conditions to enable us to predict the field photosynthesis. These measurements should preferably be made rapidly so that there is little change in the physiological status of the leaf.

We can therefore set out the requirement for leaf cuvettes:-

(1) Conditions within the cuvette should be uniform so that we can define the leaf environment accurately. This requires both a small boundary layer resistance and turbulent mixing which, unless the cuvette is very small or the air flow through the chamber very large, will necessitate a fan.

(2) We must be able to measure accurately the chamber environment with regard to radiation, humidity, temperature, carbon dioxide concentration and leaf boundary layer resistance.

(3) The materials used to make the cuvette must cause the minimum of interference with the measurement i.e. windows must be transparent, the cuvette walls must not significantly evolve or absorb carbon dioxide or water vapour, and they must not evolve materials toxic to the leaves (e.g. dibutyl phthalate used as plasticizer or PVC (Hardwick *et al.*, 1984)) or materials that affect the sensors (e.g. infra-red absorbing materials when using IRGA's). The above requirements are general to all photosynthesis measuring systems, the next two are desirable for field measurements.

(4) The system should be designed for rapid measurements so that the leaf has little time to change its physiological state. This also means that the

response times of the sensors must be fast. The
slowest responding sensor will determine the
measurement rate. This is especially important with
the closed system (see below) where the response
times for temperature, water vapour and carbon
dioxide must be effectively equal in order to derive
intercellular carbon dioxide concentrations.

(5) The leaf should not be damaged by the cuvette.

II. MATERIALS

In choosing materials for construction of cuvettes,
physical strength is important but we must also consider
transparency (for windows), permeability,
absorption/absorption of carbon dioxide and water vapour, and
effusion of noxious materials. However, the relative
importance of these factors depends to some extent on the
method of photosynthesis measurement. The most demanding is
the so called closed system method which depends on measuring
the rates of change of carbon dioxide and water vapour in a
sealed system. Ad/ab/desorption are therefore very important
and because of the extensive plumbing usually required, so is
permeability. Furthermore because the system is sealed there
is a greater possibility of the build up of noxious materials.

The accuracy of the semi-closed system in which the rates
of CO_2 addition and water vapour removal required to maintain
a steady-state, are measured, is not troubled by exchanges
with the walls. However, its equilibration time will be
influenced. The results are affected by permeability and
effusion.

The open system accuracy is not influenced by wall
exchanges, because measurements are made at a state, but the
response time of the system is affected. Permeability is
less of a problem than with the other systems because of the
limited plumbing involved and, because of the continuing flow
of fresh air, effusion can be ignored.

Tables I and II show some of the properties of materials
commonly used in leaf cuvettes. Until the mid 1970's, leaf
cuvettes were often constructed of Perspex with PVC tubing,
but the advent of rapidly responding accurate humidity
sensors, has shown these materials to be unsuitable because
of their large water absorption. In general unless we are
dealing with large, thin film windows or very long pipe runs
we can ignore permeability. Carbon dioxide doesn't present a

TABLE I. Permeabilities (at 298 K and 0.1 MPa) and absorption properties of common plastics.

Plastic Material	Common and/or Trade Name	Permeability $(m^2\ s^{-1}) \times 10^{12}$		Water Absorption $(g\ g^{-1}) \times 10^2$
		Water Vapour	Carbon Dioxide	
Cellulose acetate		1140 – 10000	1.8 – 25.1	1.8
Cellulose regenerated	Cellophane	1444 – 33000	0.001 – 0.003	20
Epoxy resins		–	0.07 – 1.06	0.2 – 1.8
Poly(6 amino caproic acid)	Nylon 6	53 – 4292	0.05 – 0.12	3.5 – 8.5
Polycarbonate		52	7.9	0.36
Polycnloroprene	Neoprene	1368	19.0	1 – 6
Polyethylene (Low density (High density	Polythene	68 – 192 9 – 19	7.6 – 50.0 1.6 – 3.3	0.024 Negligible
Polyethylene terephtnalate	Mylar,* Melinex*	86 – 200	0.076 – 0.166	0.6
Polyfluorinated ethylene propylene	Teflon FEP	38	1.3 – 7.6	Negligible
Polyisobutylene isoprene	Butyl rubber	30 – 152	3.9 – 10.6	1 – 6
Polyisoprene	Natural rubber	111 – 2280	100 – 179	1 – 6

TABLE I. Continued.

Plastic Material	Common and/or Trade Name	Permeability (m^2 s^{-1}) $\times 10^{12}$		Water Absorption (g g^{-1}) $\times 10^2$
		Water Vapour	Carbon Dioxide	
Polymethylmethacrylate	Perspex	1900 – 2837	50.2	1 – 2
Poly(4-methyl pentene-1)	PMP	61.6	–	< 0.05
Polypropylene		28.4 – 66.2	1.4 – 6.9	0.01 – 0.03
Polystyrene		454 – 1000	5.9 – 30	0.03 – 0.05
Polytetrafluorethylene	PTFE	16 – 27	1.1 – 8.3	Negligible
Polyvinyl chloride	PVC	114 – 480	0.8 – 13.7	1
Polyvinylidene chloride	PVDC, Saran**, Propafilm-C***	0.4 – 76	0.01 – 0.04	0.03
Polyurethane		270 – 2200	2.1 – 30.4	1
Silicone rubber (dimethyl)		470 – 33000	456 – 2432	0.07

*ICI Plastics Ltd., U.K. **Dow Chemical Co. Ltd. ***Propafilm-C is a PVDC/Polypropylene/PVDC film produced by ICI Plastics, Ltd. U.K.

TABLE II. Surface adsorption (g m^{-2}) by materials after 24 hours at 293 K above (a) water (Shepherd, 1973), (b) saturated sodium sulphate solution (93% RH) (Dixon and Grace, 1982).

MATERIAL	SURFACE ADSORPTION g m^{-2}	
	a	b
Soda glass	0.05	0.01
Copper	"	–
Stainless steel	"	0.09
Nickel plate on brass	"	0.06
Chromium plate on brass	0.38	–
Clean brass	"	0.07
Tarnished brass	1.05	0.11
PMP	–	0.03
PTFE	1.05	0.05
High density polyethylene	–	0.08
*Cross linked polystyrene	–	0.56

*Probably includes absorption.

problem with any of the common materials and we can base our choice on the water absorption characteristics of the materials.

For rigid windows, polymethyl pentene, polystyrene, or glass are suitable whilst for flexible windows polythene, polyethylene terephthalate and polyvinylidene chloride are good. For cuvette walls, Nylon, Perspex and polyvinyl chloride should be avoided. All the rubbers have large water absorption and whilst they may be essential for sealing, their area must be minimized. Silicone adhesives are preferable to epoxy.

All the window materials have similar transmission properties in the visible. Glass, the rigid plastics and polyethylene terephthalate absorb strongly in the far infrared and with them there will therefore be a "greenhouse" effect. With a cuvette designed for field use, we wish to maintain the cuvette conditions close to ambient, so to minimize the "greenhouse" effect we must have cuvette walls

with good thermal conductivity and couple their temperature
as close to the ambient as possible. This may require some
form of heat exchange surface and forced convection. We must
also minimize the radiation absorbed by the cuvette by using
reflective material where possible. Polyethylene is
transparent to these long wavelengths, thus eliminating the
"greenhouse" effect, and it is therefore the best material
where thin film windows are required.

III. MEASUREMENT OF CUVETTE MICRO-CLIMATE

There is no intention to give an exhaustive survey of
sensors but simply to describe the characteristics of some
that have been found suitable.

A. Humidity

The ideal humidity sensor should be small enough to fit
in the cuvette, should not affect the chamber humidity (c.f.
wet and dry bulb thermometers), should be robust and not
require frequent recalibration and should be more rapidly
responding than the cuvette so the sensor is not the limiting
factor. Until the appearance of the Vaisala sensor, nothing
approached this ideal. This sensor also offered several
bonuses; it had a linear response to humidity and a very
small temperature coefficient.
A recently introduced sensor, the CORECI, is similar to
the Vaisala both in size and performance. In both sensors,
changes in electrical capacitance with humidity are measured,
the sensors are used in bridge oscillator circuits. With the
Vaisala the capacitance is of the order of 50 pF and changes
by about 0.1 pF/% RH. The recommended oscillation frequency
is in 1-2 MHz range. The CORECI has a capacitance of 450 pF
which changes by about 1.0 pF/% RH and is measured at 50 KHz.
This lower frequency is less prone to interference.

B. Light

The ideal sensor has a quantum response between 0.4 and
0.7 μ m with a sharp cut-off at the two extreme wavelengths.
Commercially available sensors are generally based on silicon
photocells and since a silicon cell has its greatest
sensitivity in the far red part of the spectrum, considerable

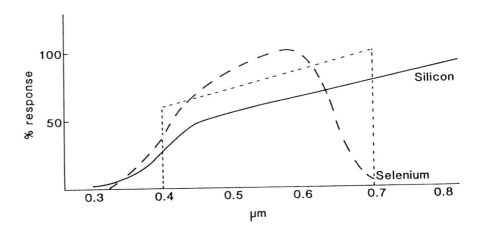

Figure 1.   The response of silicon and selenium
photocells to radiation of equal energy at wavelengths from
0.3 to 0.8 μ m.   The dotted line shows the quantum response.

TABLE III.   The readings to be expected from a selenium
photocell compared to those from an ideal quantum response
sensor when illuminated by different radiation sources.   The
values assume both sensors were calibrated in sunlight.

	Relative reading
Sun and sky	1.00
Incandescent (tungsten)	0.80
High pressure sodium	1.09
MBFR/U (mercury)	0.98
Warm white fluorescent	1.07

and expensive filtering is required (Figure 1). Selenium
photocells are also worth considering. They are only
sensitive in the visible range but they lose sensitivity too
rapidly at the red end. However, if we compare the values
that such a sensor calibrated in sunlight would read when
exposed under varying sources (Table III), Then only in
tungsten light is there a large error. In the past, selenium
cells have been criticised for having a large temperature
coefficient and suffering fatigue. Both these problems are
eliminated by modern electronics. The former by having the
cells operate into zero effective external resistance when
the temperature coefficient is negligible (Robinson, 1966)
and the latter by having a large degree of amplification so
that the cell can be used with a neutral density filter to
decrease the radiation reaching the cell to low levels.

C. Temperature

Air temperature (Ta) in the cuvette may easily be measured with thermistors, thermocouples or platinum resistance
elements. Measurement of leaf temperature is more difficult.
There are two reasons for wishing to know leaf temperature,
the first is to be able to specify the temperature of the
physiological system; for this an accuracy of $\pm 1.0°C$ is
adequate. The second is to be able to calculate stomatal
conductances from the leaf temperatures, and this requires
much greater accuracy. For leaf temperature measurements,
because of the size of the sensors, only thermocouples are
suitable. We therefore require good reference junctions and
good electronics for an accurate measure of the probe
temperature. But there will be variation in temperature
across the leaf associated with variations in boundary layer
and stomatal conductances (Hashimoto *et al.*, 1984). The
presence of the probe will disturb the boundary layer and
also there will be heat conduction along the probe wires.
Bearing in mind all these difficulties, let us therefore
consider the accuracy of the calculation of leaf temperature
from the cuvette air temperature using equations relating to
the energy balance of the leaf (Appendix 1). This calculated
temperature will be an average for the leaf.
    It can be shown that the leaf temperature (T_l) is given
by:

$$T_l = T_a - (\lambda E - \alpha I)/(0.93 \ C_p/r_b + 4 \ \sigma T_a^3) \qquad (1)$$

where E is the transpiration rate, λ the latent heat of vaporization, I the total short wave radiation energy incident on the leaf and α the fraction absorbed, C_P is the specific heat of air, r_b the boundary layer resistance to water vapour transfer (0.93 converts it to heat) and σ is Stefan's constant.

Also $r_1 = (x_1 - x_a)/E - r_b$ (2)

where x_1 and x_a are the water vapour concentration saturated

TABLE IV. Leaf temperatures and stomatal resistances calculated from the energy balance of the leaf using r_b = 0.225 $\pm$ 20% m^{2}s mol^{-1} with an atmospheric pressure of 1000 $\pm$ 25 mb. The errors in T_1 (ΔT_1) and in rs are those expected given the specification of the LI 6000 i.e. Ta $\pm$ 0.5°C, RH $\pm$ 3%, I $\pm$ 5%. The last line gives the error in rs assuming an accuracy of leaf temperature measurement of $\pm$ 1°C (as specified for LI 6000). The units of r_s are m^{2}s mol^{-1}.

T_a °C	15°C				35°C			
RH %	10		50		10		50	
I μ E m^{-2}s^{-1}	0	2000	0	2000	0	2000	0	2000
T_1 °C	14.7	17.5	13.7	16.5	34.2	36.9	31.1	33.8
T_1 °C	+0.1	+0.5	+0.3	+0.3	+0.3	+0.5	+0.8	+0.4
r_s	20.9	25.5	1.8	2.6	21.7	25.7	1.3	2.0
% Error	33	33	18	14	34	23	25	17
In r_s	(34)	(34)	(24)	(19)	(35)	(23)	(27)	(21)

at leaf temperature and that present in the cuvette air
respectively.

To determine r_1 from (2) we can either derive T_1 from (1)
or from measurements of leaf temperature. Using the
published specification of the LI-6000 portable
photosynthesis system it is possible to compare the accuracy
of the two methods (Table IV).

At high values of stomatal resistance, cuvette humidity
is low (i.e. RH = 10%) and then it is the accuracy of the
humidity measurement which predominates and both methods give
similar results. At low resistance values (RH =50%), the
accuracy of the leaf temperature measurement would have to be
considerably improved from the stated $\pm 1.0°C$ to equal that
of the energy balance method.

TABLE V.   Effects of variation in the accuracies of the
components of the energy balance equation on leaf temperature
and stomatal resistance for T_a = 35°C, RH = 50%, I = 2000 μ E
m^{-2} s^{-1}, leaf area = 5 cm^2 $\pm 5\%$ and volume flow rate = cm^3 s^{-1}
$\pm 5\%$. The standard case is for I = $\pm 10\%$, T_a = $\pm 0.3°C$, RH =
$\pm 2\%$, r_b = 2.25 m^2 s mol^{-1} $\pm 20\%$, (b) - (e) represent single
modifications to the standard case.

		T_1	$\pm$ T_1	rs	% error in rs using energy balance	% error in rs $\pm 1°C$ error in T_1
a	Standard case	33.8	0.5	2.0	12	17
b	I = $\pm$ 35%	33.8	1.0	2.0	17	17
c	r_b = 0.125 $\pm$ 20%	34.3	0.3	2.25	12	17
d	r_b = 0.325 $\pm$ 20%	33.4	0.6	1.79	16	20
e	Dewpoint $\pm$ 0.3°C	33.8	0.4	2.0	8	15

Table V shows the effects of altering the magnitude of errors of the components of the energy balance: (a) gives what we consider to be typical errors in measurement. It is only when the radiation error is very large (b) that the errors are similar in both methods. Small boundary layer resistances minimize the errors (c and d); (e) shows that smaller errors arise when a dewpoint meter is used rather than a RH sensor.

IV. TESTING OF LEAF CUVETTES

There are four tests which are essential before using a cuvette. The first is to look for leaks and consists simply of passing CO_2-free air through the cuvette and using an IRGA, confirm that the air coming out is also CO_2 free. This should be done with the cuvette fans running, as fans generate pressure differences within the cuvette. The resulting low pressure areas can, even in a slightly pressurized system be below ambient, and external air may then be drawn in through any leaks. The second test is to determine the efficiency of the mixing of the air within the cuvette and also makes use of an IRGA. Air of constant CO_2 concentration should be passed through the cuvette and the analyzer at a known flow rate. The inlet air of the cuvette should then be changed to CO_2 free gas and the subsequent changes in the outlet concentration measured. The way in which the CO_2 concentration declines gives us clues to the efficiency of mixing.

If the mixing of the inlet gas with the cuvette air was perfect then

$$\frac{C_i}{C_o} = e^{-kt}$$

where C_i is the cuvette concentration at time t after the change in the concentration, and Co is the final equilibrium concentration (=0 in this case). It can be shown that k=F/V where F is the flow rate and V the cuvette volume; k can be determined from the gradient of the logarithm of the cuvette concentration against time i.e. $-kt = \log (C_i/C_o)$.

If the mixing is not perfect the logarithmic plot will no longer be a straight line. Its exact form will then depend on the number of "compartments" that compose the exchange system, the coupling between them, their relative magnitude, and their respective response times. The "compartments" may

be physical volumes (usually true for CO_2) or particularly
with water vapour, surface adsorption or bulk absorption.
    The third test is the determination of the boundary layer
resistance in the cuvette. This is required to enable us to
determine the stomatal resistance of the leaf, and to
determine the leaf temperature from the energy balance. It
is desirable that the boundary layer resistance be small so
that errors in its calculation and differences across the
cuvette are not significant. In the past the resistance has
usually been determined from measurement of the temperature
of and the evaporation rate from wet filter paper placed in
the cuvette. Recently it has been shown that it can be
evaluated from the equilibrium cuvette humidity alone
(Appendix 2) and that this method is more accurate than
measuring the filter paper temperature (Parkinson, 1985).
    To illustrate the information provided by tests 2 and 3,
measurements were done on a number of cuvettes. The first
was a large cylindrical cuvette tested both with and without
a fan (Figure 2). Mixing was in both cases perfect as shown
by the agreement between measured chamber volume and that

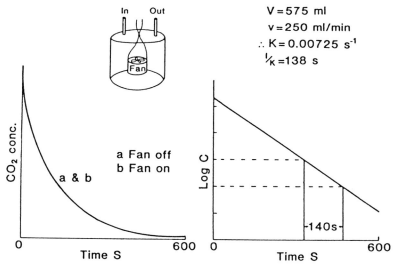

Figure 2. Concentration of carbon dioxide emerging from
the cuvette shown when the concentration of the entrant air
is changed to zero. The cuvette volume is 575 ml and the
volume flow rate 250 ml min^{-1}. The boundary layer resistance
is 1.1 m^2 s mol^{-1} with the fan off and 0.3 m^2 s mol^{-1} with
it on. The theoretical response time is 138 s and the
measured response time was 140 s.

calculated from the response time. However, with the fan, r_b was an acceptable 0.3 m² s mol⁻¹ whereas without it was much larger.

The second example (Figure 3) was the same cuvette as above but separated into two halves by porous plastic foam with the fan between the two halves. With the fan running the results were similar to before, but without the fan there was initially a more rapid decline in concentration as the upper chamber was flushed out and then a long tail that can be associated with mixing from the lower chamber. On the log plot this can be resolved into two distinct processes, this result is characteristic when mixing is incomplete or where exchange with the chamber walls is occurring. The boundary layer resistance measured in the upper part is small with the fan, but large without.

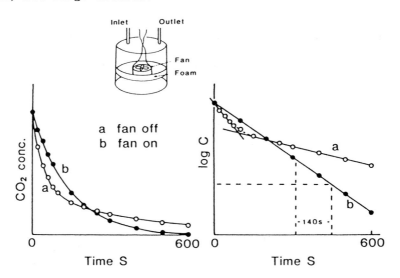

Figure 3. As for Figure 2 but with foam layer added as illustrated. The boundary layer resistance with the fan off is 1.0 m² s mol⁻¹ and with it on 0.4 m² s mol⁻¹.

The third example (Figure 4) again used the same cuvette but this time with the fan in the upper half. The foam was covered with a polythene film with a small perforation in it. We were therefore creating two distinct chambers and this is seen in the log response. There is a rapid initial fall corresponding with exchange within the top chamber followed by the steady decline dominated by exchange with the lower

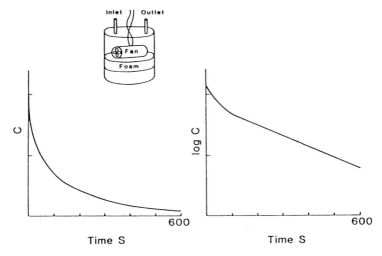

Figure 4.   As for Figure 3 but with fan moved, as illustrated.   The boundary layer resistance with the fan on is 0.08 m^2 s mol^{-1}.

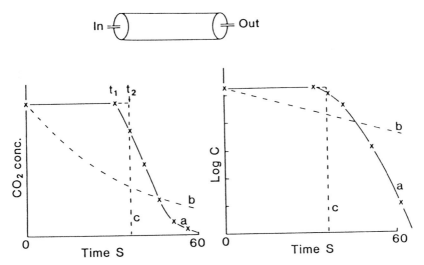

Figure 5.   The concentration of carbon dioxide emerging from the cuvette shown when the concentration in the entrant air is changed to zero.   The cuvette volume is 225 ml and the volume flow rate 350 ml min^{-1}.   The boundary layer resistance is 1.1 m^2 s mol^{-1}.   (a) is the measured value, (b) would result if mixing in the cuvette was perfect and (c) would result if there was no mixing at all.

chamber.   The boundary layer resistance in the upper chamber is small.

The fourth example (Figure 5) features an elongated chamber in which the gas could be expected to travel as a plug.   There is then a long delay before any change is recorded (t_1) (though not quite as long as would apply for a perfect front (t_2)), followed by a rapid decline.   The boundary layer resistance is large.

These examples show that measurement of the boundary layer resistance alone is insufficient to determine the suitability of a cuvette, as we have created cuvettes both with small boundary layer resistances and poor mixing and with large resistances yet good mixing.

Having satisfactorily shown that we have a leak proof chamber with acceptable mixing and small boundary layer resistance, we now test for its water-vapour properties by starting at a high humidity and measuring the humidity response as we change the air flowing into the chamber to a lower humidity.

Whereas with the CO_2 test for leaks or the mixing tests we can seek perfection, with water vapour the materials appropriate for cuvette construction are by no means ideal and we are looking for acceptable results for the intended use of the cuvette.

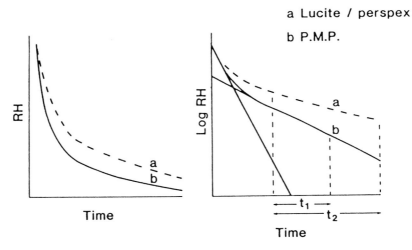

Figure 6.   Comparison of the humidity in the emergent air from similar cuvettes made with (a) perspex or (b) polymethyl pentene when the entrant humidity is changed to zero; t_2 and t_1 are their respective response times.

To illustrate the difference materials can make, Figure 6 compares the rate of decline of the humidity emerging from a long cylindrical unstirred cuvette made from either Perspex or PMP. The Perspex has nearly doubled the response time shown by the differences in shape of the graphs.

Figure 7 illustrates measurements made on a leaf cuvette constructed largely of HD 30 aluminium alloy (ADC, PLC, type 'Broad leaf'). The response times to water vapour and to CO_2 are vastly different. In fact the response to CO_2 is close to that expected from the cuvette volume and flow rate. The water-vapour response shows no indication of the initial air exchange (Appendix 2) and one can only conclude that exchanges with water adsorbed on the surfaces of the cuvette probably dominate. The long tail of the response reflects strong absorption within the rubber seals, fan blades or cuvette windows. This chamber is designed for measurements and with water vapour 90% of the response is completed within 15 seconds.

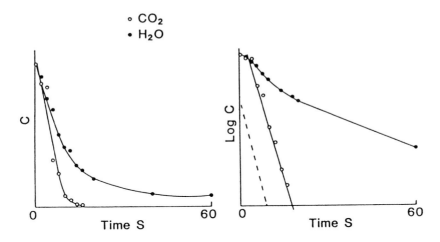

Figure 7. Changes in the carbon dioxide and water vapour content of the air emerging from a cuvette when the entrant air is changed to both dry and CO_2 free. Dotted line shows the theoretical response.

One observation on real leaves was that equilibrium appeared to be obtained more rapidly than would have been predicted from measurements made on the empty chamber. This was confirmed by putting an aluminium foil and wet filter paper "leaf" in the cuvette when the response to changes in ingoing humidity was more rapid than for the empty chamber, (Figure 8) and is supported by the analysis in Appendix 2.

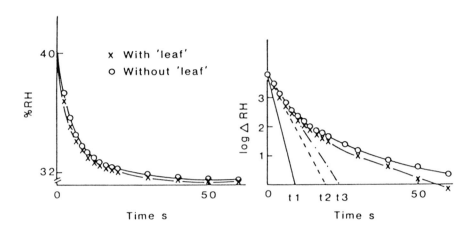

Figure 8. Changes in the relative humidity of emergent air of a cuvette similar to that used for Figure 7, when the equilibrium humidity is changed from 40% to 30% (by changing entrant air humidity), either with or without the presence of an artificial 'leaf'; t_1 is the theoretical response time, t_2 is that with the 'leaf' and t_3 without.

V. CONCLUSIONS

The accuracy of photosynthesis measurements made using leaf cuvettes ultimately rests on the design of the cuvette.
Care should be taken in the choice of suitable materials, in the design of the cuvette to ensure adequate mixing and small boundary layer resistances and in the choice and placing of sensors to monitor the chamber environment. The cuvette should then be subjected to a series of tests to confirm its design.

REFERENCES

Dixon, M., and Grace, J. (1982). Water uptake by some chamber materials. *Plant, Cell Environ.* 5:323-327.

Hardwick, R.C., Cole, R.A., and Fyfield, T.P. (1984). Injury to and death of cabbage (*Brassica oleracea*) seedlings caused by vapours of dibutyl phthalate emitted from certain plastics. *Ann. Appl. Biol.* 105:97-105.

Hashimoto, Y., Ino, T., Kramer, P.J., Naylor, A.W., and Strain, B.R. (1984). Dynamic analysis of water stress of sunflower leaves by means of a thermal image processing system. *Plant Physiol.* 76:266-269.

Parkinson, K.J. (1985). A simple method for determining the boundary layer resistance in leaf cuvettes. *Plant, Cell Environ.* 8:223-226.

Robinson, N. (Ed.) (1966). "Solar Radiation". Elsevier Publishing Company. Amsterdam/London/New York.

Shepherd, W. (1973). Moisture absorption by some instruments materials. *Rev. Sci. Instr.* 44:234.

APPENDIX 1

Consider a leaf of projected area A (m^2) at thermal equilibrium in a cuvette through which air is flowing at a mass flow rate of F (mol s^{-1}). The ingoing air has a water vapour concentration x_i (mol mol^{-1}) and because the cuvette air is vigorously stirred the vapour concentration leaving (x_a) is representative of that of the cuvette air. The transpiration rate per unit leaf area is given by: -

$$E = (x_a - x_i) \, F/A = (x_1 - x_a) \, / \, (r_1 + r_b) \qquad (1)$$

where x_1 is the saturated vapour concentration at leaf temperature T_1, r_1 and r_b are the stomatal and boundary layer resistances to water vapour transfer respectively. The energy balance of the leaf is given by: -

$$C_P(T_a-T_1)/r_H + \sigma \, (T_c-T_1) + \alpha \, I - \lambda \, (x_a-x_i)F/A = 0 \qquad (2)$$

where T_a and T_c are the cuvette air and wall temperatures, C_P is the specific heat, σ is Stefan's constant, I is the total short wave radiation flux incident on the leaf and is the fraction absorbed, λ is the latent heat of vaporization and r_H is the total resistance to heat transfer to the leaf (=0.96 r_b where r_b is the boundary layer resistance to water vapour transfer).

To close approximation $\sigma (T_c-T_1) = 4 \sigma T_a^3 (T_a-T_1)$ so

$$(T_a-T_1) = (\lambda (x_a-x_i)F/A- \alpha I)/(0.93C_P/r_b+4 \sigma {}^3T_a) \qquad (3)$$

To determine the cuvette boundary layer resistance consider a moistened filter paper ($r_1=0$), area A, suspended in a shaded cuvette ($\alpha I=0$) with dry air flowing in ($x_i=0$). From (1)

$$= (x_1/x_a-1) A/F \qquad (4)$$

and to close approximation

$$x_1 = x_s(T_a)-(T_a-T_1) \Delta \qquad (5)$$

where $x_s (T_a)$ is the saturation humidity at cuvette air temperature and Δ is $\delta x_s/ \delta T$ calculated a the average of filter paper and air temperature.

From (4) and (5)

$$r_b = (1/h-1)A/F-(T_a-T_1) \Delta A/Fx_a$$

where $h = x_a/x_s(T_a)$ = cuvette relative humidity.

Therefore, substituting from (3) with both αI and $x_i=0$

$$r_b = (1/h-1)A/F- \Delta \lambda \lambda/(0.93C_P/r_b+4 \sigma T_a^3).$$

This can be rearranged as:-

$$r_b^2(4 \sigma T_a^3)+r_b(0.93C_P+ \lambda \Delta -4 \sigma Ta^3(1/h-1)A/F)$$
$$-0.93C_P(1-(h-1)A/F = 0$$

which can be solved for r_b.

Appendix 2

Consider a cuvette made of ideal materials that absorb no water. The rate of change of water vapour concentration after closure is given by

$$V\frac{dx_a}{dt} = -F(x_a-x_i) \tag{6}$$

The concentration approaches x_i exponentially with a time constant V/F.

If a leaf of area Al and stomatal resistance r_1 is exposed in the chamber, the rate of water vapour exchange is governed by the equation

$$V\frac{dx_a}{dt} = \frac{A_1}{r_1+r_b}(x_1-x_a)-F(x_a-x_i) \tag{7}$$

The equilibrium concentration xe is given by

$$x_e = (\frac{A_1}{r_1+r_b}x_1+Fx_i)/(\frac{A_1}{r_1+r_b}+F)$$

and the concentration approaches this equilibrium value exponentially, as the difference $\delta = x_a - x_e$ is given by

$$V\frac{d\delta}{dt} = \frac{A_1}{r_1+r_b} + F\delta \tag{8}$$

The response time, τ, is $V/(\frac{A_1}{(r_1+r_b)} + F)$; this is less than that of the empty cuvette and decreases with decreasing stomatal resistance and increasing leaf area.

When the wall material of the cuvette absorbs water, there is another term in the water vapour balance corresponding to water vapour exchange with the wall. Consider walls of total area A_w that have absorbed water content W per unit area and that exchange water with an effective transfer resistance r_w so that changes in absorbed water are governed by the equation

$$\frac{dW}{dt} = - \frac{(x_w - x_a)}{r_w}$$

where x_w is the water vapour concentration in air at equilibrium with wall material of water content W. For simplicity we can assume $x_w = W$: the solutions that we now derive will be correct for small excursions of humidity such that $\frac{dx_w}{dW}$ is approximately constant.

The equation for water vapour exchange of a leaf in such a cuvette is

$$V\frac{dx_a}{dt} = \frac{A_1}{r_1 + r_b} (x_1 - x_a) - F(x_a - x_i) + \frac{A_w}{r_w} (\alpha_w - x_a) \quad (9)$$

The equilibrium water content of the walls W_e is equal to X_e/α, and, as above, the gas exchange equations can be rewritten in terms of the deviations from equilibrium $w = W - W_e$ and $\delta = Xa - Xe$, to give

$$\frac{d_w}{dt} = - \frac{(\alpha w - \delta)}{r_w}$$

$$\text{and} \quad V\frac{d\delta}{dt} = (-\frac{A_1}{r_1 + r_b} + F)\delta + \frac{A_w}{r_w} (\alpha w - \delta)$$

Solving these simultaneous differential equations leads to the following equation for :

$$V\frac{d^2\delta}{dt^2} + (F + \frac{A_1}{r_1 + r_b} + \frac{\alpha V}{r_w} + \frac{A_w}{r_w}) \frac{d\delta}{dt}$$

$$+ \frac{\alpha}{r_w} (F + \frac{A_1}{r_1 + r_b}) \delta = 0 \quad (10)$$

The solution to this second order linear differential equation is

$$\delta = \delta_1 e^{-m_1 t} + \delta_2 e^{-m_2 t}$$

SCREENING FOR STRESS TOLERANCE
BY CHLOROPHYLL FLUORESCENCE

Robert M. Smillie
Suzan E. Hetherington

CSIRO Division of Horticulture
Sydney Laboratories, P.O. Box 52
North Ryde, 2113 Australia

I.   INTRODUCTION

Chlorophyll emits a red fluorescence when excited by
visible light. In living plant tissue chlorophyll is
complexed with proteins to form part of an energy collecting
and disseminating system in the chloroplasts and as changes
occur in photosynthetic metabolism, the intensity of the
fluorescence emission varies. Consequently, chlorophyll
fluorescence has been a very useful experimental tool for
investigating photochemical mechanisms underlying
photosynthesis (Papageorgiou, 1975).
    Recently, chlorophyll fluorescence has been used to study
cellular processes other than photosynthesis, in particular,
the response of plants to environmental stress. The rationale
behind using chlorophyll as an intrinsic fluorescent membrane
probe able to detect stress-induced cellular injury is as
follows. Chlorophyll a fluorescence in vivo originates from
photosystem II which is linked to the oxygen–evolving
reaction of photosynthesis to form a complex that appears to
be especially vulnerable to cellular disturbances. With the
onset of cellular injury, there is a high probability that
this will be accompanied by changes in chlorophyll
fluorescence emission. In principle, for the purpose of
using chlorophyll fluorescence to monitor cellular stress
damage, it is unimportant whether the stress interferes
directly or indirectly with photosynthetic metabolism, as

Measurement Techniques in Plant Science
Copyright © 1990 by Academic Press, Inc.

long as a reproducible change in chlorophyll fluorescence is
elicited that is indicative of stress damage. For example,
the onset of chilling injury in leaves or other
chlorophyll-containing tissues is accompanied by a decrease
in chlorophyll fluorescence *in vivo.* Maintained under
conditions of continuous chilling, chlorophyll fluorescence
eventually declines to zero. To compare the relative
chilling tolerances of different species or cultivars, the
rate of the fluorescence decrease during chilling is
determined.

   In this chapter some properties of chlorophyll
fluorescence relevant to its use in stress physiology will be
described. The choice of the procedure adopted and the
fluorescence parameter selected for measurement is discussed
with reference to the need to devise a method capable of
rapidly handling hundreds of samples. To then illustrate the
practical application of using chlorophyll fluorescence to
quantify plant tolerance, a method of screening for chilling
tolerance will be described and examples given of its use in
physiological ecology. The principles detailed for this
screening method are also applicable to screening for
tolerance towards a number of other environmental stresses
which similarly induce a decrease in chlorophyll fluorescence.
These include heat, solar UV light, excess visible light
(photoinhibition), drought, SO_2 and NO_2.

II. CHLOROPHYLL FLUORESCENCE
AND ITS USE IN MONITORING
PLANT RESPONSE TO STRESS

   When photosynthetic tissue is irradiated with visible
light, chlorophyll fluorescence emission can be photographed
(Gibbons and Smillie, 1980), recorded on video (K. Omasa,
personal communication) or measured by a variety of light
detectors. Filters are necessary to isolate the fluorescent
light from reflected actinic light, for instance, blue light
may be used as the actinic light source and a red cut-off
filter or appropriate interference filter used to block the
blue light while allowing the red to far-red chlorophyll
fluorescence emission to pass through to the light detector
system (Gibbons and Smillie, 1980; Schreiber, 1983).

   Fluorescence emission from chlorophyll *a* contained in
living plant tissue may be separated into two components, a
fast rise occurring within 1 to 2 milliseconds usually called

the F_0 or 'constant' fluorescence and a subsequent slower
rise that reaches a peak (Fp) in a second or two (see Fig. 2).
The fluorescence rise above F_0 (Fp-F_0) is most often referred
to in the literature as either the induced or variable
fluorescence (F_v). The F_0 emission, which consists of at
least three time-resolved emissions (Green et al., 1984) is
not responsive to most physiological changes and has limited
use in measuring plant stress tolerance. However, the
development of lesions in photosystem II may cause F_0 to
increase and changes in F_0 fluorescence have been employed to
assess heat damage in leaves above about 45°C (Berry and
Björkman, 1980; Schreiber and Berry, 1977; Smillie and Nott,
1979), to detect mutants defective in photosystem II (Leto
and Miles, 1980) and, as Mn is an essential component of
photosystem II, to detect Mn deficiency (Kriedemann *et al.*,
1985).

The yield of induced chlorophyll a fluorescence, in
contrast, is highly dependent on photosynthetic metabolism
and is affected by a wide range of environmental, chemical
and biological stresses. Because of this dependence on
photosynthesis, pre-irradiation by visible light can alter
the kinetics of induced fluorescence and therefore
fluorescence measurements are performed on 'dark-adapted'
tissue, that is tissue which has been kept in darkness long
enough for reproducible fluorescence kinetics to be obtained
with repetitive irradiations. The time required for
dark-adaptation varies from a few minutes following short
exposure to low quantum flux densities typical of those used
for fluorescence measurements, to several hours following
prolonged exposure to bright sunlight. Consequently,
measurements of induced fluorescence are not easily made on
plants in the field during daylight, unless special
light-proof covers are fitted to individual leaves to allow
for a period of dark adaptation. Therefore, it is more
convenient to use harvested material which is subsequently
dark adapted. This limitation for field work may be overcome
by developing portable devices which are unaffected by
continuous ambient light and detect only modulated
fluorescence from chlorophyll excited by low-intensity pulsed
light (Schreiber, *et al.*, 1986).

The rise in induced fluorescence is thought to relate
directly to the reduction of the immediate acceptor (Q_A) of
photosystem II (Duysens and Sweers, 1963). However, this
rise is affected by a number of factors, two important ones
being the concentration of oxidized Q_A and the light-induced
energization of the chlorophyll-containing membrane system,
both of which can reduce, i.e. quench, the fluorescence

emission. Because of these interacting factors, the rise of
induced fluorescence at 'normal' physiological temperatures
is followed by a slower decline over several minutes with a
smaller second and sometimes a third peak developing before
the steady state level is reached (Papageorgiou, 1975).
(This quenching of fluorescence after Fp is largely
suppressed at 0°C and is not shown in Fig. 2). The complex
kinetics of the rise and fall of induced fluorescence is, in
part, explained by the different but interdependent metabolic
quenching factors exerting their maximum effect on
fluorescence emission at different times after the start of a
period of irradiation. Thus, under constant excitation
quantum flux density, the fluorescence signal obtained
represents an intricate function of wavelength dependent
fluorescence emission, temperature and time. This complexity
appears to have been a deterrent to using induced chlorophyll
fluorescence to reveal changes in the physiological state of
plant tissues and has given rise to uncertainties in
interpreting the precise meaning of the time-dependent
variations in fluorescence emission in terms of changes in
photosynthetic metabolism.

Even though the underlying mechanisms governing
alterations in chlorophyll fluorescence in either unstressed
or stressed plant tissues are not fully understood,
chlorophyll fluorescence can still provide valuable empirical
information about the nature and extent of damage sustained
by plants exposed to various environmental stresses. To test
the feasibility of using chlorophyll fluorescence in vivo to
monitor plant response to either single or multiple stresses,
the following approach was adopted. (1) Expose plants or
detached leaves to the particular stress or stresses of
interest and record changes, if any, in the various
fluorescence parameters. (2) Where a consistent change in an
easily measured parameter is found, establish whether it is
in proportion to the severity of the stress applied. (3)
Determine whether the rate of fluorescence change observed
under a standard stress treatment is indicative of plant
stress tolerance. Some of our studies on screening for
chilling tolerance described in the next section provide
examples of this approach.

While this direct experimental approach has the merit of
being relatively straightforward, there are nonetheless a
number of factors to be borne in mind that can influence the
results. Firstly, there are factors which can affect the
magnitude of the measured fluorescence parameter and if not
controlled can increase data variability. These include
pre-illumination, atmospheric gas composition, humidity,

temperature, optical properties of the leaves, the age of the leaves and plants and the growth conditions. As it was found that the rise in induced chlorophyll fluorescence was far less influenced by 'normal' fluctuations in environmental conditions (except for pre-illumination) than the subsequent decline in fluorescence emission after Fp, our initial studies on screening applications of chlorophyll fluorescence have been confined to those stresses preferentially affecting this rise in induced fluorescence, e.g., chilling, heat, UV irradiation and exposure to SO_2 and NO_2, rather than ones which affect more the quenching of induced fluorescence after Fp, e.g., water deficit.

Secondly, one should be aware of conditions which may alter plant stress tolerance. For instance, cool growth temperatures, exposure to water stress, salinization of root medium, mineral deprivation, and application of abscisic acid reduce injury to many plants when subsequently exposed to chilling stress (Christiansen, 1976; Riken et al., 1976; Wilson, 1976). Age of leaves is also important and young expanding leaves can be less tolerant than older leaves (Smillie, 1984). Consequently, in comparing different cultivars for chilling tolerance, the plants should be raised under the same growth conditions and leaves of comparable age used. It is especially important to be aware of the water status of the plants both before and during chilling as even mild water deficits can change chilling tolerance (Smillie et al., 1987b). Because of the complex interaction between water and chilling stresses (Wilson, 1976), all chilling treatments described in this paper were given to leaves held as close as possible to 100% RH. There does not appear to be any evidence that the level of light intensity received during growth appreciably alters chilling tolerance, but high light intensities during chilling can have profound effects, substantially increasing damage to photosynthetic metabolism (Öquist, 1983) and causing photooxidation of photosynthetic pigments (Van Hasselt, 1972). To avoid the complicating effects of light during measurements of chilling tolerance, all chilling treatments were also carried out in darkness. These considerations of course are not peculiar to measurements involving chlorophyll fluorescence but would apply to virtually any method used to assess stress damage.

III. SCREENING FOR CHILLING
TOLERANCE

A. Measurement of Chlorophyll
Fluorescence Kinetics

In screening studies on large numbers of samples, the
kinetics of chlorophyll fluorescence *in vivo* has been
measured (in relative units) with a portable fluorometer
(models SF-10 or SF-20, Richard Brancker Research Ltd, Ottawa)
designed by Schreiber *et al.*, (1975), with readout to a
potentiometric recorder with a chart speed of at least 1 sec
per cm. The compact sensor of the fluorometer contains both
a red-light-emitting diode to irradiate the immediately
underlying leaf tissue and a photodiode protected by a filter
to detect simultaneously fluoresced light above about 710 nm.
As the plant material is dark-adapted, measurements are
carried out in dim green light. After the initial rise to
the F_o level upon excitation of leaf chlorophyll, the
fluorescence emission reaches F_P in about 0.5 to 2 seconds
depending on the species of plant. In stress-damaged leaves,
the time taken may be considerably longer. In our work we
chose to measure the parameter F_R, the maximal *rate* of the
increase in induced chlorophyll fluorescence (Fig. 2), rather
than the magnitude of the variable fluorescence rise, F_V. F_R
has not generally been employed by other investigators who
have instead used F_V or the ratio of this to the constant
fluorescence (F_V/F_o). F_R was chosen in preference to F_V
mainly for technical reasons. It imparts greater sensitivity
as FR decreases faster during chilling than F_V. Also it can
be measured more quickly as it is not necessary to wait until
the induced chlorophyll fluorescence rise has reached its
maximum. In studies encompassing numerous species of
angiosperms, the kinetics of the chlorophyll fluorescence
rise were similar and were typical of the many examples
recorded in the literature. Abnormal kinetics, e.g., a
continuously curving fluorescence rise above F_o, were seen
occasionally but were uncommon and a high incidence most
likely points to unhealthy leaves, e.g., this type of curve
was observed with leaves of plants deliberately made
deficient in nitrogen.
        Since both leaf fluorescence and stress tolerance can
vary with the age of the leaf, the choice of what material to
sample becomes important when making comparisons between
different plants. In our studies comparisons were made using

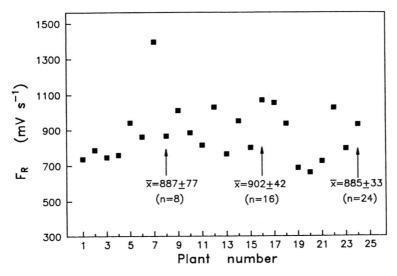

Figure 1. Chlorophyll fluorescence values, measured as F_R, for leaves harvested from 24 plants of field-grown cassava (*Manihot esculenta*). Details of sampling are given in text. The means ± S.E. for the first 8, first 16 and all 24 plants sampled are shown on the graph.

fully expanded non-senescing leaves. Figure 1 shows values for F_R obtained with 24 different plants of the same cultivar of cassava grown for 18 months in a field. Leaves were selected by picking a line of 24 plants at random and detaching from the main stem of each the eighth fully expanded leaf located below the growing point. After harvest, the leaves were kept in darkness at 100% RH for 2 hours before the chlorophyll fluorescence of each was recorded. Measurements were made on the central region, avoiding the mid-vein, of the middle lobe of each compound leaf. Values shown in Fig. 1 are respectively the mean and standard error of the mean of F_R, recorded at 0°C at the start of a chilling period, from the first 8, first 16 and all 24 plants. Variability between plants was low with this procedure and for routine comparisons between cultivars of most species, 16 plants of each were sampled.

B.   Changes in Chlorophyll Fluorescence
During Chilling and The Determination
of Chilling Tolerance

All C4 plants (Long, 1983) and most C3 plants of tropical
origin (Lyons, 1973) are susceptible to chilling injury at
low, non-freezing temperatures. These include many important
crop plants such as maize, sorghum, rice, tomato, bean,
cucumber and tropical fruits. When leaves of these plants or
leaves of chilling tolerant plants such as pea and cabbage
are dark-adapted then cooled to 0°C, all show similar
chlorophyll fluorescence curves when irradiated. However, if
fluorescence measurements are repeated at intervals while
maintaining the temperature of the leaves at 0°C, differences
between chilling tolerant and intolerant plants become
evident (Smillie, 1979; Havaux and Lannoye, 1984). Figure 2
shows that the induced fluorescence of the tropical peanut

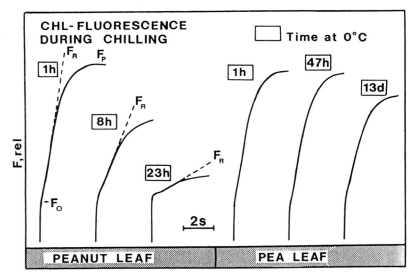

Figure 2.   Chlorophyll fluorescence in dark-adapted
leaves chilled at 0°C in darkness. The initial rise in
fluorescence to F_o was unaffected by chilling.   The
subsequent rise from F_o to F_P (the induced chlorophyll
fluorescence) decreased rapidly during chilling in peanut,
Arachis hypogaea, (chilling intolerant) compared with pea,
Pisum sativum, (chilling tolerant). F_R is the maximum rate
of increase of induced chlorophyll fluorescence. (Redrawn
from Smillie, 1979).

decreased markedly within 1 day at 0°C, whereas little change occurred in leaves of the cool temperate pea plant. Even after 13 days at 0°C the induced fluorescence of pea leaves had fallen by only about 20%. The rate of decrease in FR was exponential. This is illustrated in Fig. 3 for chilled cucumber leaves. Similar results were obtained using peanut, bean, maize, rice and other crop plants susceptible to chilling injury. In cucumber leaves the decrease in FR was chilling induced and was not due to senescence as leaves held at the higher temperature of 12°C, just above the temperature range causing chilling injury, showed little change in FR (Fig. 3). In some experiments a lag prior to the start of the decrease in FR or even a transient increase in FR upon chilling was observed. The reasons for this are unknown but could arise from inadequate dark-adaptation before chilling or because unhealthy leaves or leaves already stress injured were used.

The relationship between the chilling-induced decrease of leaf FR and chilling injury to the plant as a whole was

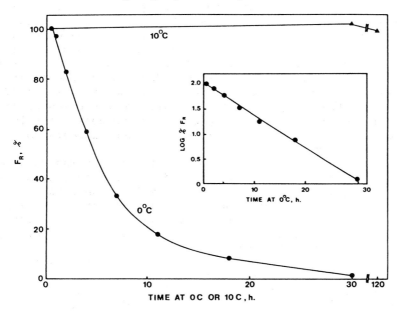

Figure 3. Time course of changes in FR of cucumber (*Cucumis sativus*) leaves held at either 0°C or 10°C. Values of FR are expressed as a percentage of the first reading (0.5 hour at 0°C or 10°C). The inset shows a logarithmic plot of the same values. From Smillie and Hetherington (1983).

investigated by chilling plants at 0°C for various periods, up to 30 hours, and comparing the degree of change in F_R during chilling with rate of plant growth following chilling. Maize seedlings grown hydroponically were used for this study and it was shown that for the 24-hour period immediately following each period of chilling, the decrease in plant growth was proportional to the extent of the chilling-induced decrease in F_R recorded at the end of the chilling period (Smillie and Hetherington, 1984).

A relationship between the rate of decrease in F_R at 0°C and chilling tolerance was established in a series of studies in which over 150 different species of plants were tested. Rates of decrease in F_R agreed with the expected relative chilling tolerances of the species tested. Values of chilling tolerance, defined as the time taken for F_R to decrease by 50% at 0°C, are shown in Table 1 for some of these different species. Values ranged from a few minutes for tropical marine algae to several hours for tropical annuals, to days for tropical perennials and up to several weeks for resistant plants originating in cool temperate regions. These results suggest that the common practice of arbitrarily grouping plants into two distinct categories, chilling sensitive and chilling resistant, is misleading. Maize, for instance, is generally categorized as a chilling sensitive plant, but it shows considerable tolerance relative to another monocotyledonous plant, the sea grass *Halophila spinulosa* (Table 1). Rather than different species being predominantly either chilling sensitive or chilling tolerant, a continuous gradient of chilling tolerance appears to exist amongst species.

Details are given below of how the method was used to assess the chilling tolerances of different races of maize (Hetherington et al., 1983a). This method can readily accommodate most terrestrial and marine higher plants and green macroalgae. To determine chilling tolerance, samples of photosynthetic tissue, usually leaves, are dark chilled at 0°C. Chlorophyll fluorescence is measured soon after the beginning of the chilling period and at one or more intervals thereafter to determine the rate of F_R decrease. Measurements can be made directly on leaves of intact plants. Alternatively, as the changes in F_R during chilling are the same for attached and detached leaves provided they are kept fully hydrated (Smillie et al., 1987b), leaves may be detached and chilled either whole or as cut pieces. For ease and speed of measurement, this latter technique is favoured. The harvested material is laid on a metal plate as described below and repetitive fluorescence measurements are

Table 1. Chilling Tolerances of Some Plants as
Determined by Chilling-induced Changes in Chlorophyll Fluorescence.

Plant	Climatic adaptation	Chilling tolerance*	
Episcia	Tropical ornamental		
	- immature leaf	0.24	min
	- semimature leaf	14.5	min
Caulerpa peltata	Tropical alga	4.2	min
Halophila spinulosa	Tropical sea grass	28	min
Cucumber	Tropical annual	4.3	h
Maize	Tropical annual	8.4	h
Guava	Tropical perennial	23	h
Custard apple	Tropical perennial	31	h
Lime	Tropical perennial	84	h
Zostera muelleri	Cool temperate sea grass	172	h
Barley	Cool temperate annual	290	h
Pea	Cool temperate annual	385	h
Kumquat	Temperate perennial	620	h

*Time for FR to decrease by 50% at 0°C, determined
mature tissues, except *Episcia*

recorded from the same portion of the leaf during chilling.
The leaves are not warmed for the fluorescence measurements
but are held at 0°C continuously. Thus the method measures

chilling induced changes as they are actually happening at
0°C.

For maize, a section about 2.5 cm long of a fully
expanded leaf (e.g., the tertiary leaf of a plant in which
the fifth leaf is still expanding), is cut from the plant at
a point one-third of the distance from the leaf tip and is
quickly placed, abaxial surface down, on wet filter paper on
an aluminium plate (46 x 31 x 0.3 cm) which can accommodate
96 leaf samples arranged in twelve columns of eight. As each
column of leaves is completed, the leaves and plate are
progressively covered with a thin polyethylene film permeable
to O_2 and CO_2 but with low permeability to water to prevent
dehydration. When all leaf samples have been positioned, a
plastic slab (46 x 31 x 1 cm) containing 96 guide holes for
the 3.3 cm-diameter sensor of the fluorometer is laid on top
of the plate. The plastic slab is fastened to the aluminium
plate so that repeated fluorescence measurements can be made
at different times on the same part of the same leaf. Prior
to recording chlorophyll fluorescence *in vivo,* the prepared
leaf samples are kept in darkness at room temperature for at
least one hour to ensure that they are completely dark
adapted and hydrated before the plate is placed onto melting
ice contained in an insulated box and stored in a darkened
cold room. One hour after the start of chilling and at
various times thereafter, the fluorescence measurements are
carried out in dim-green light. The F_R of each leaf sample
at 0°C is determined by placing the pre-chilled sensor of the
portable fluorometer on the exposed upper surface of the leaf
to irradiate it with red light of photon flux density 15 μ mol
$m^{-2}s^{-1}$ and record the rise in fluorescence yield. In practice,
irradiation time is varied between 1 and 4 seconds depending
on the speed of the fluorescence rise and measurements on 96
samples take 10 to 12 minutes to complete.

F_R can be determined directly from the fluorescence
signals recorded on a strip-chart recorder. Alternatively,
direct processing by computer can be employed. Once the
kinetics of the decrease in F_R for a particular species have
been established, the percentage decrease in F_R following a
suitable period of chilling suffices for comparisons of
chilling tolerance between different cultivars of the same
species. However, by determining the time taken for a 50%
decrease in FR, direct comparisons can be made between
different experiments and different plant species. For
applications away from the laboratory, the fluorescence
signals are monitored on a chart recorder and simultaneously
recorded onto magnetic tape via a Simtrad, a device designed
by Mr. G. Stanley of the CSIRO Division of Food Research,

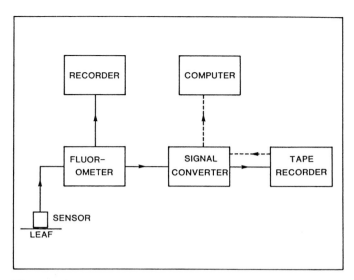

Figure 4. Instrumentation used to record chlorophyll fluorescence. Data recorded on magnetic tape is later replayed to a Hewlett Packard 9826 computer for computation of F_R.

that has an output voltage the frequency of which is dependent upon the voltage output of the fluorometer. The output signal of varying frequency from the Simtrad is recorded on the magnetic tape. Later, the tape is replayed via the Simtrad to a computer programed to compute F_R, Figure 4
outlines the arrangement of the equipment used. Compact cassette recorders are the most convenient to use but can introduce errors in values of F_R because of small fluctuations in tape speed. The use of cassette recorders has been confined to studies in which a relatively large number of replicates are taken.

IV. RELATIONSHIPS BETWEEN FR AND PHOTOSYNTHESIS DURING CHILLING

The emphasis in our studies has been to relate changes in FR *in vivo* during chilling to plant chilling tolerance and not to use chlorophyll fluorescence as a means of investigating the effects of chilling stress on photosynthesis and photosynthetic metabolism. Nonetheless, it is pertinent to ask if the data for F_R also provides information about the extent of chilling damage to photosynthesis.

Under certain circumstances, decreases in steady-state chlorophyll fluorescence have been demonstrated to parallel increases in the overall rate of photosynthesis (Walker *et al.*, 1983), but the increase in induced fluorescence obtained upon irradiating dark-adapted leaves is more likely to be related to photosystem II activity (Papageorgiou, 1975). Obviously F_R is a photosynthetic parameter, but as it has not commonly been used prior to our studies, there is little discussion of it in the literature. Since induced chlorophyll fluorescence emanates from photosystem II, we investigated the relationship between F_R and photosystem II activity during chilling. As the latter cannot at present be measured in intact tissue, maize plants were chilled at 2°C in darkness and after 0, 1, 2, 3 and 4 days of chilling, batches of leaves were rapidly homogenized and F_R and photosystem II activity determined on the homogenate. Figure 5 shows that the chilling-induced decrease in log F_R was linearly related to the decrease in photosystem II activity and that both declined linearly with respect to the time of chilling. We conclude that the exponential decrease in F_R (i.e., linear decrease in log F_R) observed in intact tissue (see Fig. 3) is most likely indicative of a concomitant linear decrease in the activity of photosystem II *in vivo*. The photooxidative (oxygen evolving) side of photosystem II has been identified as the most chilling sensitive site of inhibition of photosystem II activity (Margulies and Jagendorf, 1960) but the precise mechanism by which cellular chilling injury causes this inhibition is unknown. Some possibilities are discussed by Smillie (1984).

Since photosystem II activity is thought not to be a rate limiting step of photosynthesis in healthy leaves (Dietz *et al.*, 1984), it can be predicted that chilling-induced changes in FR will not necessarily be correlated with changes in leaf photosynthesis. Photosynthetic rates would only be expected to decrease when rate limiting steps become inhibited or when photosystem II activity has declined to the point where it becomes the new rate limiting step. To measure the changes in leaf photosynthesis, maize plants were chilled for 0, 1, 2, 3 and 4 days then warmed to 25°C and photosynthetic rates determined using a leaf disc oxygen electrode (Hansatech Ltd., King's Lynn, U.K.). Because of the high CO_2 partial pressure used in the leaf cuvette, photosynthetic rates should not be influenced by any chilling damage to stomata.

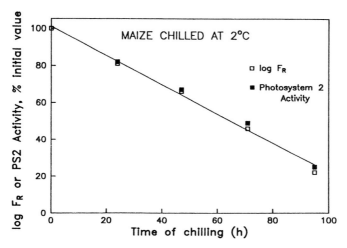

Figure 5.   Changes in log F_R and photosystem II decline in a activity in chilled maize. Seedlings were chilled in darkness at 2°C for 0,1,2,3 and 4 days, the leaves were then homogenized and rapid determinations made of F_R and photosystem II activity (spectrophotometrically by the photoreduction of ferricyanide in the presence of p-phenylenediamine).

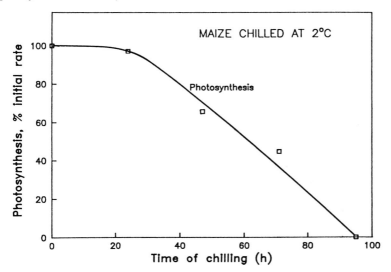

Figure 6.   Photosynthesis of chilled maize. Seedlings were chilled as described in Fig. 5.   Photosynthesis was measured by oxygen evolution using a leaf disc electrode at 25°C and quantum flux density of 650 μ mol $m^{-2}s^{-1}$.

In contrast to the decrease in F_R *in vivo* which began
immediately upon chilling, there was little change in
photosynthetic rate (Figure 6) or in the quantum yield of
photosynthesis (data not shown) after chilling for 24 hours.
Longer chilling times resulted in a decline in both rate (Fig.
6) and quantum yield and after 4 days of chilling,
photosynthetic rates were too low to be measured. These
results suggest that measurements of photosynthesis may not
give an accurate measure of chilling damage occurring in the
photosynthetic cell, at least during the early stages of
chilling injury. This may also be true for other
environmental stresses, e.g., photoinhibition, which inhibit
photosystem II preferentially.

V.   RECOVERY FROM CHILLING INJURY

As the screening by chlorophyll fluorescence measures
changes taking place at the chilling temperature, it allows a
distinction to be made between chilling injury *per se* and the
reaction to injury sustained once the temperature is raised.
These two events are likely to be governed by different sets
of genes and plant reactions to chilling injury upon
rewarming may not be specific to chilling, but may include
non-specific responses to cellular injuries promoted by any
one of a variety of agents. Further, metabolic reactions to
chilling injury are likely to be temperature dependent and
may be inhibited or take place slowly at 0°C. With the
exception of prolonged periods of chilling, plant phenotypic
expression of chilling injury will only become evident once
the temperature is raised. Most methods of assessing
chilling injury in fact require a rewarming period, e.g.,
assays of photosynthesis and other metabolic activities, ion
leakage and measurement of growth inhibition. While
recognizing that in the context of surviving a chilling
stress, both the extent of the injury incurred and the
ability to recover from the injury are important, failure to
identify these two distinct phases could lead to confusion in
interpreting results from plant physiological or plant
breeding studies. For example, while the rapidity with which
chilling injury develops at low temperature is an important
determinant of plant survival, we have noted that amongst
some related species it is the capacity to recover from
injury received during chilling and not the extent of injury
sustained that largely accounted for the different survival

rates of the species. While emphasis to date has been placed
on using leaf fluorescence to follow the onset and
development of chilling injury, it should provide a powerful
tool in future studies for following post-chilling
deterioration or recovery and for investigating presumed
complex interactions between the recovery processes and
environmental variables such as light, temperature and the
water and nutrient status of the plants.

VI. OTHER STRESSES

The monitoring of the effects of stress on green plants
by chlorophyll fluorescence has many potential applications
in physiological ecology as evidenced by the diversity of
stresses which produce changes in induced chlorophyll
fluorescence. These stresses include freezing (Klosson and
Krause, 1981), chilling (Smillie, 1979; Havaux and Lannoye,
1984), chilling in light (Baker et al., 1983) and under
anaerobiosis (Smillie and Hetherington, 1983), heat (Smillie
and Gibbons, 1981; Weis, 1982), UV light (Smillie, 1982/83),
intense visible light (Critchley and Smillie, 1981), water
deficit (Wiltens et al., 1978; Govindjee et al., 1981;
Hetherington et al., 1982), salinity (Smillie and Nott, 1982;
Downton and Millhouse, 1985), ozone (Schreiber et al., 1978),
SO_2 (Shimazaki et al., 1984), phosphorus deficiency (Conroy
et al., 1986),herbicides (Renger and Schreiber, 1985), fungal
pathogens (Ahmad et al., 1983), and genetic defects (Leto and
Miles, 1980). Changes in chlorophyll fluorescence have also
been used in studies of greening (Popovic et al., 1984), leaf
development (Nesterenko and Sid'Ko, 1980) and senescence
(Jenkings et al., 1981). Applications of fluorometric methods
to investigate the effects of several of these stresses on
plants have been reviewed by Renger and Schreiber (1986).

As mentioned previously a number of these stresses cause
a progressive decline in F_R and it ought to be feasible to
follow the development of these injuries by applying the same
experimental approach used to measure chilling tolerance.
For instance, when measuring plant heat tolerance, instead of
chilling the aluminium plate holding the leaf samples, the
plate is immersed in heated water for 10 minutes and the F_R
of each sample is measured before and after the heat
treatment. Under these conditions, leaves of heat sensitive
plants show larger decreases in F_R than plants better adapted
to withstand high temperatures (Smillie and Hetherington,
1983). It is feasible to use portions of the same leaf for

determinations of both chilling and heat tolerances. For
instance, in studies with wild and cultivated potatoes, one
half of each leaf was used for the assessment of chilling
tolerance and the other half for heat tolerance (Hetherington
et al., 1983b; Smillie *et al.*, 1983).

VII. APPLICATIONS OF CHLOROPHYLL
FLUORESCENCE SCREENING

It is only in the area of chilling and heat stresses that
screening methods like those described in this paper have
been devised and applied. Increasing the number of samples
which can be processed considerably widens the scope for
applications in physiological ecology as well as in
horticulture and agriculture including plant breeding.
Several examples are given below in which answers to the
questions posed were sought using screening by chlorophyll
fluorescence.

A. Acclimation to Chilling —
Can Maize Seedlings be
Hardened to Resist Chilling Injury?

While chilling tolerance is an inherent genetic property,
a degree of fine control exists in some species which allows
for rapid adjustment of chilling tolerance in response to
unfavourable growth conditions (Riken *et al.*, 1976; St. John
and Christiansen, 1976; Wilson, 1976). Although knowledge of
the relative abilities of different cultivars to acclimate or
harden at cool growth temperatures to become less susceptible
to chilling injury is important in agriculture, information
of this kind is scarce mainly because of the lack of easy
methods to measure chill hardening. The chlorophyll
fluorescence method would seem to be well suited as a rapid
way of determining acclimation to chilling. This was tested
using the maize cultivar, Northern Belle. Seedlings were
raised to produce plants thought to be unhardened (grown
continuously at 20/15°C), hardened (unhardened plants
transferred to 15/5°C for 4 days) and dehardened (hardened
plants returned to 20/15°C for 2 days). When chilled at 1°C,
chilling tolerances (time to 50% decrease in F_R) were found
to be 8 ± 0.3 hours for unhardened plants, 25 ± 1.9 hours
for plants given the hardening treatment and 9 ± 0.7 hours
for plants after the dehardening (Hetherington and Smillie,

1984). Hardened plants similarly showed enhanced chilling tolerance when chilled under much milder conditions, at 8°C. The increase in chilling tolerance of plants following the hardening treatment was correlated with reduced post-chilling leakage of solutes from detached leaves and higher post-chilling rates of survival, growth and chlorophyll synthesis in intact plants (S.E. Hetherington, unpublished experiments). These results indicate that seedlings of this modern maize cultivar can be chill-hardened and that the increase in chilling tolerance following hardening can be determined from measurements of chlorophyll fluorescence on chilled detached leaves.

B.   Adaptation to Chilling Along
an Altitudinal Gradient — Do
Species of Wild Potato Native
to Habitats of Increasing Elevation
Show a Progression of Chilling Tolerance?

Some groups of plants show striking genotypic adaptation to low environmental temperatures and are found growing naturally over a wide range of altitude or latitude. In the Andean region of Peru and Bolivia, different species of wild potato (*Solanum* sp.) are found at altitudes ranging from 2000m to over 4000m. Within the altitudinal cline, most individual species have specific altitudinal limits and are distributed over a fairly narrow span of altitude. The chilling tolerances of some of these species grown in an irrigated field from tubers originally collected from plants growing naturally at different altitudes were investigated with the chlorophyll fluorescence method (Smillie *et al.*, 1983). Figure 7 shows that, with two exceptions, there was a progression of increasing chilling tolerance with increasing altitude suggesting that genetic modification of cellular resistance to low temperature in potato is an important factor in adapting successfully to the generally colder conditions prevailing at high altitudes.

Altitude is only one of a number of factors influencing microclimate and the much higher cold tolerance shown by *S. olmosense* compared with the other species from low altitudes (Fig. 7) was most likely because it was native to the western side of the Andes where air temperatures are influenced by the cold Humboldt current. In contrast, the other low altitude species were all from the warmer eastern side of the Andes. The other exception to the trend of increasing cold tolerance with increasing altitude was *S. violaceimarmoratum.*

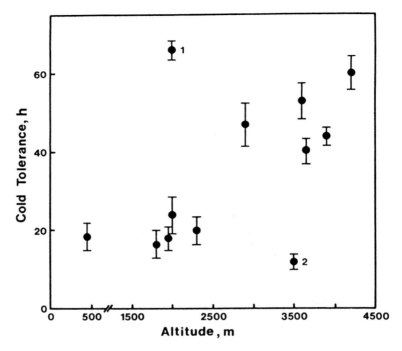

Figure 7.   Chilling tolerance of potato species from
different altitudes.   Each point represents a different
species of wild potato except the one from 450m which is a
non-tuberiferous species, *S. penelli*. Chilling tolerance is
the time in hours for a 50% decrease in F_R of leaves at 0°C.
1, S. *olmosense*; 2, *S. violaceimarmoratum* (see text).   (From
Smillie *et al.*, 1983).

This species was from the warm Yungas region of Bolivia that
is subject to warming updrafts from the Amazonian basin and
its comparatively low cold tolerance was not unexpected.
    Heat tolerance showed a reverse trend compared with
chilling tolerance, decreasing with increasing altitude, but
low cold tolerance was not closely correlated with high heat
tolerance and vice versa (Smillie *et al.*, 1983).

C. Adaptation to Chilling Along a Latitudinal Gradient — Is the Penetration of Mangroves of Eastern Australia into High Latitudes Correlated with an Increase in Chilling Tolerance?

Plants adapted to grow with their roots immersed in salt water are collectively known as mangroves and by and large are confined to tropical and mid-latitudinal coastlines. Along the entire coastline of eastern Australia at locations suitable for mangrove colonization, mangroves can be found from the tropical north (11°48'S) to the cool temperate south (to 38°52'S). However, of the 36 known species inhabiting the far northern coastline the numbers decrease with increasing latitude, dwindling to 11 species in the semitropical and warm temperate regions and to 2 species in the cool temperate south. To investigate whether the ability of mangroves to extend their range of habitation to latitudes having recurrent cold winter temperatures was correlated with an increase in chilling tolerance, 27 different species were collected and screened for chilling tolerance using chlorophyll fluorescence. Adaptation to chilling was clearly evident when species found only at northern tropical latitudes (chilling tolerances of less than 40 hours at 0°C, comparable with tropical perennials, see Table 1) were compared with the two most southerly distributed species *Aegiceras corniculatum* and *Avicennia marina* (chilling tolerances of both were greater than 120 hours). Interestingly, the manner of cold adaptation appeared to differ in these two species in that while both northern and southern populations of *A. corniculatum* were chilling tolerant, this was only true for the southern populations of *A. marina*. A Caribbean species of *Avicennia*, *A. germinans*, also shows increased chilling tolerance along a latitudinal gradient (McMillan, 1975).

For mangrove species distributed from low to mid-latitudes the overall picture was less clear; collectively they showed a wide range of chilling tolerances. This may be because mangroves are very diverse botanically and a more meaningful comparison of cold tolerance with respect to latitudinal distribution may be that made between species belonging to the same genus. Table 2 shows chilling tolerances of species from three different genera. Within genera, species found at higher latitudes had greater chilling tolerance.

Table 2.   Chilling Tolerances of Mangroves

Mangrove	Chilling tolerance*	Southern limit of distribution	
Ceriops decandra	3	18°31's	Herbert River, Qld
Ceriops tagal var. *tagal*	6	24° 1's	bustard Head,Qld
Ceriops tagal var. *australis*	27	28°11's	Tweed river, NSW
Lumnitzera littorea	20	18°32's	Herbert River, Qld
Lumnitzera racemosa	78	27°30's	Morton Bay, Qld
Sonneratia caseolaris	11	18° 5's	Murray River, Qld
Sonneratia alva	82	22°30's	Port Clinton, Qld

*Time in hours to 50% decrease in F_R at 0°C.

The results obtained with mangroves suggests that acquisition of enhanced chilling tolerance is a prerequisite for successful establishment at high latitudes. Of course chilling tolerance alone does not guarantee this as many other factors are involved in determining the distribution of a species. It should also be pointed out that in this paper chilling tolerance is determined on vegetative tissue and minimum temperatures required for reproductive processes such as flowering, fruiting and seed germination may limit distribution along a thermal gradient.

The studies with mangroves and wild potatoes illustrate advantages of utilizing chlorophyll fluorescence in screening programmes as measurements can be carried out near the field or natural growing sites and the plant material can be processed within a few hours of harvest. By comparison, other methods of determining chilling tolerance are more time consuming and less amenable for acquiring quantitative data.

D.   Selecting for Chilling Tolerance —
Do Maize Lines Regarded as Adapted
for Growth in Cool Climates Show
Enhanced Chilling Tolerance?

One of the more recent applications of chlorophyll
fluorescence screening is in plant breeding programs to
select for breeding material having increased tolerance to
chilling or other stresses. The example given here was
carried out with maize breeders and constituted our first
attempt to demonstrate the use of chlorophyll fluorescence as
a selection method for chilling tolerance. It has since been
applied to a number of other crop plants including cassava,
potato, avocado, mango and other tropical fruits. Maize was
chosen because of a world-wide interest in obtaining lines
with improved adaptation to cool temperate regions. Three
races of maize were compared, a Peruvian highland maize,
Confite Pueno, and two commonly cultivated ones, Northern
Flint and Corn Belt Dent. Confite Pueno was expected to be
the most chilling resistant as it grows at high altitudes in
Peru and has been shown to grow autotropically in controlled
environment chambers at 13°C, a temperature too low for most
other races (Hardacre and Eagles, 1980). Of the other two
races, Northern Flint is regarded as better adapted to cool
conditions and is included in hybrids used in the cooler
maize growing regions of North America and Europe (Bunting,
1978). When these populations were grown together under
non-stressful conditions and tested for chilling tolerance by
chlorophyll fluorescence, the expected rankings were obtained
(Table 3).

E.   Selecting for Heat Tolerance — Do Potato
Clones Selected for Growth in Warm Climates
Show Enhanced Heat Tolerance as Measured
by Chlorophyll Fluorescence?

Heat tolerance was the main focus of interest in a study
carried out at the International Potato Center, Lima, with
potato breeders interested in developing lines better adapted
for growth in warm climates. It was shown that five clones
of *S. tuberosum* selected for climatic adaptation to warm
growth temperatures on the basis of field trials at different
elevations had enhanced heat tolerance as determined by
chlorophyll fluorescence when compared with eight other
clones adapted for cool growth conditions only (Hetherington
et al., 1983b). Of the 13 cultivars tested, the line DTO-2

Table 3.   Chilling Tolerance of Different Maize
Populations

Population	Where grown	Chilling tolerance*
Confite Pueno	Peruvian highlands	11.8
Northern Flint	Northern latitudes	9.4
Corn Belt Dent	Warm mid-latitudes	8.4

*Time in hours to 50% decrease in F_R at 0°C.   From an
analysis of variance, values are significantly different
from each other (P = 0.05).   (From Hetherington et al.,
1983a).

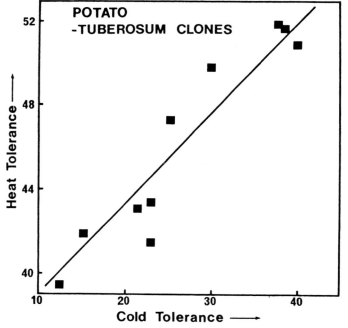

Figure 8.   Heat tolerance and chilling tolerances
compared for 10 clones of the commonly cultivated potato.
(From Hetherington et al., 1983b).

showed greatest tolerance to heat stress (Table 1,
Hetherington *et al.*, 1983b) a result which has been supported
recently in trials completed in Thailand in which up to 24
cultivars were screened for yield potential over three
successive growing seasons. The best of the cultivars for
heat tolerance as judged by the one giving the highest
consistent yield of tubers over three years was DTO-2
(Thongjiem and Chouvalitwongporn, 1985). As the
determination of heat tolerance by chlorophyll fluorescence
on say 24 clones of potato can be completed in 2 to 3 working
days, the method shows promise as an early selection
technique to be used prior to long-term field trials.

Chilling tolerances were also determined on ten of the
clones tested by Hetherington *et al.* (1983b) and as shown in
Figure 8 were found to correlate with heat tolerances
(r^2=0.82), a correlation not seen amongst species of wild
potatoes (Smillie *et al.*, 1983). This suggested that the
selection process had identified clones with a generalized
tolerance to environmental stresses rather than those having
a specific genotypic adaptation towards heat stress. Also of
interest was the finding that the two clones showing the
highest heat tolerances by chlorophyll fluorescence were the
only ones with the species *S. phureja* in their parentage.
While *S. tuberosum* is virtually the only potato cultivated
outside of Peru, S. phureja and several other species are
grown for food in Peru itself and in a study of the heat
tolerances of a number of these species, S. phureja was found
to be the most heat tolerant (Hetherington *et al.*, 1983b).
This species then may be a useful source of genes for
introgression into *S. tuberosum* to breed more heat tolerant
clones. These results demonstrate the feasibility of using
chlorophyl fluorescence to screen for either cold or heat
tolerance in potato and illustrate how the method can provide
additional information about innate characteristics of
adaptations in plants selected in field trials.

F.   Applications to Postharvest Physiology --
Can Chilling Injury in Stored Fruit
Be Monitored by Chlorophyll Fluorescence?

Storage at low, nonfreezing temperatures is one of the
most effective means of extending the postharvest life of
fruit. Tropical fruit and vegetables, however, develop
chilling injury at low temperatures. For some fruit, storage
life is improved by storing under modified atmospheres (low
oxygen, elevated carbon dioxide and removal of ethylene).

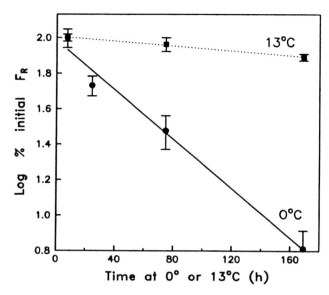

Figure 9. Time course of chilling injury development
in the peel of mature green banana fruit (n=8) kept in
polyethylene bags (poly-bags) at 0°C or 13°C. The
development of chilling injury is followed by the decrease in
log F_R. (From Smillie *et al.*, 1987a).

The combination of an optimal storage temperature, one which
favours low rates of both senescence and chilling injury,
plus a modified atmosphere may be the best way to improve
prospects of storing these fruit or transporting them to
distant markets. Precise information on the relative
chilling susceptibility and optimal storage temperature is
lacking for most tropical fruit species and in this respect
screening for chilling stress by chlorophyll fluorescence is
relevant. As shown in Figure 9, the development of chilling
injury in mature green banana fruit can be followed by the
decrease in F_R. At 0°C chilling injury developed rapidly and
at 13°C very slowly. Other experiments established that the
chilling-induced decrease in F_R measured during time of
storage at 0°C was linearly correlated with two indices of
chilling injury, the postchilling inhibition of ripening and
the postchilling development of peel darkening (R.M. Smillie,
unpublished experiments). In the experiment shown in Figure
9 the fruit prior to chilling were sealed in polyethylene
bags containing an ethylene absorbant and allowed to develop
a modified atmosphere over a period of 2 days at 20°C. An

advantage of chlorophyll fluorescence measurements over other
methods was that subsequently, chilling injurydevelopment
could not only be measured at the chilling temperature
without rewarming the fruit, but also without disturbing the
modifed atmosphere as fluorescence emission could be detected
and measured through the polyethylene bags (or similar
wrapping, e.g., clear plastic shrink wraps). Aside from
chilling stress, our recent studies have also related changes
in chloroplast fluorescence to ageing and ripening events in
green fruit held at non-chilling temperatures and chlorophyll
fluorescence monitoring appears to have the potential to be
an extremely useful and versatile technique for use in fruit
and vegetable postharvest physiology.

VIII. SUMMARY

    The notion is discussed that chlorophyll *a in vivo* can be
utilized as a naturally-occurring internal fluorescing
membrane probe to monitor plant response to environmental
stress. The use of chlorophyll in this way is feasible
because the induced fluorescence emitted by chlorophyll *a*
from within plant tissue is indicative of chloroplast
membrane functions which deteriorate in cells stressed by
exposure to a number of environmental stresses including
chilling, heat, solar UV light, photoinhibitory visible light,
drought, SO_2 and NO_2. As cellular damage develops and
chloroplast functions decline, fluorescence emission *in vivo*
decreases progressively. As a diagnostic tool for stress
injury, chlorophyll fluorescence has the advantage that the
changes can be measured rapidly and with high sensitivity on
living plant tissue well in advance of visible symptoms of
stress damage. Measurements can be made using portable
equipment and data can be stored on magnetic tape for
processing by computer.
    A method is described and some precautions given for
applying chlorophyll fluorescence to screen plants for
tolerance towards chilling stress. Modifications of the
method to measure heat and other stress tolerances are
indicated. Chilling tolerance is measured as the time taken
for F_R, the maximal rate of increase in induced chlorophyll
fluorescence, to fall by 50% at a given chilling temperature,
usually 0°C. Values for chilling tolerances at 0°C range
from a few minutes for tropical marine algae to several weeks
for chilling tolerant angiosperms. F_R was chosen for
measurement as it was the parameter of induced fluorescence

which most nearly fitted the criterion required in screening applications, namely maximum response to stress and minimum time for measurement. Evidence is given that a decline in log F_R during chilling reflects a chillinginduced inhibition of photosystem II in vivo.

Screening for chilling tolerance by chlorophyll fluorescence has a large number of applications in physiological ecology including studies of the acclimation and adaptation by plants to low temperatures. Examples are cited in which chlorophyll fluorescence screening was employed to follow chill hardening in maize and chilling injury in banana fruit and to measure chilling tolerances of potato species occurring along an altitudinal gradient and mangroves along a latitudinal gradient. Screening methods utilizing chlorophyll fluorescence have the potential to assist in achieving a major goal of plant breeders, the breeding of crop plants having enhanced tolerance towards specific environmental and man-made stresses. (Mannscript completed Sept. 1985, updated May,1987.)

REFERENCES

Ahmad, I., Farrar, J.F., and Whitbread, R. (1983). Photosynthesis and chloroplast functioning in leaves of barley infected with brown rust. Physiol. Plant Path. 23:411-419.
Baker, N.R., East, T.M., and Long, S.P. (1983). Chilling damage to photosynthesis in young *Zea mays*. *J. Exp. Bot.* 34:189-197.
Berry and Björkman (1980). Photosynthetic response and adaptation to temperature in higher plants. Annu. Rev. Plant Physiol. 31:491-54311.
Bunting, E.S. (1978). Maize in Europe. In "Forage Maize" (E.S. Bunting, B.F. Pain, R.H. Phipps, J.M. Wilkinson, and R.E. Gunn, eds.). pp. 1-13, Agricultural Research Council, London.
Christiansen, M.N. (1979). Physiological bases for resistance to chilling.HortSci. 14:583-586.
Critchley, C., and Smillie, R.M. (1981). Leaf chlorophyll fluorescence asan indicator of high light stress (photoinhibition) in Cucumis sativus L. Aust. J. Plant Physiol. 8:133-141.

Conroy, J.P., Smillie, R.M., Kuppers, M., Bevege, D.I., and Barlow, E.W.(1986). Chlorophyll fluorescence and photosynthesis responses of *Pinus radiata* to phosphorus deficiency, water stress and high CO_2. Plant Physiol. 81:423-429.

Dietz,K.-J., Neimanis, S. and Heber, U. (1984). Rate limiting factors in leaf photosynthesis II. Electron transport. *Biochim Biophys. Acta* 767:444-450.

Downton, W.J.S., and Millhouse, J. (1985). Chlorophyll fluorescence and water relations of salt-stressed plants. *Plant Sci. Lett.* 37:205- 212.

Duysens, L. and Sweers, H. (1963). Mechanism of two photochemical reactions in algae as studied by means of fluorescence. *In* "Studies in Microalgae and Photosynthetic Bacteria" (J. Ashida,ed.), pp. 353-372. Univ. Tokyo Press, Tokyo.

Gibbons, G.C., and Smillie, R.M. (1980). Chlorophyll fluorescence photograpy to detect mutants, chilling injury and heat stress. *Carlsberg Res. Commun.* 45:269-282.

Govindjee, Downton, W.J.S., Fork, D.C., and Armond, P.A. (1981).Chlorophyll a fluorescence transient as an indicator of waterpotential of leaves. *Plant Sci. Lett.* 20:191-194.

Green, B.R., Karukstis, K.K., and Sauer, K. (1984). Fluorescence decaykinetics of mutants of corn deficient in photosystem I and photosystem II. *Biochim. Biophys. Acta* 767:574-581.

Hardacre, A.K., and Eagles, H.A. (1980). Comparisons among races of maize (*Zea mays* L.) for growth at 13°C. *Crop Sci.* 20:780-783.

Havaux, M., and Lannoye, R. (1984). Effects of chilling temperatures on prompt and delayed chlorophyll fluorescence in maize and barley leaves. *Photosynthetica* 18:117-127.

Hetherington, S.E., and Smillie, R.M. (1982). Tolerance of *Borya nitida*, a poikilohydrous angiosperm, to heat, cold and high light stress in the hydrated state. *Planta,* 155:76-81.

Hetherington, S.E., and Smillie, R.M. (1984). Practical applications of chlorophyll fluorescence in ecophy-siology, physiology and plant breeding. *In* "Advances in Photosynthesis Research", (C. Sybesma, ed.), Vol. IV, pp. 447-450. Martinus Nijhoff/Dr.W. Junk Publ., The Hague.

Hetherington, S.E., Smillie, R.M., and Hallam, N.D. (1982). *In vivo* changes in chloroplast thylakoid membrane activity during viable and non-viable dehydration of a drought-tolerant plant *Borya nitida. Aust. J. Plant Physiol. 9*:611-621.

Hetherington, S.E., Smillie, R.M. Hardacre, A.K., and Eagles, H.A. (1983a). Using chlorophyll fluorescence in vivo to measure the chilling tolerances of different populations of maize. *Aust. J. Plant Physiol. 10*:247-256.

Hetherington, S.E., Smillie, R.M., Malagamba, P., and Huaman, Z. (1983b). Heat tolerance and cold tolerance of cultivated potatoes measured by the chlorophyll fluorescence method. *Planta 159*:119-124.

Jenkins, G.I., Baker, W.R., Bradbury, M., and Woolhouse, H.W. (1981). Photosynthetic electron transport during senescence of the primary leaves of Phaseolus vulgaris L. III. Kinetics of chlorophyll fluorescence emission from intact leaves. *J. Exp. Bot. 32*:999-1008.

Klosson, R.J., and Krause, G.H. (1981). Freezing injury in cold-acclimated and unhardened spinach leaves. II. Effects of freezing on chlorophyll fluorescence and light scattering reactions. *Planta 151*:347-352.

Kriedemann, P.E., Graham, R.D., and Wiskich, J.T. (1985). Photosynthetic dysfunction and in vivo changes in chlorophyll a fluorescence from manganese-deficient wheat leaves. *Aust. J. Agric. Res. 36*:157-169.

Leto, K., and Miles, D. (1980). Characterization of three photosystem II mutants in Zea mays L. lacking a 32,000 dalton lamellar polypeptide. *Plant Physiol. 66*:18-24.

Long, S.P. (1983). C4 photosynthesis at low temperatures. *Plant Cell Environ. 6*:345-363.

Lyons, J.M. (1973). Chilling injury in plants. Annu. Rev. *Plant Physiol. 24*:445-466.

McMillan, C. (1975). Adaptive differentiation to chilling in mangrove populations. In "Proceedings of the International Symposium on Biology and Management of Mangroves" (G.E. Walsh, S.C. Snedaker, and H.J. Teas, eds.), pp. 62-68. Univ. of Florida, Grainsville.

Margulies, M.M., and Jagendorf, A.T. (1960). Effect of cold-storage of bean leaves on photosynthetic reactions of isolated chloroplasts. Arch. Biochem. Biophys. 90:176-183.

Nesterenko, T.V., and Sid'Ko, F. Ya. (1980). Induction offluorescence in wheat leaves during their ontogensis. Soviet Plant Physiol. 27:262- 266.

Öquist, G. (1983). Effects of low temperature on photo-synthesis. Plant Cell Environ. 6:281-300.

Papageorgiou, G. (1975). Chlorophyll fluorescence: An
    intrinsic probe of photosynthesis. In "Bioenergetics of
    Photosynthesis" (Govindjee, ed.), pp. 319-371. Academic
    Press, New York.
Popovic, R., Fraser, D., Vidaver, W., and Colbow, K. (1984).
    Oxygen-quenched chlorophyll a fluorescence and electron
    transport in barley during greening. *Physiol. Plant.*
    62:344-348.
Renger, G. and Schreiber, U. (1986). Practical applications
    of fluorometric methods to algae and higher plant
    research. In "Light Emission of Plants and Bacteria"
    (J. Amesz, D. Fork, and Govindjee, eds.), pp. 587-619.
    Academic Press, New York.
Rikin, A., Blumenfeld, A., and Richmond, A.E. (1976).
    Chilling resistance as affected by stressing environments
    and abscisic acid. Bot. Gaz. 137:307-312.
Schreiber, U. (1983). Chlorophyll fluorescence changes as a
    tool in in plant physiology I. The measuring system.
    Photosynth. Res. 4:361-373.
Schreiber, U., and Berry, J.A. (1977). Heat-induced changes
    of chlorophyll fluorescence in intact leaves correlated
    with damage of the photosynthetic apparatus. Planta
    136:233-238.
Schreiber, U., Groberman, L. and Vidaver, W. (1978).
    Portable solid-state fluorometer for the measurement of
    chlorophyll fluorescence induction in plants. Rev. Sci.
    Instrum. 46:538-542.
Schreiber, U., Schliwa, U., and Bilger, W. (1986).
    Continuous recording of photochemical and
    non-photochemical chlorophyll fluorescence quenching with
    a new type of modulation fluorometer. Photosynth. Res.
    10:51-62.
Schreiber, U., Vidaver, W., Runeckles, V.C., and Rosen, P.
    (1978). Chlorophyll fluorescence assay for ozone injury
    in intact plants. *Plant Physiol. 61*:80-84.
Shimazaki, I., Ito, K., Kondo, N., and Sugahara, K. (1984).
    Reversible inhibition of the photosynthetic
    water-splitting enzyme system by SO_2-fumigation assayed
    by chlorophyll fluorescence and EPR signal *in vivo.*
    Plant & Cell Physiol. 25:795-803.
Smillie, R.M. (1979). The useful chloroplast: a new
    approach for investigating chilling stress in plants. In "
    Low Temperature Stress in Crop Plants" (J. Lyons, J.K.
    Raison, and D. Graham, eds.), pp. 187-202. Academic
    Press, New York.

Smillie, R.M. (1982/83). Chlorophyll fluorescence *in vivo* as a probe for rapid measurement of tolerance to ultraviolet radiation. Plant Science Lett. 28:283-289.

Smillie, R.M. (1984). A highly chilling-sensitive angiosperm. Carlsberg Res. Commun. 49:75-87.

Smillie, R.M., and Gibbons, G.C. (1981). Heat tolerance and heat hardening in crop plants measured by chlorophyll fluorescence. *Carlsberg Res. Commun.* 46:395-403.

Smillie, R.M., and Hetherington, S.E. (1983). Stress tolerance and stress-induced injury in crop plants measured by chlorophyll fluorescence *in vivo.* Chilling, freezing, ice cover, heat, and high light. *Plant Physiol.* 72:1043-1050.

Smillie, R.M., and Hetherington, S.E. (1984). A screening method for chilling tolerance using chlorophyll fluorescence *in vivo. In* "Advances in Photosynthesis Research" (C. Sybesma, ed.), Vol. IV, pp. 471-474. Martinus Nijhoff/Dr. W. Junk Publ., The Hague.

Smillie, R.M., Hetherington, S.E., Nott, R., Chaplin, G.R., and Wade, N.L. (1987a). Applications of chlorophyll fluorescence to the postharvest physiology and storage of mango and banana fruit and to screening for chilling tolerance in mango cultivars. *ASEAN J. FOOD SCI.* 3, 55-9

Smillie, R.M., Hetherington, S.E., Ochoa, C., and Malagamba, P. (1983). Tolerances of wild potato species from different altitudes to cold and heat. *Planta 157*:112-118.

Smillie, R.M., and Nott, R. (1979). Heat injury in leaves of alpine, temperate and tropical plants. *Aust. J. Plant Physiol.* 6:135-141.

Smillie, R.M., and Nott, R. (1982). Salt tolerance in crop plants monitored by chlorophyll fluorescence *in vivo. Plant Physiol.* 70:1049-1054.

Smillie, R.M., Nott, R., Hetherington, S.E., and Öquist, G. (1987b). Chilling injury and recovery in detached and attached leaves measured by chlorophyll fluorescence. *Physiol. Plant. 69*:419-428.

St. John, J.B., and Christiansen, M.N. (1976). Inhibition of linolenic acid synthesis and modification of chilling resistance in cotton seedlings. *Plant Physiol.* 57:257-259.

Thongjiem, M., and Chouvalitwongporn, P. (1985). The quest for improved potato varieties in Thailand. *CIP Circ. 13*:1-3.

Van Hasselt, P.R. (1972). Photo-oxidation of leaf pigments in Cucumis leaf discs during chilling. *Acta Bot. Neerl, 21*:539-548.

Walker, D.A., Sivak, M.N., Prinsley, R.T. and Cheesbrough, J.K. (1983). Simultaneous measurement of oscillations in oxygen evolution and chlorophyll a fluorescence in leaf pieces. *Plant Physiol*. 73:542-549.

Weis, E. (1982). Influence of light on the heat sensitivity of the photosynthetic apparatus in isolated spinach chloroplasts. *Plant Physiol*. 70:1530-1534.

Wilson, J.M. (1976). The mechanism of chill- and drought-hardening of *Phaseolus vulgaris* leaves. *New Phytol*. 76:257-270.

Wiltens, J., Schreiber, U., and Vidaver, W. (1978). Chlorophyll fluorescence induction: an indicator of photosynthetic activity in marine algae undergoing desiccation. *Can. J. Bot. 56*:2787-2794.

Chapter 4
Translocation

THE USE OF ^{11}C IN PHYSIOLOGICAL-ECOLOGICAL RESEARCH

B.R. STRAIN

Department of Botany
Duke University
Durham, North Carolina

J.D. GOESCHL
Y. FARES
C.E. MAGNUSON
C.E. JAEGER

PhytoResource Research, Incorporated
College Station, Texas

I.   INTRODUCTION

Ecologists and physiologists have long realized for many years the need for specific quantitative information on the details of carbon allocation and transport rates (Mooney, 1972).   The introduction of the infrared gas analyzer was a major technological breakthrough that enabled physiological ecologists to measure continuously and non-destructively the exchange of CO_2 by plants (Field and Mooney in this volume). This capability of "real time" measurement of photosynthetic capacity allowed the analysis of the instantaneous effect of environmental changes on photosynthetic carbon gain.   Thus, ecological analyses of different plant types in a common environment, or the reverse case of a single species in different environments, became more meaningful and more quantitative.   We now know, for example, the response time of a variety of plant types to instantaneous changes in irradiance, CO_2 concentration, temperature, ambient water vapor pressure, etc.   Our understanding of the physiological

Measurement Techniques in Plant Science

responses of plants to environmental changes can now be
coupled tightly, in time, with the underlying biochemical
changes.

Many current models of plant response require accurate
measurements of translocation rates to parallel knowledge of
net photosynthesis and respiration. Reynolds et al., (1980),
for example, utilized measurements of net photosynthesis,
respiration, and translocation rates to predict plant growth
as shown below.

$$\frac{dC_i}{dt} = \sum_j I_{ji} - \sum_h O_{ik} + A_i \tag{1}$$

Where dCi/dt is a time derivative of a growth compartment (i)
as affected by the sum of all internal inputs of carbon
($\sum_j I_{ji}$) to compartment i. Losses ($\sum_h O_{ik}$) include all
translocation losses and respiration. The effect of an
instantaneous environmental change on net photosynthesis (A_i)
can be measured on a "real time" basis. The very critical
translocation rates, however, could not be measured in "real
time" until the development of systems such as the one
described herein. Previously it was necessary to use ^{14}C as
a radioactive tracer and sample by destructive means over a
significant time span; or to measure changes in biochemical
pools or dry weight over long time spans. Thus, we were
unable to gain an understanding of the plant controlled
parameters of carbon flux to parallel our more detailed
understanding of carbon exchange with atmosphere.

We have developed methods to continuously produce ^{11}C as
a radioactive tracer (Magnuson et al., 1982). This gamma
emitting isotope can be photosynthetically incorporated as
$^{11}CO_2$ into intact plant leaves. Using externally positioned
pairs of coincidence counters, the carbon can be
instantaneously and continuously measured as it moves through
the plant. Thus, it is now possible to conduct "real time"
studies of all aspects of carbon fixation and allocation in
plants. This paper briefly reviews development of research
with ^{11}C and describes some uses of the ^{11}C facility at the
Duke Phytotron.

II. THE USE OF ^{11}C IN PLANT RESEARCH

Carbon-11 has been used for many years in studies of
plant metabolism and transport (Ruben et al., 1939; Ruben and

Kamen, 1940; Moorby *et al.*, 1963; More and Troughton, 1972, 1973; Troughton *et al.*, 1977; Fensom et al., 1977; Pickard et al., 1978 a,b; Thorpe *et al.*, 1979). There is also a very detailed review and reinterpretation of these studies by Minchin & Troughton (1980).

Major advantages of using ^{11}C have been emphasized by previous authors (Moorby *et al.*, 1963, More and Troughton, 1973; Minchin, 1978, 1979; Minchin and Troughton, 1980). These advantages are based on the decay and half-life characteristics of the ^{11}C isotope.

First, ^{11}C decays by emission of a positron (β +) which annihilates by interaction with an electron (β -) to produce a pair of gamma rays (γ) with 0.5 MeV energy. This energy level is sufficient to penetrate several centimeters of plant tissue, thus ^{11}C can be detected as it moves through live, intact plants.

Secondly, these 0.5 MeV gamma rays are emitted at essentially 180°, and continue to travel in a straight line away from each other. Thus, they can be detected by coincidence counters and/or collimated detectors to reduce background count levels and allow accurate spatial resolution. This spatial resolution has resulted in 2 and 3 dimensional gamma ray cameras developed for bio-medical use in imaging or locating metabolites in small tumors and other specialized tissues in human patients and animal subjects, (e.g. Witter *et al.*, 1979).

Third, the short half-life of ^{11}C (i.e. 20.4 min) allows repeated experiments on the same live specimens over a period of hours, days, weeks, etc. This makes it possible for each plant to serve as its own "control", especially when following a time dependent effect of some environmental input (Minchin, 1979a). Thus the genetic variability between plants in amplitude of response to environmental changes can be distinguished from differences in the base rate of each process under the initial environmental conditions.

Fourth, ^{11}C can be applied in both steady-state and pulsed or varied wave form patterns. By combining pulsed and steady-state kinetics analyses it is possible to determine the value of many dynamic parameters beyond those obtainable by either method alone. By the deconvolution of carbon-11 washout curves using Fourier transform mathematics, the soluble transport pool can be distinguished from non-transportable storage products.

III.   THE ^{11}C FACILITY AT DUKE UNIVERSITY

The Duke University integrated system for ^{11}C generation, application, detection and analysis is made up of six components including: (1) production of ^{11}C in a 4 MeV Van de Graaff accelerator in the Duke Physics Department, (2) transfer of the radioactive gas mixture to the nearby Duke Phytotron for on-line chemistry and gas conditioning, (3) measurements of isotope uptake, (4) detection of isotope movement in the plants, (5) signal processing and counting, and (6) data processing in an on-line minicomputer.   This scheme is diagramatically shown in Figure 1, and the entire system is described in detail in Magnuson *et al.*, (1982).

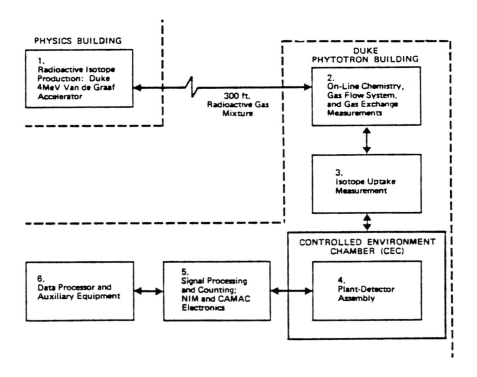

Figure 1   Outline of the integrated system for studying the carbon allocation in plants using radioactive tracers under controlled environment conditions.

The isotope production, mixing, application and detection scheme curently in use is shown in Figure 2. The symbols used for all values, flow meters and detectors in the system are defined in Table 1.

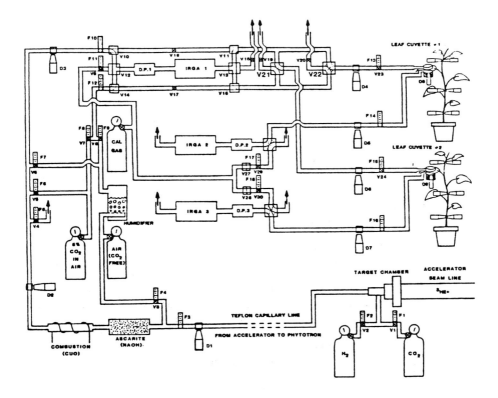

Figure 2. A diagram of the present ^{11}C tracer system for simultaneous experiments with two plants. A detailed identification of the valves, flow meters and detectors is given in the following table.

Table 1.   Definition of purpose of all values (V) flow
meters (F) and dectors (D) in Figure 2.

      A.   The Values as Numbered are Used to Control
            Flows in The System as Follows

V1	CO_2 through target chamber
V2	H_2 through target chamber
V3	Air (CO_2 free) to sweep target effluent through the processing system
V4	To vent off any excess $^{11}CO_2$ formed in the target and processing reactions to maintain constant levels of both during long experiments
V5	CO_2 (5% in air) added to ^{11}C mix if needed
V6	Air (CO_2 free) to ^{11}C mix (i.e. Air + $^{12}CO_2$ + $^{11}CO_2$)
V7	CO_2 (5% in air) added to cold ($^{12}CO_2$) mix
V8	Air (CO_2 free) to cold mix (i.e. Air + $^{12}CO_2$)
V9	Controls flow rate of calibration gases through IRGA to match flow of mixes to leaves
V10 & V11	3 way toggle valves linked mechanically to bypass ^{11}C mix around IRGA during calibration or measurment of cold mix
V12 & V13	2 way toggle valves in "T" mounting, mechanically linked to pass calibration gases through IRGA
V14 & V15	2 way toggle valves mechanically linked to send cold mix through or to bypass cold mix around IRGA
V16,V17 V18,V19 V20	These valves are adjusted in the bypass and vent lines to mimic the resistance of the appropriate IRGA, leaf cuvettes, and other components tocompensate for effects of pressure on the calibration of the various measurement devices.
V21 & V22	4 way switching valves to divert the flow of ^{11}C mix and cold mix alternately to the leaf cuvettes or vents
V23 & V24	Adjusted to balance the flow through leaf cuvettes #1 and #2
V25 & V26	4 way switching valves to divert effluent from the leaf cuvettes during calibration of the IRGA's and Dew Point Analyzers
V27 & V28	2 way toggle valves to introduce calibration gases into the IRGA's
V29 & V30	Metering valves to adjust flow of calibration gases to match effluent from leaf cuvettes

B. The Flowmeters, as Numbered, are Used to Measure
Flows in the System as Follows

F1 CO_2 into target chamber
F2 H_2 into target chamber
F3   Effluent from target chamber
F4   Air (CO_2 free) added into processing system
F5   Excess $^{11}CO_2$ and $12CO_2$ being vented
F6 CO_2 (5% in Air) added to 11C mix if need
F7   Air (CO_2 free) into ^{11}C mix (i.e. Air + $^{12}CO_2$ +
          11CO$_2$)
F8 CO_2 (5% in Air) added to Cold mix
F9   Air (CO_2 free) into cold mix (i.e. Air + $^{12}CO_2$)
F10   Final, total flow of ^{11}C mix
F11   Calibration gases
F12   Final, total flow of cold mix
F13 &   Inflow to leaf cuvettes #1 & #2
F15
F14 &   Effluent from leaf cuvettes #1 & #2
F16
F17 &   Calibration gases into IRGA systems #1 & #2
F18

Note: In addition to rotameters at positions F3, F4, F13
and F15, mass flow meters (Brooks type 5810) are positioned
to accurately monitor and/or record flows from the target
chamber air added to processing system, and flows into leaf
cuvettes #1 & 2.

C. The Detectors, as Numbered, are Used to Measure
the Activity of ^{11}C as Follows

D1   Monitors activity in the effluent from the target
          chamber after transit, but before processing
D2   Monitors activity after processing
D3   Monitors activity after dilution and mixing of
          the ^{11}C mix
D4 &   Monitor activity of the ^{11}C mix as it enters the
D6   leaf cuvettes #1 & #2
D5 &   Monitor activity in the effluent from leaf
D5   cuvettes #1 and #2
D8 &   Monitor activity in the leaves and leaf cuvettes
D9   #1 and #2
Dp1-6   Represent pairs of detectors (usually 9 pairs per
          plant) connected in coincidence and stationed at
          various positions along the plant

Abstracts of several papers describing various kinds of research done with this facility follow.

IV.   REPRESENTATIVE STUDIES COMPLETED IN THE
DUKE PHYTOTRON ^{11}C FACILITY

A.   Effects of temperature and atmospheric CO2
Enrichment on Translocation in *Echinochloa crus-galli*, a C$_4$ Grass

Potvin, C., J.D. Goeschl and B.R. Strain, 1984. *Plant Physiology 75*:1054-1057.
Potvin, C., B.R. Strain and J.D. Goeschl, 1985. *Oecologia 67*:305-309.
Plants of *Echinochloa crus-galli* from Quebec and Mississippi were grown under two thermoperiods (28°C/22°C, 21°C/15°C) and two atmospheric CO_2 concentrations (350 and 675 microliters per liter) to examine possible differential responses of northern and southern populations of this C$_4$ grass species. Translocation was monitored using radioactive tracing with short-lived ^{11}C. CO_2 enrichment induced a decrease in the size of the export pool in plants of both populations. Other parameters did not strongly respond to elevated CO_2. Low temperature reduced translocation drastically for plants from Mississippi in normal CO_2 concentration, but this reduction was ameliorated at high CO_2. Overall, plants from Quebec had a higher ^{11}C activity in leaf phloem and a higher percentage of ^{11}C exported, whereas these northern plants had lower turnover time and smaller pool size than plants from the southern population.
In addition,   low night temperature reduced translocation mainly by increasing the turn-over times of the export pool. *E. crus-galli* plants from Mississippi were the most susceptible to chilling; translocation being completely inhibited by exposure for one night to 7°C at 350 μ l $\cdot$ l^{-1} CO_2. Overall, plants from Quebec were the most tolerant to chilling-stress. For plants of all three populations, growth under CO_2 enrichment resulted in higher ^{11}C activity in the leaf phloem. High CO_2 concentrations also seemed to buffer the transport system against chilling injuries.

B.   Effect of Temperature Chilling on Translocation
In Cotton and in The Velvetleaf Weed

Strain, B.R., J.D. Goeschl, C.H. Jaeger, Y. Fares, C.E.
Magnuson, and C.E. Nelson, 1983. *Radiocarbon 25:*441-446.
A slight temperature drop (28°C to 23°C) for only 2
minutes caused translocation decline in both cotton and
velvetleaf. The weed, however, recovered much more rapidly.
Farmers have long observed that uneasonal cool nights in the
height of the growing season will impede cotton growth more
than the velvetleaf weed. This test shows a basic difference
in the way these phylogenetically closely related species
respond to temperature chilling. It is our hypothesis that
carbon allocation is more seriously impeded by cool
temperatures in cotton than in the velvetleaf. Thus,
velvetleaf recovers quickly from the occasional cool night
whereas cotton requires much longer to recover.

C.   Carbon Transport Between
Vegetative Tillers of Grasses

Welker, J.M., E.J. Rykiel, Jr., D.D. Briske and J.D.
Goeschl, 1985. *Oecologia 67*:209-212.
Labelled carbon (^{11}C) was continuously transported from
parent tillers to anatomically attached daughter tillers at a
time when morphological characteristics indicated that tiller
maturation had occurred. Steady state levels of import into
monitored daughter tillers increased with 30 min of either
defoliation or shading. Import levels decreased within 30
min of the removal of shading, but remained accelerated
throughout an 84 h observation period following defoliation.
A second defoliation further increased carbon import into
monitored tiller above the previously accelerated level
resulting from the initial defoliation. Carbon import by
vegetative tillers in the two bunchgrass species examined may
be most appropriately viewed as a series of potentially
accelerated import levels above a low level of continuous
import.

D.   Spontaneous and Induced Blockages
Of Phloem Transport

Goeschl, J.D., C.E. Magnuson, Y. Fares, C.H. Jaeger, C.E.
Nelson, and B.R. Strain, 1984. *Plant, Cell and Environment
7:*607-613.

Jaffee, MJ. and F. Telewski, 1984. *Recent Adv. in Phytochemistry,* Vol. 18. Plenum Press, NY.

Steady-state labelling with $^{11}CO_2$ was used to observe the blocking of phloem transport, induced by chilling short regions of stems or petioles of velvetleaf (*Abutilon theophrasti* Medic.) or cotton (*Gossipium hirsutum* L.). The abruptness of these blockages was evidenced by sharp decreases in ^{11}C activity below, and increases above a 2 to 3 cm region cooled from 28°C to 18°C or 13°C for a period as short as 2 min. Abrupt unblocking of transport in velvetleaf occurred a few minutes after rewarming, as evidenced by a sharp rise and overshoot in ^{11}C activity. Recovery of transport in cotton was more prolonged and was marked by occasional spontaneous blocking and unblocking of transport at various points along the petiole or stem, not necessarily in the cooled region. Similar spontaneous events were often observed in undisturbed cotton plants, but only rarely in velvetleaf.

Similar blockages of phloem transport were induced by mechanical stimulation of the stems of cotton, velvetleaf and garden bean. It was clearly shown that vibration or stem bending can induce significant declines in phloem transport.

V.   SUMMARY

The use of radioactive ^{11}C isotope to study carbon fixation and allocation in plants is described. A new system developed at Duke University is briefly described and representative studies that have been completed with the facility are presented. It is shown that the use of steady-state labelling and detection of ^{11}C is a useful analytical procedure for the study of carbon translocation in plants.

REFERENCES

Fensom, D.S., E.J. Williams, D.P. Aikman, J.E. Dale, J.
    Scobie, K.W.O. Ledington, A. Drinkwater, and J. Moorby,
    1977. Translocation of ^{11}C from leaves of Helianthus:
    preliminary results. *Can. J. Bot. 55*:1787 -1793.
Goeschl, J.D., C.E. Nelson, Y. Fares, C.H. Jaeger, C.E.
    Jaeger, C.E. Nelson, and B.R. Strain, 1984. Spontaneous
    and induced blocking and unblocking of phloem transport.
    Plant, Cell and Environ. 7:607-613.
Jaffee, M.J. and F.W. Telewski, 1984. Thigmomorphogenesis:
    Callose and ethylene in hardening of mechanically
    stressed plants. In: Timmermann, B.N., C. Steelink, and
    F.A. Loewus (eds.). Recent Advances in Phytochemistry,
    Vol. 18. Plenum Press, New York.
Magnuson, C.E., Y. Fares, J.D. Goeschl, C.E. Nelson,
    B.R.Strain, C.H. Jaeger and E.G. Bilpuch, 1982. An
    integrated tracer kinetics system for studying carbon
    uptake and allocation in plants using continuously
    produced $^{11}CO_2$. *Radiat. Environ. Biophys. 21*:51-65.
Minchin, P.E.H. 1978. Analysis of tracer profiles with
    applications to phloem transport. *J. Exp. Bot.
    29*:1441-1450.
Minchin, P.E.H., 1979. The relationship between spatial and
    temporal tracer profiles in transport studies. *J. Ex.
    Bot. 30:*
Minchin, P.E.H., 1980. Quantitative interpretation of
    phloem translocation data. *Ann. Rev. Plant Physiol.
    31*:191-215.
Minchin, P.E.H. and J.H. Troughton, 1980. Quantitative
    interpretation of phloem translocation data. *Ann. Rev.
    Plant Physiol. 31:*191-215.
Mooney, H.A., 1972. The carbon balance of plants. *Ann.
    Rev. Ecol. Syst. 3*:315-346.
Moorby, H., M. Ebert, and N.T.S. Evans, 1963. The
    translocation of ^{11}C-labeled photosynthate in soybean. *J.
    Exp. Bot. 14*:210-200.
More, R.D. and J.H. Troughton, 1972. Production of $^{11}CO_2$
    using a 3.0 MeV Van de Graaff accelerator. *Int. J. Appl.
    Rad. Isotopes. 23*:344-345.
More, R.D. and J.H. Troughton, 1973. Production of $^{11}CO_2$ for
    use in plant translocation studies. *Photosynthetica
    7*:271-274.
Pickard, W.F., P.E.H. Minchin and J.H. Troughton, 1978(a).
    Transient inhibition of translocation in *Ipomea alba* L.
    by small temperature reductions. *Aust. J. Plant Physiol.
    5*:127-130.

Pickard, W.F., P.E.H. Minchin and J.H. Troughton, 1978(b). Real time studies of Carbon-11 translocation in moonflower: I. The effects of cold blocks. *J. Exp. Bot.* *29:*993-1001.

Potvin, C., J.D. Goeschl and B.R. Strain, 1984. Effects of temperature and CO_2 enrichment on carbon translocation of plants of the C_4 grass species *Echinochloa crus-galli* (L.) Beauv. from cool and warm environments. *Pl. Physiol.* *75:*1054-1057.

Potvin, C., B.R. Strain and J.D. Goeschl, 1985. Low night temperature effect on photosynthate translocation of two C_4 grasses. *Oecologia 67*:305-309.

Reynolds, J.F., B.R. Strain, G.L. Cunningham, and K.R. Knoerr, 1980. Predicting primary productivity for forest and desert ecosystem models. In Hesketh, J.D. and J.W. Jones (eds). Predicting photosynthesis for ecosystem models, Vol. II. CRC Press, Boca Raton, Florida. pp. 279.

Ruben, S., W.Z. Hassid and M.D. Kamen, 1939. Radioactive carbon in the study of photosynthesis. *J. Am. Chem.* *61*:661-663.

Ruben, S., and M.D. Kamen, 1940. Photosynthesis with radioactive carbon. IV. Molecular weight of the intermediate products and a tentative theory of photosynthesis. *J. Am. Chem. 62:*2351-2355.

Strain, B.R., J.D. Goeschl, C.H. Jaeger, Y. Fares, C.E. Magnuson and C.E. Nelson, 1983. Measurement of carbon fixation and allocation using 11C-labeled carbon dioxide. *Radiocarbon 25*:441-446.

Thorpe, M.R., P.E.H. Minchin and A.E. Dye, 1979. Oxygen effects on phloem loading. *Plant Sci. Lett. 15*:345-350.

Troughton, J.H. and B.G. Currie, 1977. Relations between light level, sucrose concentration, and translocation of carbon-11 in *Zea mays* leaves. *Plant Physiol. 59*:808-820.

Welker, J.M., E.J. Rykiel, Jr., D.D. Briske and J.D. Goeschl, 1985. Carbon import among vegetative tillers within two bunchgrasses: Assessment with carbon-11 labeling. *Oecologia 67*:209-212.

Witter, J.P., S.J. Gatley and E. Balish, 1979. Distribution of nitrogen-13 from labeled nitrate ($^{13}NO_3^-$) in humans and rats. *Science 204*:411-413.

DEVELOPMENT OF A COMPACT MASS
SPECTROMETER FOR ANALYSIS OF
GASES DURING PLANT METABOLISM

Tadao Kaneko
Masamoto Takatsuji

Advanced Research Laboratory
Hitachi, Ltd.,
Kokubunji, Tokyo

I.   INTRODUCTION

Plants are complex systems of many simultaneous biochemical reactions responding to many interacting environmental
factors. Plant age, preconditioning environment and genetic
makeup can all affect the ultimate response of a plant to
its environment. In the past, however, we have been restricted to single or few factor measurements of single or
few reactions.
As this volume is demonstrating, modern instrumentation
is allowing increasingly accurate measurement of increasingly
complex systems. In addition, the development of portable
instruments is making it possible to make these complex
measurements in the field. This paper describes a compact
mass spectrometer with the capability of measuring several
atmospheric gasses simultaneously and under field conditions.

Measurement Techniques in Plant Science
Copyright © 1990 by Academic Press, Inc.
All rights of reproduction in any form reserved.

II.   A PORTABLE, MULTI-COLLECTOR
MASS SPECTROMETER

Through three basic physiological responses, namely photosynthesis, respiration and transpiration, plants exchange
CO_2, O_2 and H_2O gases with the atmosphere. Since these
responses are essential for plant growth, it is quite important to measure several gases simultaneously and continuously. A compact, dispersion-type mass spectrometer has
been developed for this purpose (Figure 1). By means of a
heated flexible inlet capillary 0.25mm in diameter, less

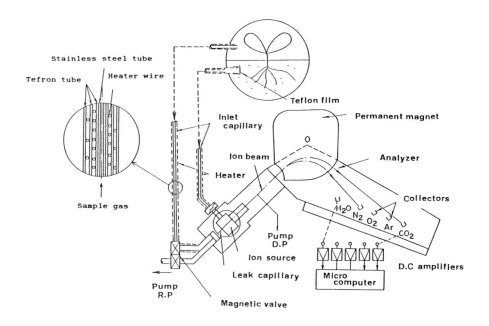

Figure 1. Schematic diagram of the multi-collector-
type mass spectrometer for analysis of gases of interest in
physiological reactions of plants.

than 15m /min of gas is directly taken from the plant-growing area using a differential pumping system. A fraction of this gas is admitted to the analyzer through a capillary tube.

A teflon membrane is put on the end of the inlet capillary to measure gases dissolved in water. Ions produced by electron impact are separated according to their masses by a specially designed permanent magnetic field and simultaneously collected by independently adjustable collectors. The effect of the temperature of the 1 m inlet capillary on rise and delay times in the mass spectrometer is shown in Figure 2 for dry air and water-vapour. Precise analysis of gases under high water-vapour pressure condition is performed under a constant capillary temperature of about 80°C. In this case rise time of this spectrometer is about 0.15 sec for air.

Figure 3 shows the new ion optical system developed for the multi-collector type mass spectrometer. The magnetic field consists of a concave entrance field and a convex exit field. Properties of second-order aberrations are determined by the radii of the concave, and convex field, r and R, respectively and the angle of incidence, , of the ion beam to the concave field. Orbits of 8,000 ion beams emitted from the entrance slit are calculated by computer according to the ion optics given by Matsuda (1964). As an example, effect of r and on second-order aberrations at focal point of M/e of 32 is shown in Figure 4. In this case the entrance slit is a straight line of 2mm in length. These aberrations are minimum when r - 31mm, R - 50mm and - 24 degree, and the three-dimensional focusing is applied clearly.

Mass spectra of the atmosphere measured by the H_2O, N_2 and CO_2 collectors are shown in Figure 5. The mass difference such that the valley between two adjacent peaks is exactly zero is less than 1 within a mass range of 4 to 50.

Also, such values as ion accelerating voltage of 850 volts, magnetic flux density of 4800 gauss and the radius of ion orbit in molecular weight of 31 of 50mm are determined. The advantages of this apparatus in comparison with the well-known infra-red gas analyzer are:

1) Such multiple gases as H_2O, O_2 and CO_2 can be measured simultaneously.

2) Several gases labeled by their stable isotopes, ^{13}C, ^{15}N and ^{18}O can be analyzed.

3) The low gas consumption and short response time enables direct measurement of the changes in gas concentration over small areas (e.g. part of a leaf).

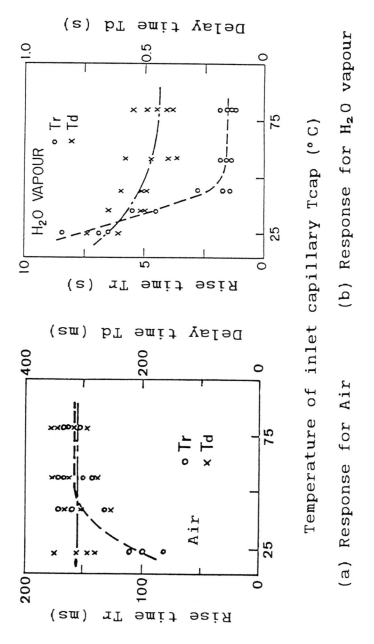

Figure 2. Effect of the temperature of the inlet capillary on rise and delay times in the mass spectrometer.

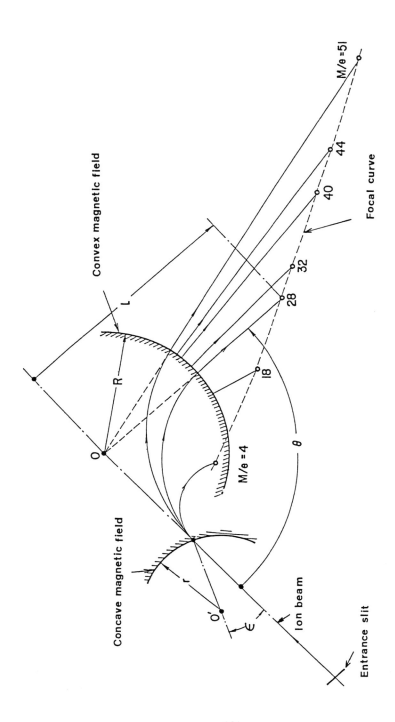

Figure 3. New ion optical system developed for the multi-collector-type mass spectrometer. r: Radius of concave magnetic entrance field. R: Radius of convex magnetic exit field. ε: Angle of incidence. θ: Deflection angle of each ion. 1: Distance between focal point of each ion and center of curvature 0.

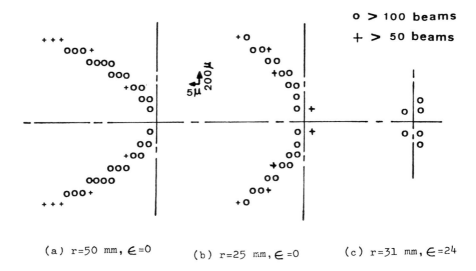

(a) r=50 mm, ϵ=0   (b) r=25 mm, ϵ=0   (c) r=31 mm, ϵ=24

Figure 4. Effect of r and ϵ on second-order
aberrations at focal point (M/e = 32). The definitions of r
and ϵ are given in Figure 3.

III. RESULTS

The above characteristics have been clarified through
the following experiments using this spectrometer:

A. Measurement of the Changes in Gas
Concentration in a Small Space

Figure 6 shows the result of the direct measurement of
distribution pattern of water-vapour pressure on the surface
of leaves of hydroponic butter-head type lettuce.
(A) shows the distribution of water-vapour pressure on
the reverse of a young leaf and an old leaf, (b) the
sampling method for water-vapour, and (c) recorder trace of
water-vapour pressure. Measurements are achieved by moving
the top of the capillary across the central vein of the
leaf. When the distance between the leaf surface and the
top of the capillary, H, is about zero, average water-vapour
pressure on the young leaf is about two and a half times
greater than that on the old leaf, whereas both water-vapour
pressures are of the same value for plant growing area of

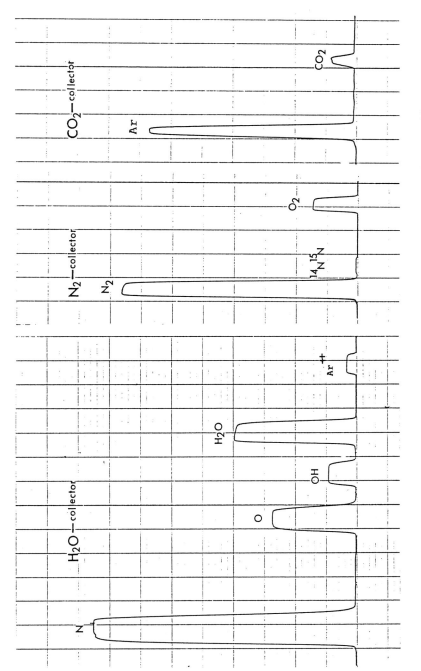

Figure 5. Mass spectra of the atmosphere measured by the H_2O, N_2 and CO_2 collectors.

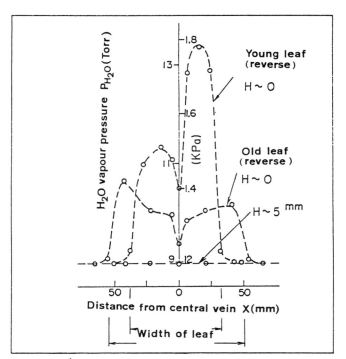

(a) Distribution of H_2O vapour pressure

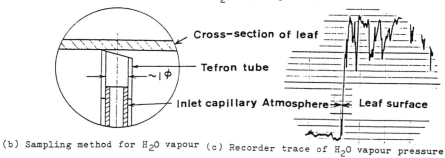

(b) Sampling method for H_2O vapour   (c) Recorder trace of H_2O vapour pressure

Figure 6. Distribution of H_2O vapour pressure on the surface of leaves of butter-head type lettuce. H is the distance between the surface and the top of the capillary.

H = 5mm. As the above experimental result shows, the difference in transpiration between the young leaf and the old one can be directly measured using a thin capillary.

B. Simultaneous Measurement of Multipule Gasses

As an example of simultaneous measurement of two gases, photosynthesis and transpiration are presented (Figure 7) against water content of cut-leaves of butter-head type lettuce, which is defined by

$$\text{water content (\%)} = \frac{W_t - M}{W_o - M} \times 100.$$

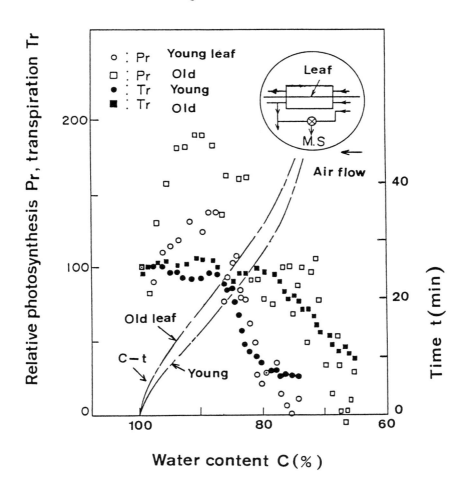

Figure 7. Effect of water content in the cut-leaves of butter-head type lettuce on photosynthesis and transpiration.

Here, W_o: leaf weight at the time of cutting the petiole
off. W_t: leaf weight at the time of measurement.
        M: weight of dry matter in the leaf
In this case a leaf chamber is used as shown in the circle.
The measurement of physiological reponses is made from the
difference of influx and efflux of gasses through the leaf
chamber. In the range of water content of 85 to 65 percent,
the two responses are highly correlated and decrease quickly
owing to the stomatal closing in both young and old
leaves. The result is that photosynthetic rate increases
temporarily during the change in water content of 100 to 85
percent. Although there are a few reports about similar phenomena, the cause still is unknown.

C. Measurement of Gas Labeled with The Stable Isotope ^{13}C

Translocation of photosynthates from leaves to a green
pepper pod using stable isotope $^{13}CO_2$ was studied against the
time after the end of $^{13}CO_2$ feeding on the leaves, as shown
in Figure 8. Stable isotope ^{13}C is introduced to a few
green pepper leaves in the form of CO_2. Subsequently,
$^{13}CO_2$ concentration emitted from the green pepper pod by
respiration after $^{13}CO_2$ feeding on the leaves is measured to
estimate translocation of photosynthates from the labeled
leaves to the pod. In figure 8, the ratio $^{13}CO_2$ to $^{12}CO_2$ is
presented against the time after the end of $^{13}CO_2$ feeding of
the leaves. The ratio $^{13}CO_2$ to $^{12}CO_2$ increases immediately
after the end of $^{13}CO_2$ feeding time for two hours on the
leaves and returns to the background level after about fifteen hours. The results may be summarized as follows:

  1) Carbon assimilated in the leaves by photosynthesis
is transferred to the pod near the labeled leaves within two
hours, and a portion of this carbon is rapidly consumed by
respiration in the pod.
  2) Part of the photosynthate transferred to the pod
change into substances that are not consumed by respiration
within about fifteen hours.

It is thus understood that analysis of the shape of respiratory loss of $^{13}CO_2$ in the pod is useful to clarify the relationships between translocation and environmental
conditions.

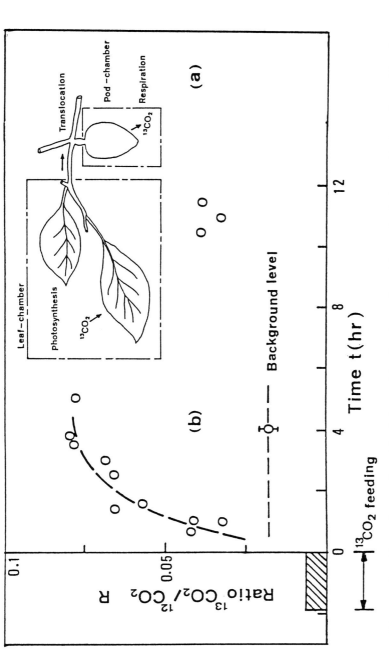

Figure 8. Experimental method to estimate translocation of photosynthates from the leaves to a green pepper pod using stable isotope $^{13}CO_2$.

IV. CONCLUSIONS

A compact, dispersion-type mass spectrometer for analysis of gases of interest in the metabolism of plants has
been developed. The advantages of this spectrometer are:
1) Such multiple gases as H_2O, O_2 and CO_2 can be
measured simultaneously.
2) Several gases labeled by their stable isotopes,
^{13}C, ^{15}N and ^{18}O can be analyzed.
3) The low gas consumption and short response time
enables measurement of the changes in gas concentration in a
small space.

REFERENCES

Matsuda H. (1964), Ion Optics (III), Second Order Aberration.
    Mass Spectroscopy 27, 165-179.

Chapter 5
Mineral Nutrition

ROOTS: NEW WAYS TO STUDY
THEIR FUNCTION IN PLANT NUTRITION

Emanuel Epstein

Department of Land,
Air and Water Resources
University of California,
Davis, California

I.   INTRODUCTION

A.   Soils, Roots, and Mineral Plant Nutrition

The mineral nutrition of plants does not loom large in the
enterprise of plant science, compared with such factors as
water relations and photosynthesis (Epstein, 1977).   There is
little point in exploring why this is so.   But I shall succeed
in this paper if I persuade you of two things.   First, mineral
nutrition is a major factor in the interplay between plants and
their   environment,   and   second,   despite   the   formidable
difficulties that in the past have stood in the way of studying
roots and mineral plant nutrition under conditions relevant to
the field, tools are now at hand, and more are in the making,
for doing just that.   These developments will give a new
impetus to the study of mineral nutrition of plants under
realistic conditions.   As we sharpen our perception of the
importance of mineral nutrition in physiological plant science
the   need   for   further   progress   in   instrumentation   and
methodology will be keenly felt.   That in turn will lead to the
development and application of ever more advanced technology
for solving the refractory problems posed by the study of roots
and their role in mineral nutrition as factors in plant growth
and development.
        I shall confine myself to terrestrial plants because for
reasons that will soon become clear, it is roots in soil and
their   function   in   mineral   nutrition   that   present   the
investigator with the most challenging problems, and by the

Measurement Techniques in Plant Science

same token, with exceptional opportunities for innovative research, especially in physiological ecology (Caldwell and Virginia, 1989).

The chief stressful soil conditions having a profound bearing on various aspects of mineral nutrition are listed in Table 1. The table includes the types of plants adapted to the several soil properties. Many of these conditions are not mutually exclusive. For example, the principal characteristic of serpentine soils is their low calcium/magnesium ratio, but many of them also are generally infertile, and in addition they may have high concentrations of heavy metals such as nickel and chromium. For another example, heavily leached, acidic soils having high concentrations of heavy metals are as a rule infertile as well. The hydrogen ions that tend to bring the metals into solution also displace exchangeable cations such as potassium and calcium from the soil cation exchange complex; this leads to the leaching of these nutrients from the root zone, and their eventual loss from the ecosystem. Finally, many of these lateritic soils, prevalent throughout the tropics, fix phosphate in forms poorly available to plants, and their nitrogen availability is generally low. Because of the inadequacy of soil surveys in many parts of the world the sum total of the area affected by one or more of these various conditions is difficult to estimate (Dudal, 1977). In any event, the areas involved are enormous, and soils fully benign in their mineral nutritional properties are probably to be found principally in carefully managed agricultural ecosystems occupied by plants bred to take advantage of these nutritionally luxurious conditions (Epstein, 1983; Epstein and Läuchli, 1983).

B. Plant Roots: Challenges for Observation and Experimentation

Many natural systems with the major function of transferring material from one compartment to another have dendritic or convoluted features (MacDonald, 1984). Both these adaptations make for a large area to accommodate and maximize the transfer of material between the compartments. Thus the branched pattern of trees and much other vegetation makes possible the exposure of a large leaf area to light and the atmosphere, maximizing the area over which light quanta can be gathered and through which essential gas exchange can take place. The bronchial tubes and alveoli of the vertebrate lung are another example of branched, convoluted structures facilitating the exchange of materials between compartments.

Table 1.   Principal Stressful Soil Conditions Affecting Mineral Plant Nutrition.

Soil condition or type	Elements, compounds, or ions of greatest importance	Types of plants adapted to the condition	References
Low fertility or nutrient avail- ability in respect to one or more nutrients	N, P	Wild species	Chapin, 1980, 1988
Salinity, sodicity	Na, Cl, SO_4	Halophytes	Albert, 1982; Reimold and Queen, 1974; Waisel, 1972
Heavy metals	Al, Mn, Ni, Cu, Zn, Cd, Pb (H^+)	Heavy metal tolerant and accumulator plants	Ernst 1982; Woolhouse, 1983
Serpentine soils	Ca, Mg (Cr, Ni)	Serpentine endemics	Kinzel and Weber, 1982; Kruckeberg, 1984; Brooks, 1987
Calcareous soils; acid soils	$CaCO_3$, H^+	Calcicoles, calcifuges	Kinzel, 1982, 1983

The root systems of plants clearly are among the types of structures I refer to. Their prime function is the transport of water and mineral nutrients from an external compartment, the soil, into the plant. As in many such situations the evolutionary response has been the elaboration of systems that expose enormous areas of absorbing surfaces to the donor compartment, the soil. Apart from that fundamental similarity, however, the soil-root system has several unique features. The principal difference between it and other such systems lies in the chemical and physical characteristics of the soil, which is the external compartment of the soil-root system and the reservoir of the materials - mainly water and nutrients - that are subject to intercompartmental transport. Whereas leaves

and lungs abut on fluids, roots exist in a matrix dominated by
an exceedingly complex, heterogeneous array of solid material,
both inorganic and organic. In most soils inorganic matter
predominates, and the total solid matter represents roughly
half of the volume of most soils. The voids among the solid
particles are occupied by the soil atmosphere and the film of
moisture which, in any soil capable of sustaining plant growth,
lines the solid particles, and contains nutrient and other ions
in solution. The concentrations of nutrient ions in this "soil
solution" usually are quite low, and vary from place to place
and from one time to another. Adding to the complexity of the
system are soil microflora and fauna ranging from bacteria to
nematodes. In addition, soil arthropods and burrowing mammals
physically and chemically alter large volumes of soil.
    It is this physically, chemically, and biologically
complex system dominated by a matrix of finely divided solid
material covered with films of water that roots invade and
inhabit. These features of the root environment present two
principal challenges to the experimentalist interested in the
growth and physiological activities of roots. First, roots in
soil are not amenable to direct observation without disturbance
of their environment, and second, their chemical medium,
especially the chemical composition of the soil solution,
cannot be accurately determined, let alone controlled. As for
the first of these challenges, viz. observation of roots _in
situ_, progress has been and is being made. I shall briefly
describe these advances. In regard to the chemical milieu of
roots and its control, I shall devote the last part of my paper
to this topic and outline instrumental systems, both existing
and yet to be built, to deal with that problem. This paper is
devoted to these two problems - observation and measurement, on
the one hand, and control of the ionic environment of roots, on
the other. Some brief description of roots and their function
in plant nutrition must be given first.

II. ROOTS: FORM AND FUNCTION

    The roots that have been most intensively studied are
those of crop plants, because of the economic interest in these
plants. Figure 1 shows roots of corn, _Zea_ _mays_, at three
stages of development. The outermost layer is the epidermis.
The next layers, four in the largest root depicted, are
cortical parenchyma cells - large, loosely arranged,
undifferentiated cells. The innermost layer of the cortex is
the endodermis; it is the outermost layer shaded in this
diagram. Finally, the center of the root consists of the stele
which contains the vessels, the large, unshaded cells. The
vessels in their mature state form long tubes of cell wall; the

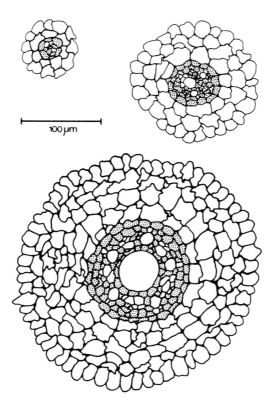

Figure 1. Tracings of corn roots at three stages of growth. The cells of the epidermis and cortex are clear, as are the xylem vessels. Stelar parenchyma cells and the endodermis are shaded, the endodermis being the outermost layer of the shaded cells. From Miller (1981).

protoplasts have been lost so that there is little obstruction to the flow of water and nutrients toward the top of the plant.

Less well known than these anatomical features is the huge size of root systems, most particularly the extent of the area that constitutes the effective interface between root and soil. In a classical investigation Dittmer (1937) grew a single rye plant, Secale cereale, for four months in soil contained in a box 30.5 cm square and 55.9 cm deep. He then carefully washed the soil from the roots, using a gentle spray of water, and systematically subsampled and measured the extent of the root system. Its surface area in contact with the soil was 639 m^2 and the total length of all the roots of this one plant was 623 km.

A recent, detailed morphometric study of roots of corn, Zea mays, was conducted by Miller (1981). Unlike Dittmer, he did not grow the plants to maturity in soil but for only about three weeks in solution culture (Miller, 1980). He expressed the results of his measurements on the basis of unit fresh weight of roots (Miller, 1981). Like Dittmer's measurements done with a different species and by different methods, Miller's results provide an impression of the enormous area that roots expose to their media. Per gram fresh weight, the length of corn roots was 4,590 cm, and the surface area, 276 cm^2.

Impressive as these numbers are, they nevertheless convey an inadequate idea of the functional interface between roots and their media, especially soil. To explain this it is necessary to consider both external and internal features of roots. Let us begin at the outside. The "surface" of the root

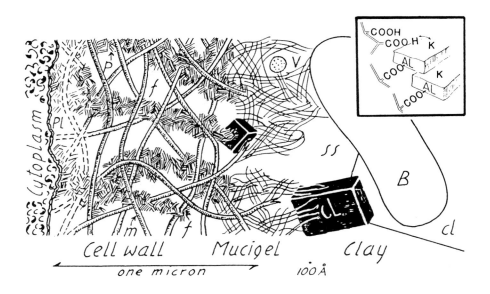

Figure 2. Highly magnified root-soil boundary region. On the left if the interior of an epidermal cell with cytoplasm and plasma membrane (Pl). The cell wall region depicts cellulose microfibrils (m), pectic (p) gel zones, and free space (f), with its channel widths exaggerated. The cell wall is fringed by the mucigel that is contacting clay (cl) particles, a bacterium (B), a virus (v) and is permeated by soil solution (ss). The 100-Å dot attests to the relative smallness of enzyme molecules and nutrient ions. From Jenny (1980).

is not easily defined; see the numerous electron micrographs of
the ultrastructure of the root-soil interface assembled by
Foster et al. (1983). Figure 2, taken from Jenny (1980), shows
the root-soil boundary region, in graphically simplified
fashion. The cell membrane lies behind the cell wall. The
wall, with its cellulose fibrils, pectic and gel substances
embedded in it, and open or free spaces, is in such intimate
contact with the soil solution, clay particles, and
microorganisms that no sharp dividing line can be drawn between
the two systems, root and soil. The soil solution and the
solution in the root cell wall are continuous.

Attention might, however, be focussed on the outer cell
membranes or plasmalemmas, beyond which lies the cytoplasm.
There is no doubt that in a functional sense these membranes
perform the acquisition of nutrient and other ions. Only after
the ions have been transported across the plasma membrane of a
root cell can they be said to have been absorbed by the plant.
Penetration into merely the cell wall is a passive process of
diffusion, mass flow, and ion exchange, and as such is readily
reversible.

But attention even to the outer cytoplasmic membranes of
the epidermal cells is not enough. Let us revert to the
tracing of the cross section of a corn root (Fig. 1). If cell
walls of epidermal cells are permeable to ions, the walls of
the cortical cells also are likely to be permeable. There is
abundant evidence that this is so (Epstein, 1972, 1973). The
question then arises how far laterally ions from the soil
solution may freely diffuse through the cell walls, thereby
gaining access to the plasma membranes of many more cells than
merely those of the epidermis. For primary, rapidly growing
roots - the ones highly active in ion absorption - the answer
to that question is probably the innermost layer of the cortex,
the endodermis. The suberized "Casparian strip" in the walls
of the endodermis forms an (imperfectly) impermeable barrier
layer that stops or at least minimizes the further centripetal
movement of water and solutes through cell wall space into the
stele. The volume diffusively accessible has been called the
"free" or "outer" space (Epstein, 1972, 1973).

As a consequence of the permeability of the cell walls of
epidermal and cortical cells the actual surface area of root
cell plasma membranes exposed to the ambient solution may be
far larger than that of the epidermal cell membranes only.
Depending on the radial extent of the cortex, cell membranes in
addition to those of epidermal cells may be reached by ions of
the external solution and participate in their absorption. As
a result the potential absorbing surface of the root is much
larger than its geometrical surface, as shown in Fig. 1.
Miller (1981), on the basis of measurements and calculation for
the three size-classes of corn roots shown in Fig. 1,

determined the total outer cell surface potentially involved in
ion absorption.   It was 2,400 cm^2 per gram root, or 8.7 times
the ostensible geometric surface. Epstein (1977) has discussed
additional features whereby roots are adapted to the soil
environment.
       The process of ion absorption by roots from solutions has
been intensively studied for many years.   The rate limiting
step for most ions is transport across the outer cytoplasmic
membrane or plasmalemma. The principal technique that has been
used is not apt to be attractive to physiological ecologists
(Chapin and Van Kleve, 1989).   Seedlings are first grown for a
few days or a week in a dilute solution of CaSO$_4$, say 0.2 or
0.5 mM.   The roots are then excised and samples of these
excised roots are immersed in an aerated solution containing
the ion under investigation in isotopically labeled form.   The
absorption period is short, usually ranging from 10 minutes to
no more than a few hours.

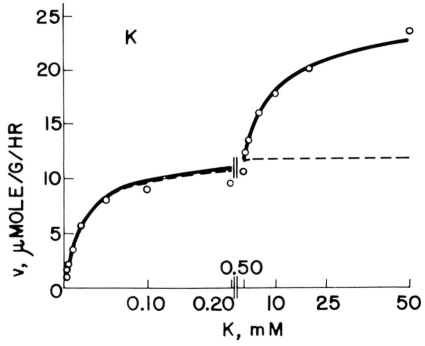

       Figure 3.   Rate, v, of potassium absorption by excised
barley roots as a function of the KCl concentration in the
solution.   Note the break in the scale between 0.2 and 0.5 mM
K$^+$, where the contribution to the rate due to mechanism 1 has
leveled off.   After Epstein (1972).

Of the many findings obtained with this and similar techniques I shall briefly describe only one, which is of particular relevance to field conditions. When the rate of absorption of an ion, e.g. K^+, is examined as a function of the KCl concentration in the solution, bimodal curves like the one shown in Fig. 3 have been observed in numerous investigations. At low K^+ concentrations the rate of its absorption increases but it levels off at concentrations of about 0.1 or 0.2 mM. At much higher concentrations, the rate of absorption rises again. There is some controversy about the interpretation of this complex isotherm, but I and many other investigators have interpreted it as indicating the operation of two transport systems, mechanism 1 and mechanism 2, respectively (Epstein, 1976). The one of greatest interest here is mechanism 1, because it is responsible for ion absorption over the range of low concentrations characteristic of many nutrient ions in soil solutions. This statement applies with particular force to soils of natural ecosystems, where the concentrations of nutrients are not artificially raised by application of mineral fertilizers.

The above discussion has brought out two features of roots that present the investigator with baffling problems, problems that require new methods and instrumentation for their solution. One is the ramification of roots through soil and the exposure of enormous root surfaces to a solid matrix where they are hidden from view and inaccessible. The other is their function of absorbing nutrients from soil solutions having low, often exceedingly low, concentrations of nutrients. Installations for observation of roots in soil are known as rhizotrons (Huck and Taylor, 1982); to apparatus in which roots can be exposed to concentrations of nutrient and other ions automatically controlled at very low concentrations, the term rhizostats has been applied (Epstein, 1984).

III. METHODS OF OBSERVATION

A. Methods in which Roots and
Soil are Separated

The oldest method for observing roots is to dig into the soil, exposing the roots, and to trace their disposition by visual observation. Figure 4 shows such a tracing of the root system of clover, <u>Trifolium</u> <u>repens</u>, taken from the root atlas

compiled by Kutschera (1960). Simple as the method is, and
limited in the contribution it can make to quantitative plant
science, it and various modifications of it nevertheless give
a vivid impression of the enormous proliferation of roots in
soil. In this 13-month old plant, the root system far exceeds
the shoot in the extent to which it occupies its medium. The
same method also shows the great diversity of root systems of
different plants (Kutschera, 1960; Weaver, 1968). Equally
important are the effects of different soil environments on the
pattern of root development in a given genotype (Pearson, 1974;
Taylor, 1974). Direct methods of observation have been
described by Böhm (1979) and Böhm et al. (1977). Although
techniques of direct observation involving the separation of
roots and soil seem straightforward enough they have the
potential for further development and refinement. To give a
recent example, Belford et al. (1986) have devised a new method
for studying intact shoot-root systems. Square metal tubes,
0.18 x 0.18 x 1.1 m were constructed of 16-gauge steel. One
side of the tube is removable. With that side in place, the
tube is forced into the soil with a hydraulic coring machine.
The location for the initial experiments were field

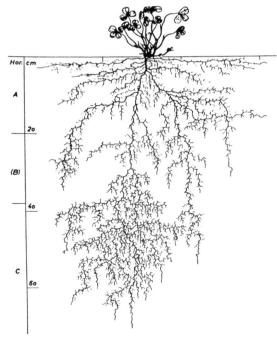

    Figure 4. Tracing of the shoot and roots of a 13-month
old clover plant. From Kutschera (1960).

plots of winter wheat, <u>Triticum</u> <u>aestivum</u> 'Stephens' and the time the tubes were installed in the soil was immediately after emergence of the plants. At intervals during the growing season tubes were extracted, the removable side was taken off, and the roots laid bare for observation and measurement by gentle washing with water. The entire plants (shoots and roots were not severed during the procedures just outlined) were laid out in shallow trays filled with water and prepared for detailed examination. Figure 5 shows a tube being extracted from the soil, and Figure 6, the extracted tube, opened side up, ready for removal of the soil with water. The great advantage of this technique is two-fold: the plant grows under near-normal conditions in the field, and it is left intact until the soil has been washed from the roots.

All methods involving the separation of roots from the soil in which they grew have in common the feature that the roots are sacrificed in the process. Observations over time require the processing of successive plants. Methods have therefore been devised for the continual observation of root systems of individual plants. Like the methods already described they have limitations of their own, but like them, also, they have their distinctive advantages.

Figure 5. Metal tube previously forced into the soil being extracted, after a wheat plant had grown in the soil. Method of Belford <u>et</u> <u>al</u>. (1986). Photograph courtesy of Betty Klepper.

B. Rhizotrons

Rhizotrons represent the most elaborate installations yet devised for the direct, sustained observation and measurement of roots growing in soil.   A full-scale rhizotron is an underground observation chamber with one or more transparent walls or windows abutting on the soil.   Some of the roots of the plants growing in the soil grow alongside these windows or

Figure 6.   One side of the metal tube of Fig. 5 has been removed; the root system is ready to be laid bare by gentle removal of the soil with water.   Method of Belford et al. (1986). Photograph courtesy of Betty Klepper.

observation panels.   Their growth can thus be observed and measured, and their development monitored by time-lapse cinematography and other means (Huck and Taylor, 1982). According to Böhm (1979, p. 3), some fifteen such installations have been constructed throughout the world. Figure 7 shows the inside of the rhizotron of the Tennessee Valley Authority's National Fertilizer Development Center at Muscle Shoals, Alabama, U.S.A., with 18 "Plexiglas"-fronted soil bins, and Figure 8, a front view of two of the bins showing the proliferation of corn roots, Zea mays, as influenced by conventional tillage vs. zero tillage.

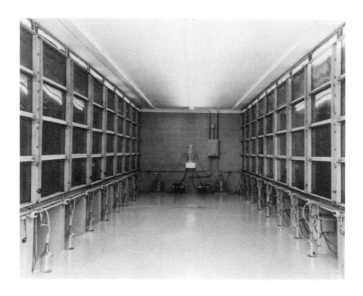

Figure 7.   The underground observation chamber of the rhizotron at Muscle Shoals, Alabama.   Photograph courtesy of M.G. Huck.

Like the methods involving separation of roots and soil, rhizotron technology is being extended.   The United States Department of Agriculture is planning a new facility at Ames, Iowa.   It will have a number of interesting features.   Instead of growing in the open, the plant shoots will be in growth chambers subject to control of light, temperature, and humidity.   The roots will be in soil with rhizotron-like facilities for observation and measurement. Ports in the walls of the soil containers will make it possible to measure or sample water potential, temperature, concentrations of gases, microbial populations, and root exudates (Tom Kaspar, private communication).

Minirhizotrons are yet another development of the basic idea.   The first such device was described by Bates (1937). Figure 9, from Böhm (1974), shows the main features.   Holes of a diameter of about 70 mm and 110 cm deep are bored into the soil.   A glass or plastic tube is lowered into the hole.   The upper part of the tube protruding above the soil is normally covered with foil to keep light out.   After seeds planted near the tube have germinated and plants have grown, their roots are periodically observed and measured as shown in the figure. Waddington (1971) had earlier described a similar arrangement

Figure 8. Two of the soil bins in the Muscle Shoals rhizotron showing corn roots as influenced by zero tillage (left) and conventional tillage (right). Photograph courtesy of M.G. Huck.

employing fiber optics (Figure 10). A more recent method using fiber optics in a minirhizotron is that of Sanders and Brown (1978). They used a medical duodenoscope, an instrument with which surgeons observe and photograph features within the human body after inserting it in a body cavity. It consists of a light source, a flexible fiberoptic tube, and an objective lens that transmits the image to an eyepiece, a camera, or a micro-TV camera (Brown, 1985). Inserted into a minirhizotron, such an instrument can be used to observe and photograph roots in situ. Figure 11 shows the instrument ready to be lowered into the soil, and Figure 12, a picture of roots of soybean, Glycine max, as transmitted by the micro-TV camera. Further refinements have been introduced. Upchurch and Ritchie (1983) have improved the techniques for observing roots through minirhizotrons by use of a video recording system, including more recently a color video camera (Upchurch and Ritchie, 1984). Stereoscopic viewing is yet another improvement (Noordwijk et al., 1985). For recent discussions of minirhizotrons and their uses, see Taylor (1987).

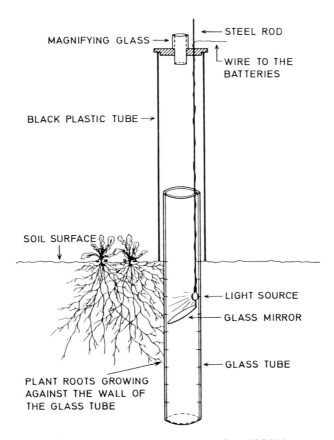

MAGNIFYING GLASS →

← STEEL ROD

└ WIRE TO THE BATTERIES

BLACK PLASTIC TUBE →

SOIL SURFACE

← LIGHT SOURCE

← GLASS MIRROR

← GLASS TUBE

PLANT ROOTS GROWING AGAINST THE WALL OF THE GLASS TUBE

Figure 9. Minirhizotron. From Böhm (1974).

All types of rhizotrons introduce some measure of disturbance of the belowground ecosystem. In addition, all roots subject to observation are the very roots also subject to an edge effect, since they are not surrounded by soil on all sides. Minirhizotrons have the advantage of affording the observer a mole's eye view of 360°; the stationary observation panels of big rhizotrons confine the viewer to essentially two-dimensional views. All types, however, provide a method for in situ, non-destructive observation of roots in soil. I believe that in the future development of rhizotron technology, emphasis will be more on minirhizotrons than the large, permanent installations. Considerations of cost, soil disturbance, versatility in application, and statistical validation of results all point in this direction, especially for ecophysiological investigations.

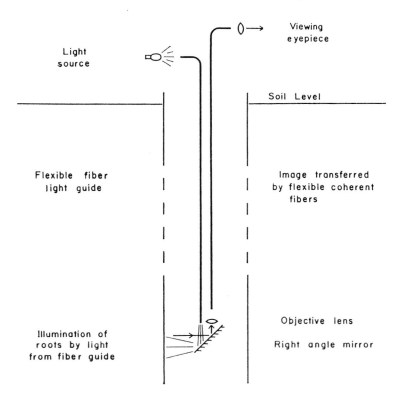

Figure 10. Minirhizotron using fiber optics. From Waddington (1971).

C.   The Perforated Soil System

Van den Tweel and Schalk (1981) and Bosch (1984) have described a method for observing roots in soil called the perforated soil system. The system consists of a soil monolith in a box with transparent front and back panels having numerous holes in exactly corresponding positions. After the soil has settled in the box holes are bored through it, these boreholes corresponding to the holes in the transparent front and back panels. For observation of the roots the cover normally kept over the panels is removed. The roots can then be observed through the holes. Their growth through these air spaces is said not the differ significantly from their growth in soil. The technique does not yet seem to have been used in the field. This may, however, be possible by digging two parallel trenches 20 to 30 cm apart (Bosch, 1984). Perforated plates would then be put in place against the soil and holes bored through it.

Figure 11. Minirhizotron of D.A. Brown ready to be lowered into a borehole in the soil. Photograph courtesy of D.A. Brown.

D. NMR Imaging

Bottomley et al. (1986) and Rogers and Bottomley (1987) applied the totally noninvasive, nondestructive technique of proton (^{1}H) nuclear magnetic resonance (NMR) to study roots and seeds in soil and other solid rooting media (vermiculite etc.). Soil is transparent to the static and radiofrequency magnetic fields employed in NMR. Using proton NMR the investigators actually imaged the water in the system. By doing the experiments just before the next watering cycle they obtained images of the distribution of the (moist) roots in situ. By virtue of its nondestructive, noninvasive nature the technique holds great promise for the study of roots in soil. Experiments of Brown et al. (1986) indicate that it is possible to differentiate among the various tissues of roots in situ and to detect changes in water content. With this technique, MacFall et al. (1990) have presented graphic evidence that water uptake may occur through the suberized region of the woody taproot of loblolly pine, Pinus taeda.

Figure 12.   TV image of soybean roots as transmitted by the apparatus shown in Fig. 11.   Photograph courtesy of D.A. Brown.

IV.   CONTROL OF THE
IONIC ENVIRONMENT

A.   What is Wrong with Conventional
Nutrient Solutions?

The physical chemical, and biological complexity of soils renders them unsuitable as media for investigations in which it is essential to have control over and knowledge of the ionic milieu of roots.   For control and monitoring of the chemical composition and the pH of the medium to which roots are exposed solution   cultures   are   therefore   indispensable.   In   a conventional   nutrient   solution   of   the   Hoagland   type (Epstein, 1972, p. 39) the concentrations of macronutrient elements range from 14 mM for nitrate to 1 mM for magnesium; that   of   dihydrogen   phosphate   is   2   mM.   A   survey   of concentrations of nutrients adequate to support normal growth of many species reveals that they are smaller than those of conventional nutrient solutions by orders of magnitude (Asher, 1978). The reason for the unrealistically high concentrations of conventional nutrient solutions is the need, not for high concentrations, but for a large supply of nutrients if plants are to grow for appreciable periods. If adequate amounts of nutrients are to be available in containers of conveniently small or moderate volume, without frequent replenishment of the solutions, use of high concentrations is inevitable. We shall see that the problem has been recognized for a long time. It is only now, however, that sophisticated new technology is at hand to devise systems of solution culture in which most nutrients can be maintained at the low concentrations that are adequate for growth. Those are also the kinds of concentrations that plants, especially wild plants in many of their natural habitats, encounter in the soil solution (Chapin, 1980, 1988; Epstein, 1985; see Table 1).

B.   Control With Solid Ion
Exchangers and Adsorbents

For many decades, solid ion exchange materials and adsorbents have been used by a number of investigators to control the pH or the concentration of (usually just one) nutrient ion in a nutrient solution in contact with the material. Absorption by the plants lowers the concentration of the ion, resulting in further release from the solid phase and the maintenance of a dynamic steady state concentration of the ion in solution. These techniques are discussed by R.M. Welch in this volume.

C.   A Once-Through Flow System

Kay and Gutschick (1984) have devised a system in which large volumes of solutions are made to flow through the plant containers in a once-through (non-circulating) pass and then discarded. Nitrate and phosphate are delivered at low to very low concentrations. The obvious advantage of this system is its relative simplicity and hence, relative freedom from malfunction. The principal difficulty is the need, in low-concentration work, of very large volumes of solution. Thus, a typical experiment requires the consumptive use of up to 3000 L of deionized water per day.

D.   Rhizostats:
Evolution of the Concept and Prospects for the Future

        Teakle (1929) devised a constant-flow system to study the
phosphate nutrition of   wheat, _Triticum aestivum_,   at low
phosphate concentrations. His experimental set-up (Figure 13)
and procedure have many of the features incorporated with
appropriate modification into later, more sophisticated
systems.   They are (i) large containers (A) with the nutrient
solution of carefully maintained composition; (ii) culture jars
(B) to the bottom of which the solution from A is delivered
through tubing; (iii) an overflow siphon and tubing through
which the solution is delivered to a carboy (C), similar to A;
(iv) periodic analyses of the solution in C for phosphate; (v)
readjustment of its concentration to the desired value; (vi)
pumping of this solution back into the delivery carboy (A),
thus maintaining a constant circulation of the nutrient
solution through the culture vessel, B.

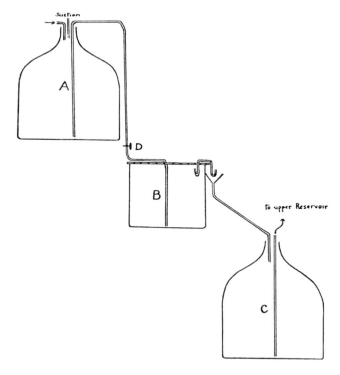

        Figure 13. The pioneering flow system of solution culture
devised by Teakle (1929). See text for explanation.

This system enabled Teakle to maintain the phosphate concentration at or close to the initially set value. The concentrations of other nutrients were sufficiently high not to become much depleted during the experimental period. Using this system Teakle (1929) found that the wheat plants grew normally at 1.0 ppm PO_4, or 10.5 μM. "Obviously the use of solutions supplying phosphate at the rate of several hundred parts per million obscures the problem rather than elucidates it." One senses a certain asperity in his comment.

Asher et al. (1965) developed a more elaborate system for control of ionic concentrations and pH, as well as of the temperature of the circulating nutrient solution. Instead of readjusting the concentration of the ion under investigation at intervals they installed a drip-feed system of stock solution to re-supply the nutrient on a continual basis. In accordance with the results of frequent analyses of the solution in the storage tank (corresponding to Teakle's carboy A; see Fig. 13), the drip rate was adjusted to maintain the concentration of the ion being studied close to the rate at which it was being withdrawn through absorption by the plants. Their systems is shown diagrammatically in Figure 14. To give an impression of the flow rates needed, these investigators found it necessary to supply solution at rates of 1000 to 1300 L per plant container per day, the particulars depending on the number of plants per pot and their stage of growth, the particular nutrient and its concentration, and the decrease in nutrient concentration upon passage of the solution through each pot

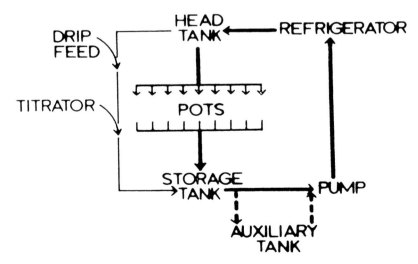

Figure 14. The flow system of Asher et al. (1965). © by Williams and Wilkins, by permission. See the text.

that the investigators considered permissible. The system made possible adequate maintenance of concentrations as low as about 2 and 1 x 10^{-8} M zinc and manganese, respectively. The system of Asher et al. (1965) and other systems patterned after it have furnished much information on nutrient absorption and mineral nutrition of plants at low concentrations of nutrients in the medium. Clement et al. (1974) built an installation incorporating the automation of several functions. They used ion-sensitive electrodes to control pumps for delivery of the necessary volumes of stock solution to maintain the concentration of the ion studied (nitrate initially) in the flowing nutrient solution. Breeze et al. (1982) made an important addition to the same system by incorporation of continuous-flow colorimetry for phosphate analysis and a computer to control nutrient pumps for addition of the required volumes of phosphate solution; see Breeze et al. (1985). Phosphate concentrations can be maintained at a level as low as 0.1 uM, or 0.01 ppm H_2PO_4. At that concentration, the dry weight of the plants 29 days after sowing was 14 percent of that of plants grown at 6.4 uM H_2PO_4. Thus the concentration of this nutrient was controlled at so low a level as to cause a severe deficiency.

Because of the particular importance of pH in affecting the nitrate and ammonium absorption by plants, Moritsugu and Kawasaki (1983) built a flow system for automatic control of pH and the two nitrogen ions, using ion sensitive electrodes. For a current review of flowing nutrient solution cultures and their use, see Wild et al. (1987).

Two installations currently being used are those of A.D.M. Glass and collaborators at the University of British Columbia (Glass et al., 1987) and A.J. Bloom at the University of California (A.J. Bloom, private communication; also see Bloom, 1989). They have two features in common. Unlike those described so far they are not continuous-flow systems. Instead, the plant roots are in a container with nutrient solution, as in a conventional solution culture; the solution however, is monitored automatically for its concentration of several nutrient ions and the pH, and readjusted by appropriate additions, this also being done automatically. The second feature these two installations have in common is exclusive reliance on ion sensitive electrodes for analysis of the ions being controlled.

I have no intention of compiling a bibliography of low nutrient concentration technology. I have mentioned some pioneering work to make a more general point. Only now, about 130 years after the introduction of the solution culture method in the 1860s, technology is finally at hand to devise sophisticated systems of solution cultures whose concentrations are truly relevant to research on mineral nutrition in contexts

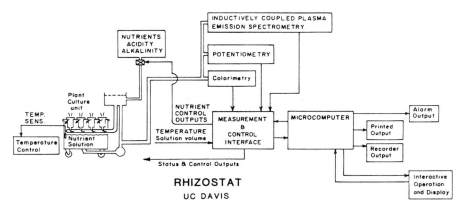

Figure 15. Diagram of rhizostat under development at the University of California, Davis. The plant culture unit is at the left. The flowing nutrient solution is automatically sampled at intervals and analyzed for the controlled ions by one of several methods. The data are processed by a computer, which directs the necessary additions of ions.

such as agriculture, forestry, and physiological ecology.

Sophisticated growth chambers and phytotrons have been in existence for a long time. They permit control of many if not all of the environmental conditions to which the shoot is exposed. The ionic milieu of the roots of a growing plant consists of no less than about 15 ions of nutrient elements (including the pH). Even in the most advanced existing installations, only a few of these ions are under simultaneous control, and the degree of automation varies. The thesis I submit to you is that with existing technology it is now possible to control the concentrations of the majority of the nutrient elements at realistically low concentrations, and do so in a fully automated fashion. The obstacles to accomplishing this are no longer scientific or technical; they are "only" economic. The scheme for an elaborate rhizostat as envisaged by a number of members of my department is shown in Figure 15. The system is not limited to any one analytical technique but meant to use whatever method is best suited to the element in question. As the diagram shows, the operation is fully automatic and computer controlled. Amounts of nutrients supplied, corresponding to those absorbed, will be monitored on a continual basis and stored in the computer, which will be programmed to calculate rates of absorption. Such a rhizostat will make possible a large variety of experiments now difficult or impossible to do. They include the effects on the growth and mineral nutrition of plants at suboptimal concentrations of one or more nutrients, mutual interactions of ions at very low concentrations, the mineral nutrition of wild plants at concentrations common in their natural habitats, rhizobial and mycorrhizal symbioses under controlled conditions of the medium, the effects of low concentrations of heavy metal and other pollutants, and selection and breeding of nutrient-efficient genotypes; the list is by no means exhaustive. In conclusion, I think that the development of rhizostat technology will add a new dimension to research in mineral plant nutrition and nutritional (edaphic) ecophysiology.

ACKNOWLEDGMENTS

In writing this paper I have been greatly aided by the generous collaboration of many colleagues. They willingly answered questions, supplied papers, graphic material, and information not yet published, and discussed various aspects of their research with me. These colleagues and friends are named individually in connectin with the specific information they furnished. I must, however, acknowledge in particular my debt to my colleagues in this department, R.G. Burau and G.S. Pettygrove, and to A.J. Bloom in the Department of Vegetable Crops.

REFERENCES

Albert, R. (1982). Halophyten. In: Pflanzenökologie und Mineralstoffwechsel (H. Kinzel), pp. 33-204. Verlag Eugen Ulmer, Stuttgart.

Asher, C.J. (1978). Natural and synthetic culture media for spermatophytes. In: CRC Handbook Series in Nutrition and Food. Section G: Diets, Culture Media, Food Supplements. Vol. III. Culture Media for Microorganisms and Plants. (M. Rechcigl, Jr., ed.), pp. 575-609. CRC Press, Inc., Cleveland.

Asher, C.J., P.G. Ozanne, and J.F. Loneragan (1965). A method for controlling the ionic environment of plant roots. Soil Sci. 100:149-156.

Bates, G.H. (1937). A device for the observation of root growth in the soil. Nature 139:966-967.

Belford, R.K., R.W. Rickman, B. Klepper, and R.R. Allmaras (1986). Studies of intact shoot-root systems of field-grown winter wheat. I. Sampling techniques. Agron. J. 78:757-760.

Bloom, A.J. (1989). Continuous and steady-state nutrient

absorption by intact plants. In: Application of Continuous and Steady-State Methods to Root Biology. (J.G. Torrey and L.J. Winship, eds.), pp. 147-163. Kluwer Academic Publishers, Dordrecht.

Böhm, W. (1974). Mini-rhizotrons for root observations under field conditions. Z. Ackerpflanzenbau 140:282-287.

Böhm, W. (1979). Methods of Studying Root Systems. Springer-Verlag, Berlin.

Böhm, W., H. Madakor, and H.M. Taylor (1977). Comparison of five methods for characterizing soybean rooting density and development. Agron. J. 69:415-419.

Bosch, A.L. (1984). A new root observation method: the perforated soil system. Acta Oecol./Oecol. Plant.5:61-74.

Bottomley, P.A., H.H. Rogers, and T.H. Foster (1986). NMR imaging shows water distribution and transport in plant root systems in situ. Proc. Natl. Acad. Sci. USA 83:87-89.

Breeze, V.G., R.J. Canaway, A. Wild, M.J. Hopper, and L.H.P. Jones (1982). The uptake of phosphate by plants from flowing nutrient solution. I. Control of phosphate concentration in solution. J. Exp. Bot. 33:183-189.

Breeze, V.G., A.D. Robson, and M.J. Hooper (1985). The uptake of phosphate by plants from flowing nutrient solution. III. Effect of changed phosphate concentrations on the growth and distribution of phosphate within plants of Lolium perenne L. J. Exp. Bot. 36:725-733.

Brooks, R.R. (1987). Serpentine and its Vegetation. A Multidisciplinary Approach. Dioscorides, Portland.

Brown, D.A. (1985). Methods for measuring root development and activity of crop plants. 1985 Yearbook of Science and Technology, pp., 366-373. McGraw-Hill Book Company, New York.

Brown, J.M., G.A. Johnson, and P.J. Kramer (1986). In vivo magnetic resonance microscopy of changing water content in Pelargonium hortorum roots. Plant Physiol. 82:1158-1160.

Caldwell, M.M. and R.A. Virginia (1989). Root systems. In: Plant Physiological Ecology. Field Methods and Instrumentation. (R.W. Pearcy, J. Ehleringer, H.A. Mooney, and P.W. Rundel, eds.), pp. 367-398. Chapman and Hall, London.

Chapin, F.S., III (1980). The mineral nutrition of wild plants. Ann. Rev. Ecol. Syst. 11:233-260.

Chapin, F.S., III (1988). Ecological aspects of plant mineral nutrition. Adv. Plant Nutr. 3:161-191.

Chapin, F.S., III and K. Van Kleve (1989). Approaches to studying nutrient uptake, use and loss in plants. In: Plant Physiological Ecology. Field Methods and Instrumentation. (R.W. Pearcy, J. Ehleringer, H.A.

Mooney, and P.W. Rundle, eds.), pp. 185-207. Chapman and Hall, London.

Clement, C.R., M.J. Hooper, R.J. Canaway, and L.H.P. Jones (1974). A system for measuring the uptake of ions by plants from flowing solutions of controlled composition. J. Exp. Bot. 25:81-99.

Dittmer, H.J. (1937). A quantitative study of the roots and root hairs of a winter rye plant (Secale cereale). Amer. J. Bot. 24:417-420.

Dudal, R. (1977). Inventory of the major soils of the world with special reference to mineral stress hazards. In: Plant Adaptation to Mineral Stress in Problem Soils. (M.J. Wright, ed.), pp. 3-13. A special publication of Cornell University Agricultural Experiment Station, Ithaca, N.Y.

Epstein, E. (1972). Mineral Nutrition of Plants: Principles and Perspectives. John Wiley and Sons, New York.

Epstein, E. (1973). Mechanisms of ion transport through plant cell membranes. Intern. Rev. Cytol. 34:123-168.

Epstein, E. (1976). Kinetics of ion transport and the carrier concept. In: Encyclopedia of Plant Physiology New Series Vol. 2. Transport in Plants II. Part B. Tissues and Organs. (U. Lüttge and M.G. Pitman, eds.), pp. 70-94. Springer-Verlag, Berlin.

Epstein, E. (1977). The role of roots in the chemical economy of life on earth. BioSci. 27:783-787.

Epstein, E. (1983). Crops tolerant of salinity and other mineral stresses. In: Better Crops for Food. (J. Nugent and M. O'Connor, eds.), pp. 61-82. Ciba Foundation Symposium 97, Pitman, London.

Epstein, E. (1984). Rhizostats: controlling the ionic environment of roots. BioSci. 34:605.

Epstein, E. (1985). Salt-tolerant crops: origins, development, and prospects of the concept. Plant and Soils 89: 187-198.

Epstein, E. and A. Läuchli (1983). Mineral deficiencies and excesses. In: Challenging Problems in Plant Health. (T. Kommedahl and P.H. Williams, eds.), pp. 196-205. The American Phytopathological Society, St. Paul, Minnesota.

Ernst, W.H.O. (1982). Schwermetallpflanzen. In: Pflanzenökologie und Mineralstoffwechsel. (H. Kinzel), pp. 472-506. Verlag Eugen Ulmer, Stuttgart.

Foster, R.C., A.D. Rovira, and T.W. Cock (1983). Ultrastructure of the Root-Soil Interface. The American Phytopathological Society, St. Paul, Minnesota.

Glass, A.D.M., M. Saccomani, G. Crookall, and M.Y. Siddiqui (1987). A microcomputer-controlled system for the automatic measurement and maintenance of ion activities in nutrient solutions during their absorption by intact plants in hydroponic facilities. Plant Cell and Environment 10:375-381.

Huck, M.G. and H.M. Taylor (1982). The rhizotron as a tool for root research. Adv. Agron. 35:1-35.

Jenny, H. (1980). The Soil Resource: Origin and Behavior. Springer-Verlag, New York.

Kay, L.E. and V.P. Gutschick (1984). Solution culture method for studying nutrient uptake and stress. VIe Colloque International pour l'Optimisation de la Nutrition des Plantes. Actes: 3:1003-1007.

Kinzel, H. (1982). Pflanzenökologie und Mineralstoffwechsel. Verlag Eugen Ulmer, Stuttgart.

Kinzel, H. (1983). Influence of limestone, silicates and soil pH on vegetation. In: Encyclopedia of Plant Physiology New Series Vol. 12C. Physiological Plant Ecology III. Responses to the Chemical and Biological Environment. (O.L. Lange, P.S. Nobel, C.B. Osmond, and H. Ziegler, eds.), pp. 201-244. Springer-Verlag, Berlin.

Kinzel, H. and M. Weber (1982). Serpentin-Pflanzen. In: Pflanzenökologie und Mineralstoffwechsel (H. Kinzel), pp. 381-410. Verlag Eugen Ulmer, Stuttgart.

Kruckeberg, A.R. (1984). California Serpentines: Flora, Vegetation, Geology, Soils, and Management Problems. University of California Press, Berkeley.

Kutschera, L. (1960). Wurzelatlas mitteleuropäischer Ackerunkreuter und Kulturpflanzen. DLG-Verlag, Frankfurt.

MacDonald, N. (1984). Trees and Networks in Biological Models. John Wiley and Sons, New York.

MacFall, J.S., G.A. Johnson, and P.J. Kramer (1990). Observation of a water-depletion region surrounding loblolly pine roots by magnetic resonance imaging. Proc. Natl. Acad. Sci. USA 87: 1203-1207.

Miller, D.M. (1980). Studies of root function in Zea mays. I. Apparatus and methods. Can. J. Bot. 58:351-360.

Miller, D.M. (1981). Studies of root function in Zea Mays. II. Dimensions of the root system. Can. J. Bot. 59:811-818.

Moritsugu, M. and T. Kawasaki (1983). Effect of nitrogen source on growth and mineral uptake in plants under nitrogen-restricted culture condition. Ber. Ohara Inst. Landwirtsch. Biol., Okayama Univ. XVIII(3):145-15.

Noordwijk, M. van, A. deJager, and J. Floris (1985). A new dimension to observations in minirhizotrons: a stereoscopic view on root photographs. Plant and Soil 86:447-453.

Pearson, R.W. (1974). Significance of rooting pattern to crop production and some problems of root research. In: The Plant Root and Its Environment. (E.W. Carson, ed.), pp.

247-270. University Press of Virginia, Charlottesville.
Reimold, R.J. and W.H., Queen, eds. (1974).   Ecology of
    Halophytes. Academic Press, New York.
Rogers, H.H. and P.A. Bottomley (1987).   In situ nuclear
    magnetic resonance imaging of roots: influence of soil
    type, ferromagnetic particle content, and soil water.
    Agron. J. 79:957-965.
Sanders, J.L. and D.A. Brown (1978).   A new fiber optic
    technique for measuring root growth of soybeans under
    field conditions. Agron. J. 70:1073-1076.
Taylor, H.M. (1974).   Root behavior as affected by soil
    structure and strength.   In: The Plant Root and Its
    Environment. (E.W. Carson, ed.), pp. 271-291. University
    Press of Virginia, Charlottesville, Virginia.
Taylor, H.M., ed. (1987).   Minirhizotron Observation Tubes:
    Methods and Applications for Measuring Rhizosphere
    Dynamics. American Society of Agronomy, Madison.
Teakle, L.J.H. (1929). The absorption of phosphate from soil
    and solution cultures. Plant Physiol. 4:213-232.
Tweel, P.A. van den and B. Schalk (1981).   The horizontally
    perforated soil system: a new root observation method.
    Plant and Soil 59:163-165.
Upchurch, D.R. and J.T. Ritchie (1983).   Root observations
    using a video recording system in mini-rhizotrons. Agron.
    J. 75:1009-1015.
Upchurch, D.R. and J.T. Ritchie (1984).   Battery-operated color
    video camera for root observations in mini-rhizotrons.
    Agron. J. 76:1015-1017.
Waddington, J. (1971).   Observation of plant roots in situ.
    Can. J. Bot. 49: 1850-1852.
Waisel, Y. (1972). Biology of Halophytes. Academic Press, New
    York.
Weaver, J.E. (1968).   Prairie Plants and their Environment.
    University of Nebraska Press, Lincoln.
Wild, A., L.H.P. Jones, and J.H. Macduff (1987).   Uptake of
    mineral nutrients and crop growth: the use of flowing
    nutrient solutions. Adv. Agron. 41:171-219.
Woolhouse, H.W. (1983).   Toxicity and tolerance in the
    responses of plants to metals. In: Encyclopedia of Plant
    Physiology New Series Vol. 12C.   Physiological Plant
    Ecology III. Responses to the Chemical and Biological
    Environment. (O.L. Lange, P.S. Nobel, C.B. Osmond, and H.
    Ziegler, eds.), pp. 245-300.   Springer-Verlag, Berlin.

MODERN TECHNOLOGIES FOR STUDYING THE REQUIREMENTS AND FUNCTIONS OF PLANT MINERAL NUTRIENTS

Ross M. Welch

USDA, Agricultural
Research Service
U.S. Plant, Soil &
Nutrition Laboratory
Ithaca, New York

I. INTRODUCTION

During the past several decades, there has been a veritable explosion of new technologies which aid scientists in studying natural phenomena. Life scientists have taken advantage of these modern technologies to study many complex biological processes which allow life to exist on earth. Within the plant science research community, there are several research areas where the application of these new scientific tools could prove to be advantageous in helping solve some important and long-standing research problems. Plant mineral nutrition is such an area.

Research concerning the relationship between mineral nutrients and their functions in plant growth and metabolism and their role in natural selection and competition within plant communities has received relatively little attention in the recent past. Moreover, research directed toward discovering new essential micronutrients and determining their possible roles in plant growth has been de-emphasized (Bieleski and Läuchli, 1983). Besides inadequate research funding, a primary factor responsible for this neglect has been the lack of suitable experimental techniques, instrumentation, or training needed to study many of the difficult problems facing plant scientists interested in of mineral nutrition.

As a result of recent developments in analytical instrumentation and methods, conditions are ripe for significant advances in several difficult mineral nutrition research areas including : 1) determining the metabolic function of known essential micronutrients and their interactions with other nutrients, 2) discovering new essential micronutrients and identifying their metabolic roles, 3) localizing and characterizing the sites and mechanisms responsible for the translocation and deposition of mineral nutrients within higher plant tissues including seeds, and 4) opening the root-soil interface "black box" to discover the processes controlling root growth in soil in situ and identifying and delineating those mechanisms which regulate the movement of mineral elements to root surfaces and their transport across root-cell membranes.

I have selected some new technologies which I feel have great potential to provide plant scientists with the ability to make significant advances in some of these challenging research areas. A brief summary of these technologies will be followed by a more detailed discussion of specific applications of two techniques which can be used to study trace element essentialities and plant growth in solution culture at nutrient levels similar to those occurring naturally in soil solution.

I have not attempted to present a comprehensive discussion of these modern technologies. Rather, I hope my brief overview will result in an increased awareness among plant scientists of the potential usefulness of these techniques in solving many important mineral nutrition research problems.

Table 1 lists some new techniques and a few of their applications (or potential applications) in various aspects of plant mineral nutrition research. The analytical instrumentation listed has not been widely used by plant scientists because of: 1) limited instrument availability and/or high costs associated with their use, 2) the lack of special training required in order to reliably interpret the complex data generated or 3) a lack of awareness of the potential usefulness of certain instruments in helping to solve various mineral nutrition research problems. Encumbrances to the use of these instruments could be alleviated if various private and government granting agencies provided more grants to a greater number of plant science research institutions for the purchase, operation, and maintenance of modern, expensive research equipment.

Table 1. Examples of modern techniques having applications in various areas of mineral nutrition research.

Technique	Examples of applications
Nuclear Magnetic Resonance	1) Noninvasive, *in vivo* studies concerning ionic fluxes of P, Na, K, H, etc. from nutrient media to root surfaces and transport of ion across cell membranes and between cell organelles; 2) Determinations of shifts in chemical forms of essential mineral elements within cells and between cells *in situ* 3) Studies concerning the function of micronutrient elements in biochemical pathways involving macronutrient metabolites that can be analyzes by NMR techniques.
Electron Spin Resonance	1) Determinations of *in vivo* shifts in oxidation states of Fe, Mn and Cu ions in whole tissues, sells, organelles, and metabolites; 2) Studies concerning the effects of Ca, B and Zn on the structure of cellular membranes by using spin-label probes to measure changes in membrane fluidity induces by variations in the level of these nutrients supplies to plant cells.
Inductively couples, Argon-plasma, mass spectrometry	1) Studies concerning the movement and cycling of nutritionally important elements through the soil-plant-animal and human food chain by using stable isotopes in place of radioactive isotopes in various types of tracer experiments; 2) Quantitative analyses of nearly all elements in the periodic table in environmental samples and biological tissues in nanogram levels.

Table 1 – continued

Mixed chelate and ion exchange resin nutrient solution systems	Maintaining and controlling the ratios of free metal,H and P ionic activities in plant nutrient solutions at concentrations normally found in soil solutions.
Chelation column chromatography	Rapid removal of trace metal contaminants from concentrates macronutrient solutions used to prepare ultrapure nutrient solutions to study the essentiality of trace metals for plant growth.

II. NUCLEAR MAGNETIC RESONANCE

Clearly, nuclear magnetic resonance (NMR) spectroscopy
has a wide range of potential applications within mineral
nutrition research. NMR spectroscopy is already one of the
most potent, non-destructive methods available to chemists
(Gadian, 1982). The technique has been used to determine the
distribution of various isotopes within molecules, determine
molecular structures, characterize interactions between
molecules and measure steady state reaction rates.
Applications of NMR spectroscopy in plant physiology research
were reviewed by Roberts (1984). He summarized the benefits
of NMR for plant physiologists into four categories: 1)
observe and quantitate mobile metabolites *in vivo*, 2)
non-destructively study cellular compartmentation of
metabolites *in vivo*, 3) trace the metabolism of atoms and
chemical bonds in plant tissues and cells, and 4) measure
the rates of ATP synthesis and degradation in plant cells *in
vivo*. While Roberts (1984) stressed the use of NMR methods
to study certain isotopes of carbon, nitrogen, hydrogen and
phosphorus, isotopes of a number of other mineral elements
are also amenable to NMR analysis (Harris and Mann, 1978).
All isotopes which possess a nucleus with a magnetic dipole
moment (i.e. nonzero spin angular momenta) can be studied
using NMR methods. Usually these isotopes have an odd atomic

number. Furthermore, the benefits of NMR in mineral nutrition research could be extended by artificially enriching the abundance of various NMR analyzable isotopes having low natural abundance in plant tissues (e.g. magnesium-25). NMR spectroscopy should play an increasingly important role in the development of our understanding of various mineral nutrient functions in plant metabolism, ion transport processes in plant cells, and the role of certain mineral nutrients in the maintenance of cellular membrane integrity (e.g. calcium). The use of NMR technology is discussed in more detail by Kramer and Johnson in this volume.

III. ELECTRON SPIN RESONANCE

Electron spin resonance (ESR) spectroscopy is based on transitions between energy levels of an unpaired electron in an atom produced when the atom is subjected to a magnetic field. The technique is applicable to many paramagnetic transition metal atoms which contain unpaired electrons in their d or f atomic orbitals (e.g. the Fe^{3+} cation), various free organic radicals (e.g. nitroxide spin-labels), as well as several other groups of substances (e.g. active chlorophyll). Free radical nitroxide spin labels have been used in ESR investigations to study several cellular properties of interest to biologists including studies of protein structure, cytoplasmic water viscosity, and membrane structure and function. Bioenergetic parameters have also been studied using spin labels including cell volumes, pH and electrical gradients across membranes, one-electron oxidation/reduction potentials within cellular compartments, and electrical surface and boundary potentials at membrane surfaces (Mehlhorn and Packer, 1983). Certainly greater use of ESR techniques in studies concerning the effects of various nutrient deficiencies (e.g. calcium and boron) on metabolic processes and membrane function in plant cells is warranted.

IV. INDUCTIVELY COUPLED ARGON-PLASMA/MASS SPECTROMETRY

Many important advances in our knowledge of plant mineral
nutrition can be attributed to the use of radioisotopes of
mineral elements as tracers in soil-plant systems. However,
there are many research situations where radioisotope tracers
are not appropriate or are not available. For example, there
are no long-lived radioisotopes of some mineral nutrients
(e.g. copper, boron, magnesium and molybdenum). The lack of
long-lived radioisotope of these elements has precluded their
use in prolonged experiments with plants. Additionally,
while the use of radioactive tracers has yielded important
information, the shortcomings associated with their use under
field situations can be prohibitive (i.e., long-term soil
contamination, very expensive radioactive waste disposal
costs, etc.). The use of stable isotope tracers, in lieu of
radioactive isotopes, presents a viable alternative in many
situations.

Highly enriched stable isotope salts of mineral nutrients
have been available from Oak Ridge National Laboratories for
some time. However, their use in mineral nutrition studies
has been very restricted because of limitations in analytical
procedures. Previous analytical options included either
neutron activation analysis or mass spectrometry methods.
Neutron activation analysis requires a costly neutron source
(i.e. a nuclear reactor or nuclear particle accelerator) to
activate the sample to be analyzed. Mass spectrometry methods
have been dependent on either relatively low temperature
thermal ionization techniques or the use of electron and
chemical ionization of volatile metal chelates. Both of
these mass spectrometry techniques are restricted to the
analysis of relatively few stable isotopes.

The recent development of inductively coupled
argon-plasma mass spectrometers (ICP/MS) (Houk *et al.*, 1980)
should revolutionize the use of stable isotopes as tracers of
mineral elements and trace metals in mineral nutrition
research. The ICP/MS combines the high ionization efficiency
of inductively coupled argon-plasma (commonly used as an
excitation source for atomic emission spectroscopy) with the
spectral simplicity and very high sensitivity of a mass
spectrometer. Not only can the technique perform
multielement analyses of most elements (i.e. over 90% of the
elements in the periodic table) in concentrations as low as
0.1 to 10 ppb, but also can determine isotope ratios and
isotopic abundance on the same samples at the 0.5% precision
level. The method has a large practical dynamic
concentration range of at least six orders of magnitude and
is rapid; at least 30 elements per minute can be analyzed
when their concentrations exceed ten times the detection

limit (information obtained from SCIEX Elan™ 250 ICP/MS
System sales literature).[1] Thus, the ICP/MS technique has
great potential in the areas of determining element
concentrations and isotope ratios in various types of mineral
nutrition studies.

V.   TECHNIQUES FOR STUDYING
POTENTIALLY ESSENTIAL
TRACE METALS

Progress in discovering new essential trace metals for
higher plant life has been limited by the lack of methods for
removing subnanogram amounts of the trace metal under study
from nutrient culture media. Recent developments in
purification and preconcentration techniques now make it
possible to effectively eliminate trace metal contaminates
from high ionic strength macronutrient stock solutions and
from the highly purified deionized water used to prepare
nutrient solutions and irrigate plants. Styrene
divinylbenzene copolymer ion exchange resins, containing
iminodiacetate functional groups which efficiently bind trace
metals (such as Fe, Cu, Mn, Zn, Pb, Cd, Ni, and Co), are
commercially available (e.g. Chelex 100[1]). They have been
used to concentrate and recover subnanogram levels of trace
metals from various aqueous biological and environmental
samples and to ultrapurify buffers and concentrated ionic
reagents (for examples, see Baetz and Kenner, 1975; Samsahl
et al., 1968; Shmuckler, 1965). These resins should be very
useful in the ultrapurification of concentrated macronutrient
stock salt solutions for use in trace metal essentiality
studies. Another commercially available (Pierce Chemical
Company) chelating ligand immobilized on controlled pore
glass beads (CPG) is 8-hydroxyquinoline (8HQ).

---

[1]Mention of a trademark, proprietary product, or vendor
does not constitute a guarantee or warranty of the product by
the U.S. Department of Agriculture and does not imply its
approval to the exclusion of other products or vendors that
may also be suitable.

We have employed CPG-8HQ chelation column chromatography to study the essentiality of nickel for higher plant growth (Eskew *et al.*, 1983). The CPG-8HQ method proved to be a very rapid and efficient procedure for removing nickel contaminants from concentrated salt solutions used to make ultrapure nutrient solutions (Eskew *et al.*, 1984a). Table 2 shows the efficiency of nickel removal from highly concentrated macronutrient solutions used to prepare nickel deficient nutrient solutions. By using this technique, we

Table 2. Efficiency of Ni removal from concentrated salt solutions by chelation column chromatography on CPG-8HQ. Solutions were labeled with 63Ni and then passed through columns of CPG-8HQ. Recovery was measured by eluting retained ions with 1.2 N HCl (from Eskew *et al.*, 1984a).

Salt	Concentration	Retention	Recovery
	M	percent	percent
$Ca(NO_3)_2$	2	99.9	99.8 (3)[a]
KNO_3	2	99.9	100.8 (3)
NH_4NO_3	2	99.9	97.5 (2)
$MgSO_4 \cdot 7H_2O$	2	99.9	104.1 (1)
$CaCl_2 \cdot 2H_2O$	2	99.9	101.1 (2)
Na Mes[c]	2	99.9	98.8 (2)
K phosphate	2	99.9	92.3 (1)
$NH_4H_2PO_4$	1	ND[b]	96.3 (1)

[a] Numbers in parentheses indicate the number of observations.
[b] N.D. = not determined.
[c] Mes = 2-(N-morpholino) ethanesulfonic acid.

Table 3.   Efficiency of Zn removal from stock solutions used to prepare *Chlorella* media.   Solutions were labeled with ^{65}Zn and then passed through columns of CPG-8HQ.   Recovery was measured by eluting retained ions with 1.2 N HCl (from Eskew *et al.*, 1984a).

Salt	Concentration	Retention	Recovered
	M	percent	percent
$MgSO_4$	1	100	99.6
K phosphate	1	100	N.D.[a]
Urea	1	99.8	92.7
Glucose	1	99.5	102.3
$CaSO_4$	0.01	99.9	100.3

[a]N.D. = not determined.

were able to show that nickel was essential for the growth of those legumes that accumulate ureides (Eskew, *et al.*, 1983; 1984b).   Table 3 provides an example of the removal of zinc from highly concentrated macronutrient solutions used in Chlorella growth studies.

Traditionally, nutrient solutions prepared for plant nutrition studies have been purified by tedious methods such as multiple recrystallizations of nutrient salts and liquid-liquid solvent extraction techniques employing organic solvents to extract complexed metal ions from aqueous solutions.   CPG-8QH chelation chromatography has several advantages over the former techniques.   The primary advantages are the rapidity of purification and the high trace metal removal efficiency from high ionic strength aqueous solutions.   The method is simple to use and the columns can be readily regenerated and reused repeatedly. Furthermore, no potentially carcinogenic, or explosive organic solvents are required.   Finally, CPG-8QH columns have the advantage that trace metals (e.g. Co, Ni, Cu, Fe, Al, Zr,

Ti, V, and Zn) can be concentrated on and quantitatively
recovered from the columns (Eskew, *et al.*, 1984a). Thus,
trace metals can be preconcentrated from dilute acid digests
of biological tissues or from deionized water for subsequent
determination of element concentrations.

VI.   USING ION EXCHANGE AND CHELATING
RESIN CULTURE SYSTEMS TO MIMIC
SOIL SOLUTION

For over a century, nutrient solution culture has been
one of the most important techniques used in studies
concerning plant mineral nutrition (Asher and Edwards, 1983;
Epstein, 1972; Gauch, 1972). The technique has been used in
experiments concerning: the essentiality of mineral elements
and establishing characteristic deficiency symptoms in crops,
interactions among nutrient ions, the physiological roles of
essential elements, and mechanisms of both mineral ion
absorption and translocation. In modern, well-stirred
solution culture systems, nutrient ion concentrations, pH,
pO_2, Eh (redox potential) and root temperature, can be kept
relatively constant or these parameters can be varied
independently of each other. Unlike heterogenous and complex
soil systems, growth media factors affecting plant growth can
be accurately controlled in homogeneous nutrient solutions.
This allows scientists the opportunity to study experimental
variables precisely.
Unfortunately, traditional types of simple nutrient
solution cultures are not ideal for several reasons. First,
they do not provide constant nutrient ion concentrations in
solution at levels commonly occurring in soil solution. In
order to prevent nutrient depletion, mineral nutrients are
usually supplied at initial concentrations much greater than
those found in soil solution. As a result, large
fluctuations in mineral ion concentrations typically occur in
the culture during the course of an experiment. Further,
they have limited hydrogen ion buffering capacity
particularly when nutrient treatments produce large
imbalances in plant uptake of either nutrient cations or
anions or when phosphorus concentrations are supplied at low
levels. Excessive levels of phosphorus are necessary in
conventional nutrient solutions to provide adequate
hydrogen-buffering capacity. However, these high phosphorus
levels (e.g. 2mM) can result in plant growth problems under

certain circumstances including: phosphorus toxicity in
phosphorus sensitive plants (Loneragan & Asher, 1967) and in
plants grown under zinc-deficiency conditions, (Loneragan *et
al.*, 1982), inhibition of nodulation and nitrogen fixation in
legumes dependent on symbiotic dinitrogen fixation for their
nitrogen supply (Imsande & Ralston, 1981), and prevention of
root hair development (Asher and Edwards, 1983; Foehse and
Junke, 1983).

Other limitations of conventional nutrient solutions have
been summarized in a recent editorial (see Epstein, 1984).
Continuously flowing nutrient solution cultures are one means
by which these difficulties may be circumvented. Detailed
examples of these systems are reviewed by Epstein in another
chapter of this book. They permit a high degree of control
of the root environment (Asher, 1981; Asher and Edwards,
1983). However, sophisticated flowing culture systems
require a large initial capital expense and their maintenance
can be costly and labor intensive. Hence, most plant science
laboratories do not have this equipment available to them.

Currently, an attractive alternative system to
continuously flowing nutrient culture techniques is being
developed in several laboratories, (Checkai *et al.*, 1987a; b).
It is based on the use of mixed synthetic ion exchange and
chelating resins to precisely control the activities of free
metal (both macronutrient and micronutrient), hydrogen, and
phosphate ions in nutrient solutions at concentrations
similar to these occurring in natural soil solutions.

The concept of using ion exchange resins to buffer
nutrient media ions is not new. Since the 1940s, various
investigators have grown plants in "exchanger cultures," but
the technique has not had wide acceptance among plant
scientists for several reasons (Epstein, 1972). Typically,
ion exchange resin techniques used in the past have consisted
of single cation or anion exchange resins. They were usually
employed to buffer pH and/or to supply a few nutrients in
comparatively undefined matrices (Checkai *et al.*, 1987a).
Thus, some plant scientists believe that the use of this
technique results in a fairly complex plant growth medium
with a nutrient ion composition which is difficult to
accurately define and control. However, several developments
and improvements in ion exchange resin technology now make it
likely that mixed bed ion exchange resin culture systems may
become more generally accepted. For example, current use of
various types of commercial metal chelating resins (e.g.
Chelex 100) permit the accurate control of cation activities
of various micronutrient metals (i.e. cooper, manganese and
zinc) independently of macronutrient metals (e.g. calcium,

potassium and magnesium), and at very low solution
concentrations (i.e. comparable to soil solution levels).
Additionally, the activity of phosphate ions in nutrient
solutions can now be continuously buffered at very low levels
by using a partially neutralized cation-exchange resin (such
as Dowex 50W X4) containing adsorbed polynuclear
hydroxyaluminum to which appropriate amounts of exchangeable
phosphate ions are initially bound (Checkai et al., 1987a;
Robarge and Corey, 1979).

Finally, the development and general availability of
sensitive instruments for the determination of element
concentrations (e.g. heated graphite furnace atomic
absorption spectrophotometers and inductively coupled plasma
emission spectrophotometers) now permit routine
determinations of metal ions in aqueous solutions at
concentrations in the ppm to ppb range. These analytical
tools have made it possible to accurately prepare and test
the ability of ion exchange and chelating resins to provide
nutrient ions in amounts and ratios sufficient to meet the
growth requirements of plants.

Recently, Checkai et al., (1987a; b) have developed a
mixed ion exchange and chelating resin nutrient solution
culture system to study the uptake of cadmium and
micronutrient metals by tomato (*Lycopsersicon esculentum*)
plants. Their method attempted to mimic soil solution ionic
activities in a nutrient solution culture system. The types
of ion exchange and chelating resins they employed, the ion
binding functional groups, and the specific ions buffered by
each type of resin are shown in Table 4. Table 5 shows the
manner in which other mineral nutrients, not buffered by the
resins, were maintained in their culture system. Membrane
filter bags, constructed of a hydrophilic acrylic copolymer
material with non-woven nylon support (Versapor-800[1]), were
used to enclose the ion exchange and chelating resins within
the dilute nutrient solutions contained in six 1 plastic pots
(see Figures 1 and 2). Figure 3 shows the entire nutrient
culture system operating in a growth chamber during their
experiment with tomato plants. The nutrient cultures were
continuously swirled during plant growth by plastic coated
magnetic stirring bars placed in each pot. The magnetic bars
were driven by water actuated magnetic stirrers situated
under each plastic pot.

Table 4. Types of exchange and chelating resins, their functional ion binding groups and specific ions buffered in nutrient solution employed by Checkai *et al.*, (1987a; b).

Resin type	Ion binding functional groups	Specific ions buffered[a]
Chelating cation exchange resin (Chelex-100)	iminodiacetic acid on styrene-divinylbenzene copolymer lattice	Fe^{2+}/Fe^{3+}, N^{2+} Ni^{2+}, Cd^{2+} (Ca^{2+}, Mn^{2+})
Strongly acidic cation exchange resin (Dowex 50W-X4)	sulfonic acid on styrene divinylbenzene copolymer lattice	Ca^{2+}, Mg^{2+}, K^+, Mn^{2+}
Weakly acidic cation exchange resin (Bio-Rex 70)	carboxylic acid on macroreticular acrylic polymer lattice	H^+, (Ca^{2+}, Mg^{2+}, Mn^{2+})
Partially neutra-lized Al-strong acid resin pH 6 (Al saturated Dowex 50W-X4)	polynuclear hydroxy-Al surface on sulfonic acid functional groups in styrene divinylbenzene copolymer lattice	H_2PO_4, (HPO_4, PO_4, H^+)

[a]Ions in parentheses are buffered to a limited extent by resin type.

Table 5. Procedure for maintaining macronutrient anions and other micronutrients for culture system developed by Checkai *et al.*, (1987a).

Nutrient	Methods of Maintenance
Nitrate[a], sulfate, water, (Phosphate)	Continuously supplied via Mariotte bottle as a very dilute nutrient solution at a rate equivalent to the plant's water transpiration rate during their growth.[b]
Iron	Initially added to the nutrient solution as FeEDDHA (10 μM).[c]
Chlorine	Furnished initially, with iron, at a concentration of 30 μM in dilute nutrient solution.
Molybdenum	Furnished initially as Na_2MoO_2 at a concentration of 0.1 μM in dilute nutrient solution.
Boron	Maintained by adding borosilicate spun-glass to each pot (0.17 g glass per 1).

[a]Solutions received 0.55 moles/1 KNO_3 after 20 days of plant growth.

[b]Dilute nutrient solution contained: 104.4 μM nitrate, 42.2 μM potassium, 20.6 μM calcium, 17.0 μM magnesium, 8 μM phosphate, and 5.0 μM sulfate.

[c]EDDHA - ethlenediaminedi-0-hydroxyphenylacetic acid. EDDHA buffered with iron via exchange with iron from the iron-saturated chelating resin.

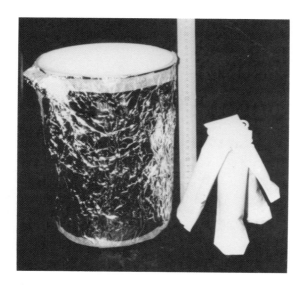

Figure 1.   Picture of a six-l plastic pot and four membrane filter bags containing various types of resins used in a ion exchange and chelating resin culture system developed by Checkai *et al.*, (1987a; b).

Figure 2.   Picture of the arrangement of resin filter bags in plastic pot used to grow tomato plants (Checkai *et al.*, 1987a; b).

Figure 3.   Picture of entire nutrient culture system
developed by Checkai *et al.*, (1987a; b) operating in a growth
chamber during an experiment with tomato plants.

The ability of this system to maintain nutrient ion levels while maintaining various cadmium ion activities in solution during the growth of tomato plants is shown in Tables 6 and 7. The ionic strength of the nutrient solution culture system declined less than 10% from 0.030M initially to 0.028M at the termination of the experiment. The average pH values of the nutrient solutions slightly increased from 6.00 initially to 6.26 on day 25 of the experiment.

The experiment of Checkai *et al.*, (1987a; b), presented above, demonstrates the feasibility of using mixed ion exchange and chelating resin nutrient solution systems in plant mineral nutrition research. Using this technique, they were able to maintain the activity ratios of various nutrient metal ions in solution while varying the ionic activity of cadmium by three orders of magnitude. All nutrient ionic activities were maintained at realistic soil-solution levels, and the system supplied sufficient nutrients to meet the growth requirements of tomato plants for at least 25 days.

Table 6. Variations in the average concentrations of some nutrients in mixed-resin nutrient solutions during the growth of tomato plants over a 25-day period. (Data from Checkai, 1987a)

Nutrient	Day_1	Day_7	Day_14	Day_21	Day_25
			μ M		
Calcium	6040	5901	5950	6150	5680
Magnesium	1780	1840	1810	1760	1750
Potassium	4100	4020	3720	4090	3780
Sulfur	2110	2260	2300	2390	2410
Phosphorus	9.4	7.3	4.8	4.1	4.0
Iron	8.6	8.0	5.4	2.9	3.4

Table 7.   Average ionic activities, expressed as pM
values, of some micronutrient cations and cadmium in mixed
resin nutrient solutions during the growth of tomato plants
over a 25 day period (data from Checkai *et al.*, 1987a).

Metal	Day 1	Day 7	Day 14	Day 21	Day 25
			pM value[a]		
Zinc	6.82	7.1	7.6	7.7	7.7
Copper	9.6	9.9	10.5	10.5	10.7
Manganese	6.52	6.50	7.00	6.90	7.20
Nickel	9.7	10.1	10.6	10.5	10.6
Cadmium - Low[b]	9.3	9.6	9.7	9.6	9.8
Cadmium - Medium	8.0	8.3	8.3	8.4	8.4
Cadmium - High	6.9	7.2	7.4	7.4	7.4

[a]pM value is defined as the negative log of metal ion
activity in solution.
[b]Low, medium and high refer to the concentrations of the
cadmium treatments supplied to the tomato plants on the
chelating resin.

Table 8 lists some beneficial characteristics of mixed ion
exchange and chelating resin systems which are applicable to
plant mineral nutrition studies.   However, there are still
some limitations to this type of system.   First, various
nutrient anions (i.e. nitrate, sulfate, chloride, molybdate
and borate) are not buffered by the mixed resin system and
must be supplied continuously.   Secondly, the rates of
release of some nutrient ions from this mixed resin system
may be limited by diffusion, which may not be high enough to
sustain maximum plant growth rates, especially when plants
approach maturity.   Finally, the technique is labor intensive
requiring a great deal of preliminary elemental analysis and
resin preparation before an experiment can be performed.
Further development of the mixed resin system may eliminate

some of these undesirable limitations. Possibly, future
developments of a mixed resin system, in conjunction with a
continuously flowing nutrient culture system, will provide
the ideal nutrient culture system for plant studies involving
mineral nutrients.

Table 8.   Beneficial characteristics of nutrient solution
cultures buffered via mixed ion exchange and chelating resin
techniques.

---

1.   The activity ratio of selected free ions can be
maintained at very low levels similar to those found in soil
solutions.
2.   Nutrients and pH can be buffered in solution at
levels sufficient to meet plant growth requirements without
the need for systematic replacement of nutrient solution to
counter nutrient depletions and pH changes resulting from
plant nutrient uptake.
3.   Synthetic metal chelates can be used without
affecting the ionic activities of various metals in nutrient
solutions.
4.   Transition metal ionic activities can be maintained
independently of other metal ions.
5.   Trace metal contaminants can be scavenged from
nutrient solution cultures in experiments designed to test
the essentiality of various transition metals for higher
plant growth.
6.   Naturally occurring compounds exuded by plant roots
are allowed to accumulate (e.g. metal complexes, reductions,
oxidants, etc.) in the nutrient media.

VII. CONCLUSIONS

Hopefully, this brief review of selected modern
analytical techniques and methods in plant mineral nutrition
research will stimulate their application to challenging
problems facing plant scientists.   Wider use of these
techniques should lead to i) major advances in our knowledge
of the processes controlling mineral nutrient absorption,
translocation, deposition, and remobilization in plants, ii)

greater understanding of the functions of essential mineral nutrients in plant metabolism, and iii) possibly, the discovery of some new essential micronutrients and their function(s) in cellular metabolism. These advances are predicated on whether or not plant scientists have access to these instruments and, furthermore, that plant scientists are trained to operate and interpret the data obtained. However, this can only happen if granting agencies and educational institutions make funds available for their purchase and maintenance, and, if universities begin to train their majors in plant science in the use and applications of these modern technologies.

REFERENCES

Asher, C.J. (1981). Limiting external concentrations of trace elements for plant growth: use of flowing solution culture techniques. *J. Plant Nutrition 3*:163-180.

Asher, C.J. and Edwards, D.G. (1983). Modern solution culture techniques. In: Inorganic Plant Nutrition. (A. Läuchli and R.L. Bieleski, eds.). pp. 94-119. Springer-Verlag, New York.

Baetz, R.A. and Kenner, C.T. (1975). Determination of trace metals in food using chelating ion exchange concentration. *J. Agr. Food Chem. 23*:41-45.

Bieleski, R.L. and Läuchli, A. (1983). Synthesis and outlook. In: Inorganic Plant Nutrition. (A. Läuchli and R. L. Bieleski, eds.). pp. 745-759. Springer-Verlag, New York.

Checkai, R.T., Hendrickson, L.L., Corey, R.B., and Helmke, P.A. (1987a). A method for controlling the activities of free metal, hydrogen, and phosphate ions in hydroponic solutions using ion exchange and chelating resins. *Plant and Soil 99*:321-334.

Checkai, R.T., Corey, R.B., and Helmke, P.A. (1987b). Effects of ionic and complexed metal concentrations on plant uptake of cadmium and micronutrient metals from solution. *Plant and Soil 99*:335-345.

Epstein, E. (1972). Mineral Nutrition of Plants: Principles and perspectives. John Wiley and Sons, Inc., New York.

Epstein, E. (1984). Rhizostats: controlling the ionic environment of roots. *Bioscience 34*:605.

Eskew, D.L., Welch, R.M., and Cary, E.E. (1983). Nickel: an essential micronutrient for legumes and possibly all higher plants. *Science 222*:621-523.

Eskew, D.L., Welch, R.M., and Cary, E.E. (1984a). A simple plant nutrient solution purification method for effective removal of trace metals using controlled pore glass-8-hydroxyquinoline chelation column chromatography. *Plant Physiol. 76*:103-105.

Eskew,D.L., Welch, R.M., and Norvell, W.A. (1984b). Nickel in higher plants: further evidence for an essential role. *Plant Physiol. 76*:691-693.

Foehse, D. and Jungk, A. (1983). Influence of phosphate and nitrate supply on root hair formation of rape, spinach and tomato plants. *Plant and Soil 74*:359-368.

Gadian, D.G. (1982). Nuclear Magnetic Resonance and Its Applications to Living Systems. Clarendon Press, Oxford.

Gauch, H.G. (1972). Inorganic Plant Nutrition. Dowden, Hutchinson, and Ross, Inc., Stroudsburg, Pa.

Harris, R.K. and Mann, B.E. (1978). NMR and The Periodic Table. Academic Press, London, New York, and San Francisco.

Houk, R.S., Fassel, V.A., Flesch, G.D., Svec, H.J., Gray, A.L., and Taylor, C.E. (1980). Inductively coupled argon plasma as an ion source for mass spectrometric determination of trace elements. *Anal. Chem.* 52:2283-2289.

Imaande, J. and Ralston, E.J. (1981). Hydroponic growth and the nondestructive assay for dinitrogen fixation. *Plant Physiol. 68*:1380-1384.

Loneragan, J.F. and Asher, C.J. (1967). Response of plants to phosphate concentration in solution culture: II. Rate of phosphate absorption and its relation to growth. *Soil Science 103*:311-318.

Loneragan, J.F., Grunes, D.L., Welch, R.M., Aduayi, E.A., Tengah, A., Lazar, V.A., and Cary, E.E. (1982). Phosphorus accumulation and toxicity in leaves in relation to zinc supply. *Soil Sci. Soc. Am. J. 46:*345-352.

Mehlhorn, R.J. and Packer, L. (1983). Bioenergetic studies of cells with spin probes. *Ann. N.Y. Acad. Sci. 414*:180-189.

Roberts, J.K., (1984). Study of plant metabolism in vivo using NMR spectroscopy. *Ann. Rev. Plant Physiol. 35*:375-386.

Robarge, W.P. and Corey, R.B. (1979). Adsorption of phosphate byhydroxy-aluminum species on a cation exchange resin. *Soil Sci. Soc. Am. J. 43:*481-487.

Samsahl, K., Wester, P.I., and Landstrom, O. (1968). An automatic group separation system for the simultaneous determination of a great number of elements in biological material. *Anal. Chem. 40*:181–187.

Shmukler, G. (1965). Chelating resins- their properties and
    applications. *Talanta 12*:281–190.

Chapter 6
Image Processing

REMOTE–CONTROL LIGHT
MICROSCOPE SYSTEM

Kenji Omasa
Ichiro Aiga[1]
Jiro Kondo[2]

National Institute for Environmental Studies
Tsukuba, Ibaraki 305, Japan

I. INTRODUCTION

Porometers are generally used to monitor the average movement of a great many stomata of attached leaves (Meidner and Mansfield, 1968; Burrows and Milthorpe, 1976). A thermography system with a digital computer can measure spatial distributions in stomatal resistance (or conductance) (Omasa et al., 1981a,b,c; 1983a), as well as leaf temperature (Schurer, 1975; Hashimoto and Niwa, 1978; Horler et al., 1980; Hashimoto et al., 1984; Omasa et al. 1990). Although these devices are satisfactory for many purposes, they cannot be used to observe responses of individual stomata and their neighboring cells to environmental stimuli.

Direct observation of the stomatal movement of attached leaves had been very difficult under the plant's actual growing conditions (Meidner and Mansfield, 1968; Meidner, 1981). Although the scanning electron microscope (Turner and Heichel 1977; Shiraishi et al., 1978) and the light microscope in which a piece of leaf is immersed in water or liquid paraffin (Monzi, 1939; Stålfelt, 1959; Meidner, 1981) can provide a clear image at high magnification, observation of intact stomata under their growing conditions is

---

[1]Present address: College of Agriculture, University of Osaka Prefecture, Sakai, Osaka 591, Japan.
[2]Present address: Science Council of Japan, Roppongi, Tokyo 106, Japan.

Measurement Techniques in Plant Science

impossible. Observation with an ordinary light microscope
under a plant's growing conditions poses some problems
(Heath, 1959; Meidner and Mansfield, 1968). First, visual
observation under weak light is very difficult. Second, the
environment of the lower side of the leaf cannot be
controlled because the leaf is directly held on the
microscope stage; the environment is also affected by human
manipulation of the microscope. And third, the working
distance, that is, the distance between the leaf and the
objective during the observation is very small at high
magnification, thus subjecting the leaf to the danger of
sticking to the objective during focusing and destroying the
environment between the leaf and the objective.

Recently, Omasa et al., (1983b) and Kappen et al.
(1987) developed new light microscope systems for direct
observation of the guard and epidermal cells of attached
leaves under actual growing conditions. These systems
solved the above problems of the ordinary light microscope.
In this paper we introduce our remote-control light
microscope system and show several observations.

II. OUTLINE OF THE SYSTEM AND ITS PERFORMANCE

Figure 1 shows our remote-control light microscope
system for direct observation of stomatal movements in
attached leaves. This system has a light microscope with a
wide working distance (ca. 13 mm) at high magnification (a
50x objective, 1.5x and 2x amplifiers and a TV adapter lens;
ca. 1,600-fold magnification on a TV monitor), an SIT
(Silicon intensifier target) video camera with high
sensitivity S20 type spectral response, image resolution of
over 600 TV lines, distortion within 2% and shading within
20% as a detector of the microscope image, a monochromatic
TV monitor with image resolution of ca. 1,000 TV lines and
distortion within 3% and remote controllers for adjusting
camera sensitivity, microscope focus, and movement of the
microscope stage in a separate room. The micoscope images
are projected on the TV monitor in a separate room and
recorded photographically on black-and-white (B&W) film or
by a video tape recorder (VTR) (horizontal resolution, 340
TV lines at B&W mode; S/N, ca. 50 dB) with a digital time
base corrector and a time code generator/reader. The image
processing for evaluating stomatal aperture and cell damage
are carried out by an image processing system composed of a
high speed video processor, graphic displays and a host
computer (Omasa and Onoe, 1984; Omasa and Aiga, 1987).

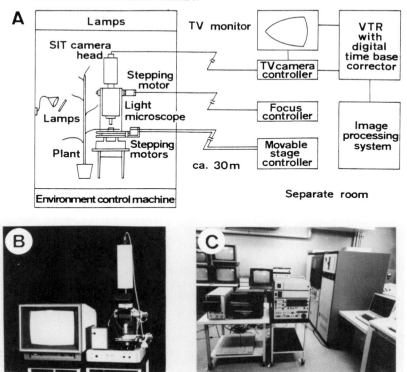

Fig. 1 Remote-control light microscope system for direct observation of stomatal movements in attached leaves. A, block diagram; B, light microscope system; C, VTR and image processing system. (Adapted from Omasa _et al._, 1983b and Omasa 1990).

Figure 2 shows a schematic cross-sectional view of the microscope stage for holding the leaf of an intact plant. The leaf (C) is held on a ring (F, 30 mm in inner diameter, 10 mm wide and 10 mm high) fixed to a remote-control movable stage (G) by a holding ring (E, the same diameter and width as F, 3 mm in height) in order to have the conditioned air pass under the surface of the leaf. Furthermore, since the center of the movable stage is cut out to a circle 30 mm in diameter and the distance between the mobvable stage and plate (H) fixed on base (K) is kept at 10 mm, the same temperature and humidity can be maintained on both sides of the leaf. The movable stage and the plate are made of

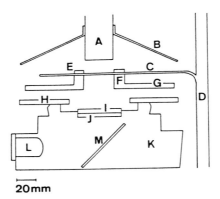

20mm

Fig. 2 Schematic cross-sectional view of the microscope stage for holding an intact leaf. A, objective; B, shade cover; C, leaf; D, stem; E, holding ring; F, ring fixed to remote-control movable stage; G, remote-control movable stage; H, plate; I, heat absorbing glass filter; J, diffusing filter; K, base; L, halogen lamp; M, mirror (Omasa et al., 1983b)

transparent acrylic resin in order to allow light from the environment to enter. The shade cover (B) is not used except for observation with transmitted light. Although observation is usually carried out with light from the environment, a halogen lamp (L) is sometimes used as a supplementary light source for observation with transmitted light.

Figure 3 shows photomicrographs of an intact stoma observed with reflected and transmitted light using the remote-control light microscope system. The stomatal image was clear at high magnification (ca. 1,600-fold magnification on the TV monitor). The stoma was observed with reflected light and then rapidly observed with transmitted light, and the stomatal aperture was found to be the same for both. Although this system could provide stomatal images with a mixture of reflected and transmitted lights, the clearest image was obtained with either reflected or transmitted light alone. The stomata could be observed with either reflected or transmitted light above ca. 0.1 mW cm^{-2} (environment illumination; ca. 2 klx (0.5 mW cm^{-2}) with reflected light, ca. 0.5 klx (0.5 mW cm^{-2}) with transmitted light). If the observation with a single light of 0.1 mW cm^{-2} is made by naked eye through the eyepiece instead of the SIT camera, the eye must be

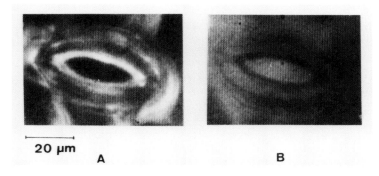

Fig. 3 Photomicrographs of an intact sunflower stoma observed with reflected or transmitted light using the light microscope system. A, reflection image; B, transmisssion image. (Omasa et al., 1983b)

sufficiently acclimatized to the dark room. The attachment of an image intensifier to the SIT camera produces a further increase in sensitivity although the image quality becomes poor.

Figure 4 shows a photomicrograph of a test chart measured using the light microscope system. Judging from the photomicrograph, we could tell that the resolution of the microscope image was within 1 μm. The resolution is improved by a digital image processing technique. Omasa and Onoe (1984) were able to evaluate the stomatal aperture

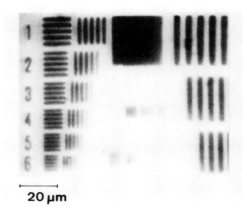

Fig. 4 Photomicrograph of test chart measured with the light microscope system. (Omasa et al., 1983b)

within 0.3 μm standard error using this technique, even when
the microscope image was of poor quality.

III. CONTINUOUS OBSERVATION OF GUARD AND EPIDERMAL CELLS

1) Response to illumination change

When the light microscope system was used for continuous
observation of the stomatal movement of an intact growing
plant, we obtained photomicographs similar to those in
Figure 5 which show the response of an intact stoma of the

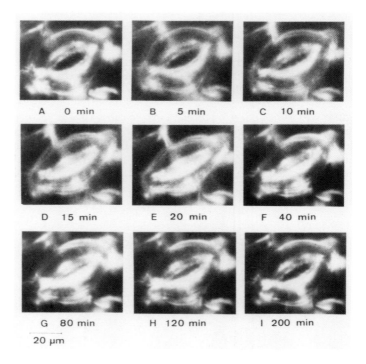

A   0 min   B   5 min   C   10 min

D   15 min   E   20 min   F   40 min

G   80 min   H   120 min   I   200 min

20 μm

Fig. 5 Photomicrographs of responses of an intact stoma of
an adaxial epidermis of a broad bean plant to illumination
change. The time after the first illumination change is
shown under the photographs. The illumination was changed
from 30 to 2 klx at 0 min (A) and from 2 to 20 klx at 20 min
(E). Other environmental conditions: air temperature, 20.0
$^{\circ}$C; humidity, 70% RH (Omasa et al., 1983b)

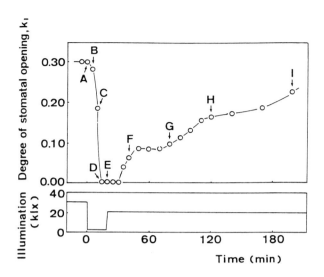

Fig.6 Changes in k_1 of the stoma in Fig.5; k_1 is the degree of stomatal opening expressed by the ratio l_a/l_{bmax}, where l_a is the width of the stomatal pore and l_{bmax} is the maximum length of the fully opened stomatal pore. "A" to "I" in Fig. 5 correspond to time points from "A" to "I". (Omasa et al., 1983b)

adaxial epidermis of a broad bean plant to illumination change. The illumination was changed from 30 klx (11.9 mW cm^{-2}) to 2 klx (0.5 mW cm^{-2}) at 0 min (A) and from 2 to 20 klx (7.7 mW cm^{-2}) at 20 min (E). The movement of the central pore of the stoma could be continuously observed. Figure 6 shows changes in the degree of opening (k_1) of the stomatal pore shown in Fig. 5; k_1 is expressed by the ratio l_a/l_{bmax}, where l_a is the width of the stomatal pore and l_{bmax} is the length of the fully opened stomatal pore. The stoma began to close within 5 min (B) after lowering the illumination (30 to 2 klx) and became completely closed after ca. 15 min (D). It began to reopen within 15 min after raising the illumination (2 to 20 klx), and after 180 min (I), had recovered to ca. 75% of aperture before the illumination change.

2) Response to water deficit

Figure 7 shows changes in k_1 of an intact sunflower stoma to water deficit caused by ice water perfusion to roots and Fig. 8 shows photomicrographs of the stoma and subsidiary cells at time points (A to F) in Fig. 7. The

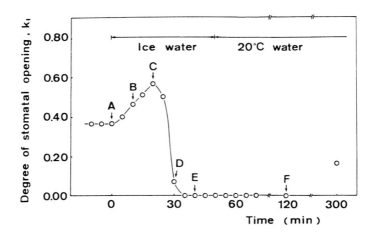

Fig.7 Changes in k_1 of an intact sunflower stoma to water deficit. The water deficit was caused by ice water perfusion to roots. Similar results were obtained in other experiments. Environmental conditions: air temperature, 25.0 °C; humidity, 60% RH; illumination, 600 μmol photons $m^{-2}s^{-1}$ (Omasa and Maruyama 1990).

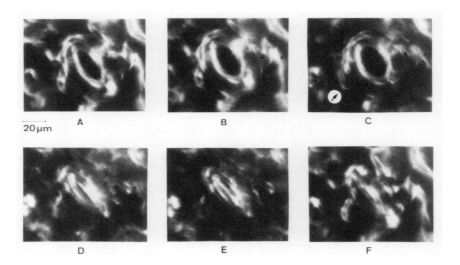

Fig.8 Photomicrographs of the stoma shown in Fig.7. "A" to "F" correspond to time points ("A" to "F") in Fig. 7. The arrow ($\rightarrow$) in "C" shows first hollow of subsidiary cell (Omasa and Maruyama 1990).

stoma increased aperture within a few minutes after the perfusion and reached a maximum opening at 20 min (C). Thereafter, it began to close and reached complete closure at 35 min (E). Subsidiary cells observed in the photomicrographs began to hollow at ca. 20 min (C) and then the hollow expanded from the epidermal cell to the guard cell (C to E). The transient opening from "A" to "C", therefore, may be caused by the rapid decrease in subsidiary cell turgor in comparison to guard cell turgor due to a decrease in water uptake from the root. After 20 °C water perfusion, the cell form slowly recovered (F) and the stoma began to reopen ca. 3 hrs later.

3) Response to air pollutants

Figure 9 shows changes in k_1 of an intact sunflower stoma to mixtures of SO_2, NO_2 and O_3 and Figure 10 shows photomicrographs of the stoma at time points (A to F) in Fig. 9. After 0.1 ppm exposure (A), the stoma began steady closure within 45 min although a slight opening related to acclimatization from dark to light was found at 10 min, and reached constant aperture (C to D) after ca. 90 min. However, the increase in concentration from 0.1 ppm to 0.2 ppm produced rapid and complete closure (D to F). The stoma recovered to the same aperture as before the exposure at ca. 5 hrs after the end of the exposure. The reopened

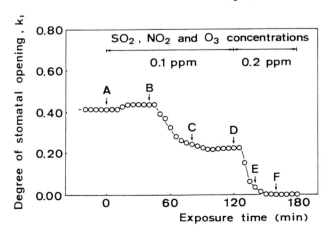

Fig.9 Changes in k_1 of an intact sunflower stoma to mixtures of SO_2, NO_2 and O_3. Similar results were obtained in other experiments. Other environmental conditions: air temperature, 25.0 °C; humidity, 60% RH; illumination, 600 μmol photons $m^{-2}s^{-1}$ (Omasa 1990).

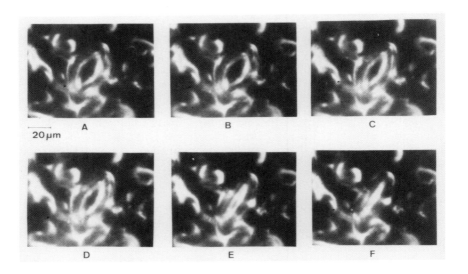

Fig.10 Photomicrographs of the stoma shown in Fig.9. "A" to "F" correspond to time points ("A to "F") in Fig. 9 (Omasa 1990).

stoma responded normally to change in illumination. During the experiment, the guard and epidermal cells maintained normal form.

Figure 11 shows changes in the k_1's of six neighboring stomata in a small leaf region (ca. 300 x 300 μm^2) of an intact sunflower plant to 2 ppm SO_2 exposure. These stomata showed almost uniform and constant k_1 until ca. 20 min after the start of the exposure, and then a wide variety of stomatal movements began; the largest k_1 value was about twice as large as the smallest value at 45 min and become about three times as large at 90 min. Water-soaking and wilting (cell collapse) began to appear in the subsidiary cells at about 55 min, when k_1 was a local maximum value, and then all the stomata began to close. This phenomenon may be caused by increased water loss from the subsidiary cells due to SO_2, which affects the membrane and osmotic pressure, resulting in a difference in the turgor between the guard cell and the subsidiary cell. At the end of the exposure, the stomata did not completely respond to change in illumination because both the guard and epidermal cells were disorganized. The variety of responses of neighboring stomata to SO_2 increased at the border of the injured region (Omasa <u>et al.</u>, 1985).

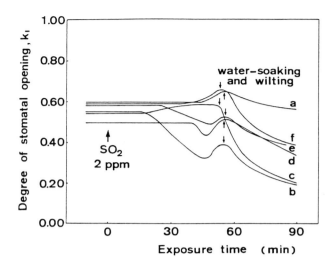

Fig.11 Changes in k_l's of six neighboring stomata in a small leaf region (ca. 320 x 320 μm^2) of an intact sunflower plant to 2 ppm SO_2 exposure. Small arrows ($\uparrow$ or $\downarrow$) show when water-soaking and wilting (cell collapse) began to appear. Other environmental conditions: air temperature, 25.0 °C; humidity, 60% RH; illumination, 600 μmol phptpons $m^{-2}s^{-1}$. (Omasa and Onoe 1984)

IV. RELATIONSHIP BETWEEN STOMATAL APERTURE AND CONDUCTANCE

The light microscope system was used to analyze the relationship between the stomatal aperture and conductance. The experiments were carried out according to the following procedure. An intact leaf of the test plant was held on the microscope stage under constant illumination. After the leaf had been sufficiently acclimatized to the new conditions, ca.30 stomata in an area ca.15 mm in diameter of the leaf were randomly observed with the reflected light, and then the stomatal conductance of the area was quickly measured with a porometer. This procedure was repeated for the same area after the illumination was changed.

Figure 12 shows relationships between the l_a's of the stomata of the attached middle leaves of various plants and their stomatal conductances (g_s). There was a positive correlation between l_a and g_s measured in the same area of the leaf. However, the regression curves varied with the kind of plant and epidermis. The maximum values of l_a and

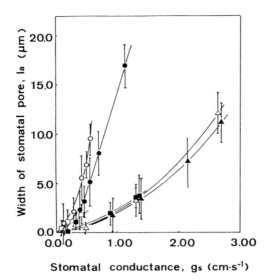

Fig. 12 Relationships between l_a's of the stomata of the attached middle leaves of various plants and their stomatal conductances (g_S). Symbols represent mean values of the l_a and vertical bars indicate $\pm$ standard error.○,broad bean adaxial epidermis; ●,broad bean abaxial epidermis; △, sunflower adaxial epidermis;▲,sunflower abaxial epidermis; □ ,tomato adaxial epidermis;■,tomato abaxial epidermis. Environmental conditions: air temperature,25.0 ℃ (sunflower and tomato plants), 20.0 ℃ (broad bean plants); humidity, 70% RH; illumination, 0 to 40 klx. (Omasa et al., 1983b)

g_S also varied. Since all regression curves were concentrated near the origin of the coordinate axes, the transpiration from the cuticle of these plants was negligible in comparison with that from the stomata. Table 1 shows the density of the stomata (n_S) and the mean value of l_{bmax} [$E(l_{bmax})$] in the same areas as Fig. 12. From Fig. 12 and Table 1, we could see that g_S was dependent not only upon l_a but also upon n_S. The maximum values of g_S of the sunflower plants with large n_S and $E(l_{bmax})$ were larger than those of the broad bean and tomato plants with small n_S or $E(l_{bmax})$. Figure 13 shows the relationships between the degree of stomatal opening ($\bar{k}_l$) and g_S obtained from Fig. 12 and Table 1. From the ratio $E(l_a)/E(l_{bmax})$, where $E(l_a)$ was the mean value of l_a, $\bar{k}_l$ could be calculated and was found to decrease at a given value of g_S in the order of: tomato adaxial epidermis > broad bean adaxial epidermis >

Table 1 Density of stomata (n_s) and mean value of l_{bmax} [$E(l_{bmax})$] in the same area as Fig.12. (Omasa et al.,1983b)

Plant species	Kinds of epidermis	Density of stomata (n_s) (pieces/mm^2)	Mean value of l_{bmax} [$E(l_{bmax})$] (μm)
Sunflower	adax	86.6	30.1
	abax	76.8	35.9
Broad bean	adax	17.7	31.1
	abax	34.7	31.9
Tomato	adax	24.8	11.6
	abax	77.8	14.5

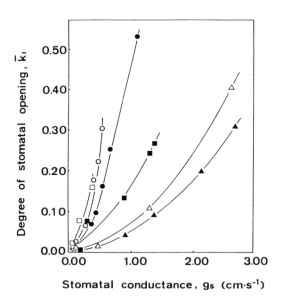

Fig.13 Relationships betwwn $\bar{k}_l$ and g_s obtained from Fig. 12 and Table 1; $\bar{k}_l$ is degree of stomatal opening expressed by the ratio $E(l_a)/E(l_{bmax})$, where $E(l_a)$ is the mean value of l_a. Symbols are the same as those in Fig. 12. (Omasa et al., 1983b)

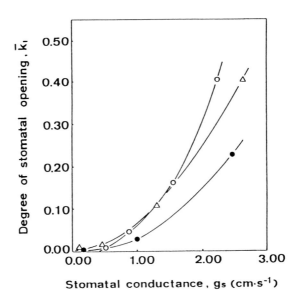

Fig.14 Relationships between $\bar{k}_1$'s measured at different leaf positions of sunflower plants and their stomatal conductances (g_S). O ,young leaf; Δ ,middle leaf; ● ,old leaf. Environmental conditions: air temperature, 25.0 °C; humidity 70% RH; illumination, 0 to 40 klx (Omasa 1990).

broad bean abaxial epidermis > tomato abaxial epidermis > sunflower adaxial epidermis > sunflower abaxial epidermis.

Figure 14 shows relationships between $\bar{k}_1$'s measured at different leaf positions of sunflower plants and their stomatal conductance (g_S) and Table 2 is n_S and $E(l_{bmax})$ in the same area as Fig. 14. The growth decreased n_S and increased $E(l_{bmax})$. As a result, the regression curve changed; the $\bar{k}_1$ of a young leaf was about twice as large as that of an old leaf at each given value of g_S. The transpiration from the cuticule was negligible in the mature and old leaves but the young leaf had some transpiration.

The stomatal conductance is dependent upon the structure of stomata as well as the aperture (Meidner and Mansfield, 1968). For example, the conductance of stomata with long effective length is smaller than that with short length in spite of the same aperture. Since the stomatal structure varies with plant species, age, and injury,we cannot exactly evaluate the aperture from the conductance (Meidner and Mansfield, 1968; Omasa et al., 1985).

Table 2 Density of stomata (n_s) and mean value of l_{bmax} $[E(l_{bmax})]$ in the same area as Fig.14 (Omasa 1990).

Leaf age	Leaf area (cm^2)	Density of stomata (n_s) (pieces/mm^2)	Mean value of l_{bmax} $[E(l_{bmax})]$ (μm)
Young	158	115.2	19.4
Middle	197	86.6	30.1
Old	67	53.0	43.0

V. CONCLUSIONS

Porometry or gravimetric measurements of transpiration do not provide the real aperture of stomata and the information for cell form and surface conditions. They also involve the risk of losing important information about the stomatal response because they average the behavior of many stomata. Therefore, we need to directly observe the individual stomata and their neighboring cells in order to study stomatal movement.

In this paper we described our remoto-control light microscope system for continuous observation of the guard and epidermal cells of attached leaves. This system was effectively used to observe changes in stomatal aperture and epidermal cells to environmental stimuli, such as illumination change, water deficit and exposure to air pollutants. Since this system is also effective for observing many intact stomata because of its easy and rapid operation, we could use it to analyze the relationship between stomatal aperture and conductance under the plant's growing conditions.

We gratefully acknowledge the helpful suggestions of Professors P.J. Kramer, B.R. Strain, M. Onoe and Y. Hashimoto.

REFERENCES

Burrows, F.J., and Milthorpe, F.L. (1976) Stomatal

conductance in the control of gas exchange. In "Water Deficits and Plant Growth Vol. 4" (T.T. Kozlowski, ed.), p. 103 Academic Press, New York.

Hashimoto, Y., Ino, T., Kramer, P.J., Naylor, A.W., and Strain, B.R. (1984) Dynamic analysis of water stress of sunflower leaves by means of a thermal image processing system. Plant Physiol. 76:266.

Hashimoto, Y., and Niwa, N. (1978) Image processing of leaf information. Proc. Joint Conf. Image Technol. Jpn. 9:51. (in Japanese).

Heath, O.V.S. (1959) The water relations of stomatal cells and the mechanisms of stomatal movement. In "Plant Physiology Vol. 2" (F.C. Steward, ed.), p.193. Academic Press, New York.

Horler, D.N.H., Barber, J., and Barringer, A.R. (1980) Effects of cadmium and copper treatments and water stress on the thermal emission from peas (Pisum sativum L.): Controlled environment experiments. Remote Sens. Environ. 10:191.

Kappen L., Andresen G., and Losch R. (1987) In situ observations of stomatal movements. J. Exp. Bot. 38:126.

Meidner, H., (1981) Measurements of stomatal aperture and responses to stimuli. In "Stomatal Physiology" (P.G. Jarvis and T.A. Mansfield, eds.), p. 25. Cambridge University Press, Cambridge.

Meidner, H., and Mansfield, T.A. (1968) "Physiology of Stomata". McGraw-Hill, London.

Monzi, M. (1939) Die Mitwirkung der Stomata-Nebenzellen auf die Spältoffnungsbewegung. Jpn. J. Bot. 9:373.

Omasa, K. (1990) Study on changes in stomata and their surrounding cells using a nondestructive light microscope system: Responses to air pollutants. J. Agr. Met. 45:251. (in Japanese and English summary)

Omasa, K., Abo, F., Aiga, I., and Hashimoto, Y. (1981a) Image instrumentation of plants exposed to air pollutants -Quantification of physiological information included in thermal infrared images. Trans. Soc. Instrum. Control Eng. Jpn. 17:657 (in Japanese and English summary) also in Res. Rep. Natl. Inst. Environ. Stud. Jpn. 66:69 (1984) (in English translation).

Omasa, K., and Aiga, I. (1987) Environmental measurement: Image instrumentation for evaluating pollution effects on plants. In "Systems and Control Encyclopedia. Vol 2" (M.G. Singh, ed.), p. 1516. Pergamon Press, Oxford.

Omasa, K., Aiga, I., and Hasahimoto, Y. (1983a) Image instrumentation for evaluating the effects of air pollutants on plants. In "Technological and Methodological Advances in Measurement Vol. 3" (G.

Striker, K. Havrilla, J. Solt, and T. Kemeny, eds.). p. 303. North Holland Publishing Co., Amsterdam.

Omasa, K., Hashimoto, Y., and Aiga, I. (1981b) A quantitative analysis of the relationships between SO_2 or NO_2 sorption and their acute effects on plant leaves using image instrumentation. Environ. Control Biol. 19:59.

Omasa, K., Hashimoto, Y., and Aiga, I. (1981c) A quantitative analysis of the relationships between O_3 sorption and its acute effects on plant leaves using image instrumentation. Environ. Control Biol. 19:85.

Omasa, K., Hashimoto, Y., and Aiga, I. (1983b) Observation of stomatal movements of intact plants using an image instrumentation system with a light microscope. Plant Cell Physiol. 24:281.

Omasa, K., Hashimoto, Y., Kramer, P.J., Strain, B.R., Aiga, I. and Kondo, J. (1985) Direct observation of reversible and irreversible stomatal responses of attached sunflower leaves to SO_2. Plant Physiol. 79:153.

Omasa, K., and Maruyama, S. (1990) Study on changes in stomata and their surrounding cells using a non-destructive light microscope system: Responses to changes in water absorption through roots. J. Agr. Met. 45:259 (in Japanese and English summmary)

Omasa, K., and Onoe, M. (1984) Measurement of stomatal aperture by digital image processing. Plant Cell Physiol. 25:1379.

Omasa, K., Tajima, A., and Miyasaka, K. (1990) Diagnosis of street trees by thermography: Zelkova trees in Sendai city. J. Agr. Met. 45:271 (in Japanese and English summary)

Stålfelt, M.G. (1959) The effect of carbon dioxide on hydroactive closure of the stomatal cells. Physiol. Plant. 12:691.

Schurer, K. (1975) Thermography in agricultural engineering. Bibl. Radiol. 6:249.

Shiraishi, M., Hashimoto, Y., and Kuraishi, S. (1978) Cyclic variations of stomatal aperture observed under the scanning electron microscope. Plant Cell Physiol. 19:637.

Turner, N.C., and Heichel, G.H.. (1977) Stomatal development and seasonal changes in diffusive resistance of primary and regrowth foliage of red oak (Quercus rubra L.) and red maple (Acer rubrum L.). New Phytol. 78:71.

DIGITAL PROCESSING OF PLANT IMAGES SELECTED BY SPECTRAL CHARACTERISTICS OF REFLECTANCE FOR EVALUATION OF GROWTH

Hiromi Eguchi

Biotron Institute
Kyushu University 12,
Fukuoka 812, Japan

I. INTRODUCTION

Digital image processing of plants has been used for non-destructive and on-line measurement of plant growth and vigor (Eguchi and Matsui, 1977; Eguchi *et al.*, 1979, 1982; Hashimoto *et al.*, 1980; Matsui and Eguchi, 1978). Information from these measurements might be usable for optimizing the environment. However, the image taken in the glass room or in the greenhouse contains a background of many kinds of light from reflecting surfaces. This background radiation complicates the use of imagery for routine control of plant growth. Therefore, it is necessary to develop a more reliable method for image processing in order to clearly separate plant images from background images.

This paper describes a technique of digital processing for selecting plant images on the basis of spectral characteristics of reflected light. Plant growth may then be evaluated by using binary images.

Measurement Techniques in Plant Science

II. MATERIAL AND METHODS

Cucumber plants (*Cucumis sativus* L. var. Hort. Chojitsu-Ochiai) were potted in Vermiculite moistened with nutrient solution and grown under air conditions of $22\pm1°C$ and $25\pm1°C$ at $70\pm5\%$ relative humidity in a phytotron glass room. Six cucumber plants were set in an array in the phytotron glass room. A TV camera (Silicon vidicon camera, HV-13287, Hitachi Electronics, Ltd.) was fixed at a constant distance from plants and at a camera angle of $55°$ to the normal (Figure 1). Camera operation was controlled by the computer signal and synchronized with illumination to the plants: The plants were illuminated for 3 min once a day at 4:00-4:03 a.m. (before sunrise) by tungsten light (5 incandescent lamps of 500W) with an intensity of 10 W m^{-2} (in visible light region), and at the same time, the plant image was taken by the TV camera. The image was displayed on the TV monitor and digitized into video memory (Video densitometer 520A, NAC Inc.) with 240 (lines) x 256 (pixels) meshes. The digitized image was transmitted to CPU (TOSBAC-40C, Tokyo Shibaura Electric Co., Ltd.) and filed in disk memory (BLK 4201Z, Tokyo Shibaura Electric Co., Ltd.). The digitized image was denoted as a matrix (M), where the arrangement of the elements (x_{ij}) was expressed as:

$$M = \begin{bmatrix} x_{11} & x_{12} & \cdots x_{1n} \\ x_{21} & x_{22} & \cdots x_{2n} \\ & \cdots\cdots \\ x_{m1} & x_{m2} & \cdots x_{mn} \end{bmatrix}, \; m = 240, \; n = 256 \tag{1}$$

Each element (x_{ij}) was evaluated as reflectance (%) of the light and displayed as a digital symbol.

III. RESULTS AND DISCUSSION

Figure 2 shows a photograph taken by a still camera with panchromatic film. In this panchromatic image, there are many kinds of backgrounds brighter than the plants. The spectral reflectances of the background media appeared in

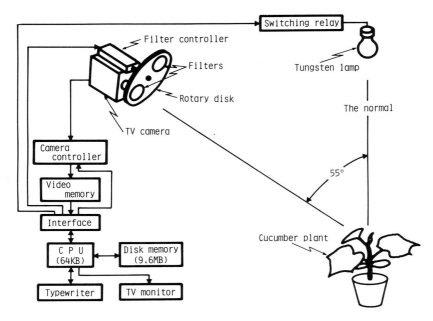

Figure 1. Schematic diagram of image processing system.

Figure 2. Panchromatic photograph of 25 day old plants taken by still camera in a phytotron glass room (air temperature of 22℃).

different patterns as shown in Fig. 3.   Most   of   them   were
flat in wave length regions of 400-1000 nm.   The brighter
background media of aluminium, stainless steel and the   steel
painted   in   olive   reflected   50-70%   of   the   light.   The
reflectance of concrete was almost constant at about 30%   in
this wave length region. In black polyethylene film, sandy
and clay soils, the reflectances were remarkably lower   than
others, being less than 20%. On the other hand, the spectral
reflectance of the cucumber leaf was characterized by a small
peak at about 15% at about 550 nm, a dip to about 6% in   670-
680 nm and a broad peak at 60-70% in   the   infrared   region.
Thus, the leaf reflectance appeared in a specific spectrum,
but   this   reflectance   and   the   background   reflectances
overlapped   each   other.   These   patterns   indicate   that
selection of   the   plant   image   could   be   confused   by   the
background reflectances in any of the wave length regions   of
light.   Therefore, for clear selection of   the   plant   image
from the background, it is   necessary   to   use   two   or   more
images taken in   different   wave   length   regions   for   image
processing. In this experiment, two wave bands were employed
for   selecting   the   plant   images,   on   the   basis   of   the
characteristics of spectral reflectance of the leaf; one   was
the red band (R; 670nm, absorbing region), and the other   was
the infrared band (I; 900nm highly reflecting region).

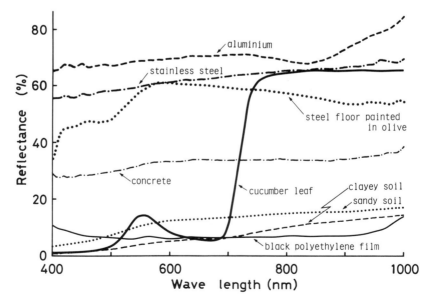

Figure 3. Patterns of spectral reflectances of the
cucumber leaf and background media in a phytotron glass room.

To take two images in respective wave bands of R and I, two kinds of interference filters of F_R and F_I (Eguchi *et al.*, 1979) were used, where the peaks of transmittances were at 670 nm and at 900 nm, respectively. Figure 4 shows the spectral characteristics of the filters, the tungsten light, the camera sensor and neutral test card with 90% reflectance (Eastman Kodak Co.). As observed in the spectral transmittances, the filters of F_R and F_I made it possible to take the two images reliably in the respective wave bands of R and I. These filters were attached to a rotating disk in front of the camera lens and alternated with each other by turning the disk which was manipulated by the computer signal synchronizing with the camera drive, as illustrated in Fig.1. Thus, R and I images were automatically taken in R and I bands, respectively.

Figure 5 shows photographs of R (a) and I (b) images displayed on the TV monitor. In the R image, the plants appeared darker with the partially brighter background. In the I image, the plants were brighter, but there were many brighter backgrounds. These R and I images were digitized

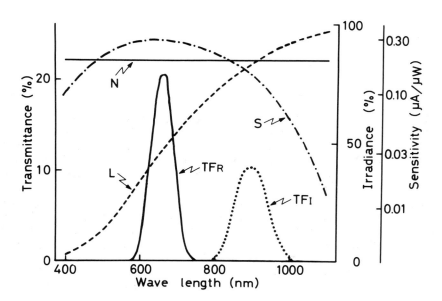

Figure 4. Spectral characteristics of transmittances of filters (*TF*$_R$ and *TF*$_I$), light energy (*L*), camera sensitivity (*S*) and reflectance (*N*) of neutral test card (90% reflectance).

and denoted as M_R = $[a_{ij}]$ and M_I = $[b_{ij}]$, respectively, where respective elements of a_{ij} and b_{ij} were denoted as the reflectances (%) on the basis of the calibration by using the reflectances (18% and 90%) of the neutral test cards taken by the TV camera in the optical system used: The brightness in each image was affected by spectral characteristics of light energy, filter transmittance and camera sensitivity, and was calibrated to denote the reflected light intensity as the reflectance (%) by using two neutral test cards with 18% and 90% reflectances. Figure 6 shows digital displays of M_R (a) and M_I (b). As found in the displays, digits in the plant area appeared lower in M_R and higher in M_I, but the plant images were not clear in either M_R or M_I with the complicated backgrounds. Thus, it was difficult to specify the plant image in either image.

On the other hand, the difference in reflectance between R and I bands was less than 25% in the backgrounds and was remarkably larger in the plants as shown in Figs. 3 and 6. Therefore, a matrix (M_S) was derived from M_I - M_R = $[b_{ij} - a_{ij}]$ for selecting the plant image and given as

$$M_S = [c_{ij}] \tag{2}$$

where

$$c_{ij} = \begin{cases} 0 & (b_{ij} - a_{ij} \leq 25\%) \\ 1 & (b_{ij} - a_{ij} > 25\%) \end{cases} \tag{3}$$

In M_S, elements were denoted as '1' in the plants and '0' in backgrounds by setting a threshold level of the reflectance at 25% as defined in Eq. 3. Figure 7 shows the digital display of M_S derived from M_R and M_I in Fig. 6, where the element, c_{ij} = 0 is displayed as blank, as illustrated on the scale. The background were completely eliminated, and the images of six cucumber plants were clearly found in arrays of '1'. Thus, M_S made it possible to select the plant image separated from background noises in the glass room. Figure 8 shows digital displays of M_S taken at different growing stages. The plant image enlarged with growing stages, and the plant growth was found in increase in number of the digits of '1'. So, the sum (P) of the elements of M_S was calculated from

$$P = \sum_{i=1}^{m} \sum_{j=1}^{n} c_{ij} \tag{4}$$

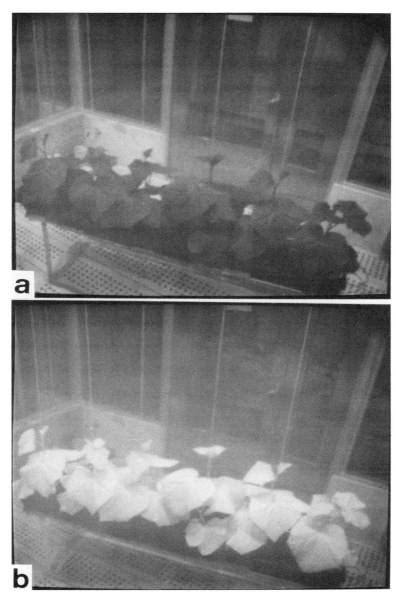

Figure 5. Photographs of R image (a) and I image (b) displayed on a TV monitor (the plants were grown at air temperature of 22℃, and the images were taken when the plants were 25 days old).

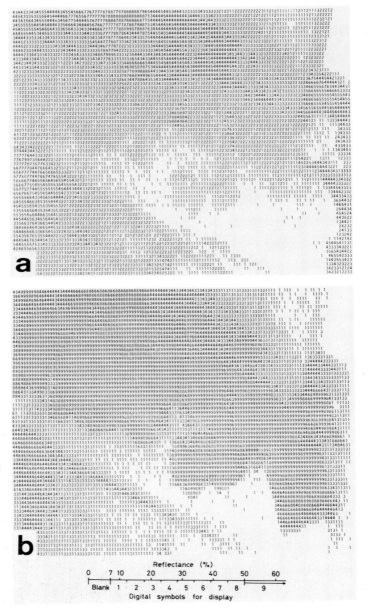

Figure 6. Digital displays of M_R (a) and M_I (b) derived from R and I images shown in Fig. 5; reflectances are symbolized as illustrated on the scale (odd rows and columns of matrix elements are omitted).

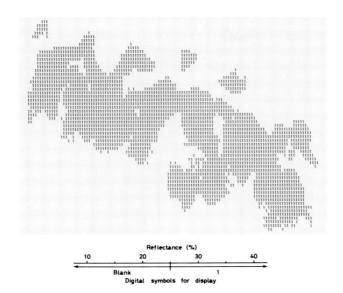

Figure 7. Digital display of M_S calculated from M_R and M_I displayed in Fig. 6; reflectances are symbolized as illustrated on the binary scale (odd rows and columns are omitted).

Figure 9 shows relationship between P and measured fresh weight of a plant. The distribution of the measured fresh weights was represented by a regression curve of P at 0.1% significance level even when the plants were grown at different air temperatures of 22℃ and 25℃, as reported in the previous papers (Eguchi and Matsui, 1977; Matsui and Eguchi, 1978). The regression equation of fresh weight (F_W, g) of a plant on P was given by

$$F_W = 1.03 \times 10^{-7} P^2 + 6.21 \times 10^{-3} P - 5.43 \qquad (5)$$

Figure 10 shows patterns of cucumber plant growth evaluated by the fresh weight calculated from Eq. 5 and the measured fresh weights. The calculated fresh weights were near the measured ones. The calculated fresh weight of a plant grown at 25℃ began to increase faster at 27 days and finally became larger than that at 22℃. Thus, the effect of air temperature on plant growth was clearly observed in

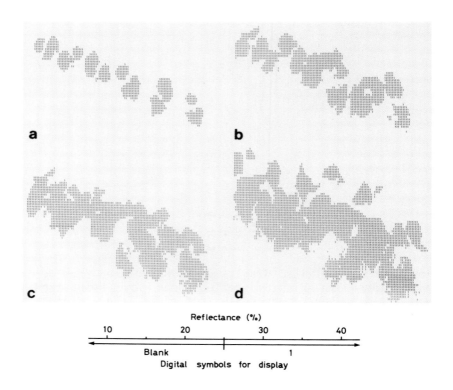

Figure 8. Digital displays of M_S, where the plants were grown at air temperature of 22°C, and the images were taken when the plants were 17 (a), 20 (b), 23 (c) and 26 (d) days old, respectively; reflectances are symbolized as illustrated on the binary scale (odd rows and columns of matrix elements are omitted).

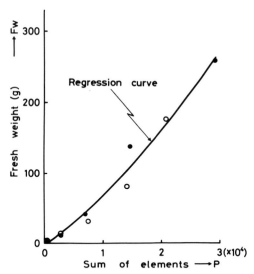

Figure 9. Relationship between *P* and measured fresh weight of a plant, where the plants were grown at the respective air temperatures of 22°C (○) and 25°C (●) and sampled at different growing stages.

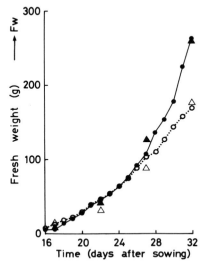

Figure 10. Time course of measured and calculated fresh weights of each plant grown at different air temperatures; the measured fresh weights were obtained by weighing the plants grown at 22°C (△) and 25°C (▲), and the calculated fresh weights were obtained from the images of plants grown at 22°C (○) and 25°C (●).

the patterns of the calculated fresh weights. This fact indicates that P can be used as a reliable parameter for evaluation of plant growth through the mathematical representation.

Thus, the digital image processing in two wave bands of red (R) and infrared (I) made it possible to select clear plant images separated from the background in the glass room and to represent the plant growth through time. In the previous paper (Eguchi *et al.*, 1979), the author reported that the images taken in respective bands of R and I are useful for analysis of vigor of plants. From these results, it could be estimated that digital processing of R and I images can be used for evaluation of plant growth and also used for estimation of plant vigor in an on-line system.

REFERENCES

Eguchi, H . and Matsui, T. (1977). Computer control of plant growth by image processing. II. Pattern recognition of growth in on-line system. *Environ. Control in Biol*. 15:37-45.

Eguchi, H., Hamakoga, M. and Matsui, T. (1979). Computer control of plant growth by image processing. IV. Digital image processing of reflectance in different wave length regions of light for evaluating vigor of plants. *Environ. Control in Biol*. 17:67-77.

Eguchi, H., Hamakoga, M. and Matsui, T. (1982). Digital image processing in polarized light for evaluation of foliar injury. *Environ. Exp. Bot*. 22:277-283.

Hashimoto, Y., Ioki, K. Kaneko, S., Funada, S. and Sugi, J. (1980). Process identification and optimal control of plant growth (VIII). Relationship between distribution of leaf temperature and stomatal aperture. *Environ. Control in Biol*. 18:57-65 (in Japanese with English summary).

Matsui, T. and Eguchi, H. (1978). Image processing of plants for evaluation of growth in relation to environment control. *Acta Hort*. 87:283-290.

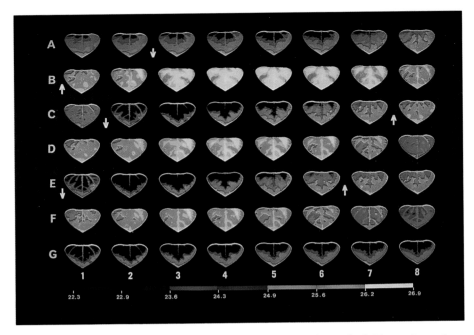

Figure 6 Distributional patterns of leaf temperature in sunflower leaf. Picture imaged at 2-min intervals. For details see the text discussion on page 382.

LEAF TEMPERATURE BASED ON
IMAGE PROCESSING

Yasushi Hashimoto

Department of Biomechanical System
Ehime University
Tarumi, Matsuyama, Ehime, Japan

I . INTRODUCTION

A number of methods have been used to measure plant water
status under several stress conditions (Kramer, 1983).
These include measurements of leaf water content, relative
water content, and leaf water potential by means of the
pressure chamber and the phychrometric methods (Boyer, 1969).
Stomatal aperture generally is measured by porometers (Hsiao
and Fisher, 1975). The drawback, however, is that there is
no satisfactory method of monitoring changes in leaf water
status of all parts of a leaf blade. Localized water
deficits may well foreshadow the necrosis and localized
spotting often observed following imposition of stress.
It was reported that the rates of photosynthesis and
respiration were significantly higher at the base than at the
tip of tobacco leaves, and the osmotic potential was lower
(Slavik, 1963). Differences in water status in various parts
of large leaves were also observed (Rawlins, 1963).
Differences have been noted as well in basal, mid-, and
distal segments of maize leaves (Michelena and Boyer, 1982).
Only a few attempts have been made to measure variation in
temperature across a leaf. The observed differences have
been attributed to differential transpiration rates (Cook et
al., 1964). Many studies on leaf temperature have been
undertaken from the standpoint of the physical energy budget

Measurement Techniques in Plant Science

(Gates, 1980), but variations in leaf temperature caused by differential transpiration as a result of localized stomatal behavior differences have rarely been taken into consideration. It is very desirable to know the temperature of various regions of leaves because temperature can serve as an indicator of transpiration. If temperature measurements were combined with measurements of stomatal conductance or aperture, useful information would be provided concerning leaf water status as the leaf responds to environmental stresses.

Leaf temperature can be measured continuously and non-destructively with infrared (IR) thermometers (Fuchs and Tanner, 1966; Takiuchi and Hashimoto, 1977). Distributional patterns of surface temperature also can be measured with a scanning IR thermal camera (Cetas, 1978) and some physiological ecological information was obtained by thermography (Pieters, 1975). Described herein is an improved method that permits non-destructive monitoring of temperature changes over an entire leaf. This is done by means of an IR mechanical scanning camera connected to an image interface and a computer that processes the signals to produce a thermal image of a leaf (Hashimoto and Niwa, 1978; Hashimoto et al., 1982, 1984).

II. MEASUREMENT OF LEAF TEMPERATURE

We can measure the surface temperature of leaf by an IR thermometer. Its principle is fundamentally based on Plank's laws. Let's suppose that the outward flux of thermal energy to the infrared thermometer is described by $L(\lambda,T)$ as follows:

$$L(\lambda,T) = \varepsilon(\lambda,T)L_b(\lambda,T) + \rho(\lambda,T)E(\lambda,T_s)/\cos\beta \quad (1)$$

where $\varepsilon(\lambda,T)$ is the spectral emissivity of the leaf surface at the surface temperature (T), $L_b(\lambda,T)$ is the value of Planck's energy distribution law for the surface temperature, $\rho(\lambda,T)$ is spectral reflectance of the leaf, $E(\lambda,T_s)$ is the radiant energy flux incident to the surrounding surface temperature (T_s), and the angle β is formed between the normal of the leaf and measuring line. The characteristics of the filter of the infrared thermometer may be approximated as follows.

$$f(\lambda) = 0 \text{ , for } \lambda < \lambda_1 \text{ , } \lambda > \lambda_2$$

$$f(\lambda) \neq 0 \text{ , for } \lambda_1 \leq \lambda \leq \lambda_2$$

Then, output voltage of the infrared thermometer is described by V as follows.

$$V = A \omega \int_{\lambda_1}^{\lambda_2} f(\lambda)s(\lambda)L(\lambda ,T) \, d\lambda \qquad (2)$$

where $A\omega$ is the amplification factor involved in the amplifier of the thermometer, is the geometric coefficient involved in the optical characteristics of the input, and $s(\lambda)$ is the efficiency of transducer.

On the other hand, we may define $\bar{\varepsilon}(T)$ and $F(T)$ as follows.

$$\bar{\varepsilon}(T) = \frac{\displaystyle\int_{\lambda_1}^{\lambda_2} \varepsilon(\lambda ,T)f(\lambda)s(\lambda)L_b(\lambda ,T) \, d\lambda}{\displaystyle\int_{\lambda_1}^{\lambda_2} f(\lambda)s(\lambda)L_b(\lambda ,T) \, d\lambda} \qquad (3)$$

$$F(T) = A \omega \int_{\lambda_1}^{\lambda_2} f(\lambda)s(\lambda)L_b(\lambda ,T) \, d\lambda \qquad (4)$$

If we substitute for $L(\lambda ,T)$ given in Eq. (1) into Eq.(2), and use he relation given in Eq. (3) and (4), then we have

$$V = \bar{\varepsilon}(T)F(T)$$

$$+ A \omega \int_{\lambda_1}^{\lambda_2} \rho(\lambda ,T)f(\lambda)s(\lambda)E(\lambda ,Ts)/\cos \beta \qquad (5)$$

Let's suppose that the measured material is the black body, then we have easily

$$\bar{\varepsilon}(T) = 1 \text{ , } \rho(\lambda ,T) = 0$$

Therefore, we can obtain

$$V = F(T) \qquad (6)$$

In the general case, we can know the characteristics of F(T) based on the calibration curve obtained from actual observation of the instrument.

On the other hand, E(λ,Ts) may be an unknown factor. But ρ (λ,T) is found to be nearly zero for $8 \mu m \leqq \lambda \leqq 14 \mu m$ in Figure 1 obtained from the actual observations of tobacco leaves. Therefore, the second term of Eq. (5) may be negligible. Thus, we obtain V as follows.

$$V = \bar{\varepsilon} (T)F(T) \tag{7}$$

In general, leaves are not black bodies. Therefore $\bar{\varepsilon}$ (T) should be decided from another experiment. In case of tobacco leaves, $\bar{\varepsilon}$ (T) is about 0.99. Most leaves are found to be from 0.95 to 0.99. Thus, we can obtain the temperature of· leaf surface based on Eq. (7). By using the infrared thermometer (Matsushita ER2005), we can easily measure the leaf temperature with the accuracy of 0.1°C (Takiuchi and Hashimoto, 1977).

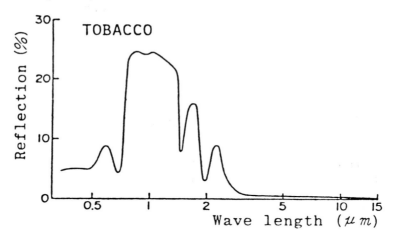

Figure 1.   Spectral reflectance of tobacco leaves.

III.   IMAGE PROCESSING SYSTEM

As explained in the previous section, the leaf temperature can be measured by an IR thermometer. Although we can obtain the accurate and noncontact temperature of the leaf surface by the IR thermometer, the sensor can monitor not the distribution of the leaf temperature, but the average temperature of a circle of 30 mm in diameter on the leaf.

The distribution of surface temperature is obtained by a thermography based on the same IR sensor as in the IR thermometer. In the thermography, a thermal image is obtained by scanning numerous spots over the surface as if they were obtained by numerous IR thermometers. The detector for the thermography is the thermal camera which was confirmed to accurately detect the temperature on the surface (Cetas, 1978). The maximum resolution in the thermography (for example, JTG-MD) is 0.05 C. One weak point, however, was that it produced a black and white photograph which was difficult to distinguish exactly. An attempt was made to estimate photosynthesis and stomatal aperture from leaf temperature measured by thermography (Pieters, 1975). However, the distributional patterns could not be printed clearly.

Thus, we designed a digital image processing system which was composed of a thermography and a computer system (Hashimoto and Niwa, 1978). The thermal camera (JTG-MD, Japan) is drivel with a mechanical scanner based on a moving mirror. It has a field of 10x10cm. The signals were fed to a computer (MELCOM 70; core, 64 Kbytes; disc, 5 Mbytes; and magnetic tape; Japan) through an image interface. Two s is enough time to obtain one image composed of 256x240 picture elements, and 4 s are required to store the information on

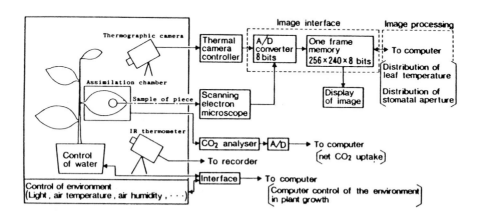

Figure 2. Schematic diagram of the thermal image processing system.

the disc via the image interface. Each picture element has 8
bits and the resolution of the thermal camera is 0.05°C.
Therefore, the integrated accuracy of one picture element in
the system is guaranteed to be no less than 0.1°C.

Figure 2 shows the system used for measurement and
control in the research. It consists of the image processing
system for measuring the pattern of leaf temperature
distribution, a scanning electron microscope to measure
stomatal aperture (Hashimoto et al., 1982), a system for the
measurement of photosynthesis, and a computer controlled
environmental system for the leaf cuvette and root (Hashimoto
et al., 1981).

IV. RELATION BETWEEN LOCALIZED LEAF
TEMPERATURE AND STOMATAL APERTURE

Localized leaf temperature is closely related to the
localized transpiration caused by stomatal behavior.
Therefore, it seems that there may be some relation between
leaf temperature and stomatal aperture.

Figure 3 shows the time course changes in average
temperature of lighted tobacco leaves following 5 hours of
darkness (Hashimoto et al., 1982). Oscillation of the
temperature after the start of irradiation may be explained
as follows: When the plant is exposed to light (570 μ E
$m^{-2}s^{-1}$), leaf temperature rises at first because of a sudden
gain in radiant energy. Later, the temperature begins to
fall as transpiration increases following stomatal opening.
However, transpiration later reached a plateau and leaf
temperature ceases to fall at C. Since water content of the
leaf decreases, transpiration is suppressed some time later,
and leaf temperature again starts to rise. Such oscillation

Table 1 shows the average value and standard deviation of
the opening area of the central pore of stomata, which is
obtained statistically from 30 stomata in different points of
the leaf. At C, leaves suffer from water deficit caused by
much transpiration, and almost all stomata seem to close.
But, except C, we can observe the distribution of stomatal
aperture. Furthermore, at D, it can be observed that the
stomata localized at lower temperature are more open than
those at higher temperature.

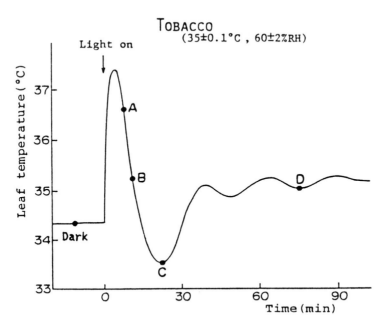

Figure 3. Time course changes of leaf temperature.
is damped gradually and reaches a stationary state at a
certain temperature. The time points in the chart shown as
Dark and A to D give the stomatal aperture obtained with the
image processing system based on scanning electron microscope
observation.

Table 1. Opening area of stomata obtained by image
processing system.

Point	Opening area (μm^2)
Dark	1.13 + 0.51
A	4.02 + 1.17
B	3.89 + 1.29
C	0.96 + 0.37
D	2.33 + 0.77

Figure 4 shows the relation between localized leaf
temperature and stomatal aperture obtained from sunflower
plant at the steady state just like D shown in Figure 3
(Hashimoto et al., 1982). In the left side of Figure 4, the
distributional pattern of the leaf temperature is marked with
the quaternary pattern composed of one set of (N, I, -,
blank), for N ≪ I ≪ - ≪ blank. In the middle of Figure 4,
images of stomata sampled from the localized sub-area marked
with A, B and C in the left quaternary pattern are seen. In
the right side, the opening areas of the central pore are
obtained from the images of stomata shown in the middle based
on the digital image processing. Therefore, it may be
evident that localized leaf temperature is closely related to
the stomatal aperture at the steady state under such the
controlled environment as in constant radiation and constant
wind.

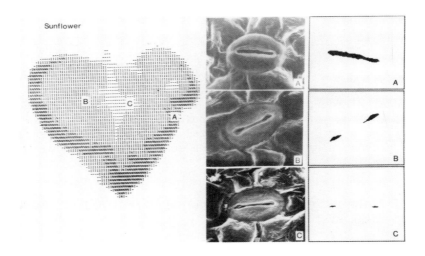

Figure 4. Relation between localized leaf temperature
and stomatal opening. Leaf temperatuare is shown by
localized temperataure regions of A, B, and C for A < B < C.

V. DISTRIBUTIONAL PATTERNS OF
LEAF TEMPERATURE

The experiments with the digital image processing system

indicate that in the steady state of leaves that are well
watered, the pattern of leaf temperature is closely related
to the stomatal aperture patterns over the leaf surface as
observed with a scanning electron microscope.

Furthermore, it might be clear that variation in
temperature over the leaf surface declined, when stomata were
kept open by treatment with kinetin (Hashimoto et al., 1980);
but that variation in temperature across the leaves increased,
when the stomata were closed with ABA (Hashimoto, 1982).

Distributional patterns of leaf temperature obtained
every 2 minutes from leaves of sunflower plants subjected to
increasing water stress showed that water deficit developed
first at the margins of leaves, accompanied by stomatal
closure and increase in temperature (Hashimoto et al., 1984).
Therefore, it should be kept in mind that the distributional
patterns of leaf temperature may be valuable information for
studying dynamic behavior in physiological ecology.

Now, let's attempt to improve the resolution of detecting
distributional patterns of leaf temperature and stomatal
opening. Figure 5 shows the average time course leaf
temperature obtained from the experiments of sunflower plants.

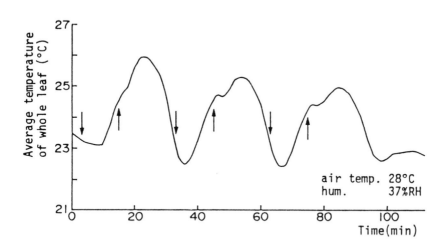

Figure 5. Cyclic changes of leaf temperature caused by
repeated water stress.
     ↑ : root zone is filled with water of hydroponics after
irrigating water on.
     ↓ : root zone is exposed to the air after draining
water off.

In the chart, it may be seen that the cyclic phenomena are caused by square wave manipulations of water supply for the roots in hydroponics. It seems evident that the cyclic patterns of the average leaf temperature reappear subjected to the cyclic water stress caused by the manipulations.

Figure 6 shows 56 patterns of leaf temperature distribution obtained from the thermal images every 2 min from the starting point in Figure 5. Let's suppose that A, B, .., and G mean the row of the figure and 1, 2, ..., and 7 in the line means the time course. For example, the pattern described with (A, 1) in Figure 6 means the pattern at the starting point shown in Figure 5; the pattern with (B, 3) means that at the 20 min point after the starting point in Figure 5; the pattern with (F, 1) means that at the 80 min point after the starting point in Figure 5; and so on. The distributional pattern is displayed in eight colors: white (26.9 - 26.2°C), yellow (26.2 - 25.6°C), pink (25.6 - 24.9°C), red (24.9 - 26.2°C), blue (24.3 - 23.6°C), green (23.6 - 22.9°C), dark blue (22.9 - 22.3°C), and black. Black denotes temperatures either over 26.9°C or under 22.3°C. (See the color plate version of Figure 6 facing page 373.) It is

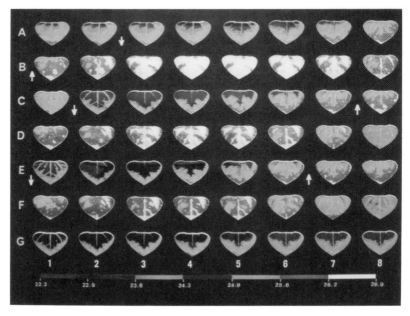

Figure 6.   Distributional patterns of leaf temperature in sunflower leaf.   Picture imaged at 2-min intervals.

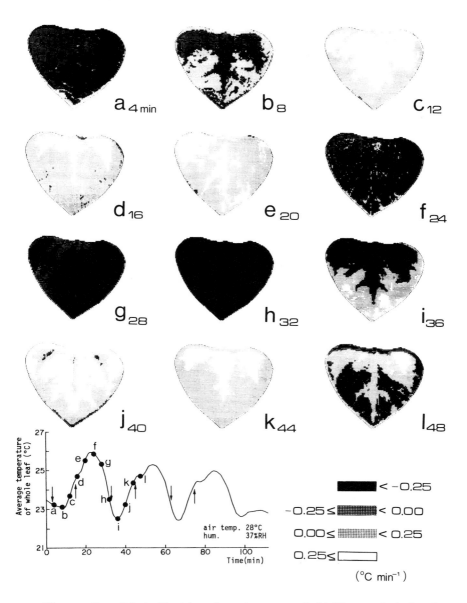

Figure 7. Distributional patterns of differential leaf temperature in sunflower leaf. The picture is shown by quaternary pattern of four differential values as depicted in the lower part of the right side in the figure. Frame a, b, ... to l correspond to the same character as in the chart shown in the lower part of the left side in the figure.

especially clear from (A, 3) to (B, 5) in Figure 6 that there
is a progressive increase in leaf temperature from the edge
inward. These findings can be interpreted to mean that a
water deficit develops initially at the edge of the leaf,
resulting in stomatal closure and a decrease of transpiration
with a concomitant rise in temperature. Next, the patterns
from (B, 6) to (C, 1) show that leaf temperature decreased
from near the edge of leaf as transpiration began to increase
following an increase in leaf water content. From (C, 2) and
(C, 3), we can observe that almost all parts of the entire
leaf are kept at lower temperature due to preceding adequate
water supply.

These data show that significant differences arise in
temperature patterns over the whole leaf during the
development of water stress. Thus, stomatal aperture and
transpiration within a single leaf may show considerable
variation over a short time span. Furthermore, we emphasize
that these patterns reappear when the leaves are subjected to
the cyclic water stress shown in Figure 5.

In order to discriminate the rate of changing temperature,
differential operation was done by the digital image
processing system. Figure 7 shows distributional patterns of
the differential value in leaf temperature obtained from the
images given in Figure 6. Frames a, b, c, ..., and l
correspond to a, b, c, ..., and l marked on the lower chart
which is the same chart as shown in Figure 5. At b in Figure
7, we can easily understand that water deficit develops from
the edge of the leaf, because the changing rate of the
temperature in the sub-area is a positive value. We can
infer a lot of dynamic behavior from these patterns. The
results of these experiments and their image processing
indicate that there are significant differences in the
temperature and water status of different parts of a leaf.
This makes it important to determine causes of such behavior
and decide in what part of a leaf the temperature and water
status should be measured. The thermo-imaging method can be
useful in monitoring short term temperature changes occurring
in leaves undergoing water, chilling, and other stresses.

REFERENCES

Boyer, J.S., (1969). Measurement of water status of plants.
    Ann. Rev. Plant Physiol. 20:351-364.

Cetas.D.M. (1978). Practical thermometry with a thermographic camera. Rev.Sci. Instrum. 49:245-254.

Cook, G.D., Leopold, A.C., and Dixon, J.R. (1964). Transpiration: Its effects on plant leaf temperature. Science 144:546-547.

Fuchs, M., and Tanner, C.B. (1966). Infrared thermometry of vegetation. Agron. J. 58:597-601.

Gates, D.M. (1980). Biophysical Ecology. Springer-Verlag, New York.

Hashimoto, Y. (1982). Dynamic behavior of leaf temperature. Biol. Sci. (Tokyo). 34:68-75 (in Japanese).

Hashimoto, Y. and Niwa, N. (1978). Image processing of leaf information. Joint Conf. Image processing. 9:51-54 (in Japanese).

Hashimoto, Y. Morimoto, T., and Funada, S. (1980). Computer processing of speaking plant for climate control and computer aided cultivation. Acta Hortic. 115:317-325.

Hashimoto, Y. Morimoto, T., Funada, S., and Sugi, J. (1981). Optimal control of greenhouse climate by the identification of water deficiency and photosynthesis in short-term plant growth. In "Control Science and Technology for The Progress of Society" (H. Akashi, ed.) pp. 3621-3626. Pergamon Press, Oxford.

Hashimoto, Y. Morimoto, T., and Funada, S. (1982). Image processing of plant information in the relation between leaf temperature and stomatal aperture. In Technological and methodological Advances in Measurement" (G. Striker, K. Havrilla, J. Solt and T. Kemeny, eds.). Vol. 3, p. 313-320. North-Holland, Amsterdam.

Hashimoto, Y. Ino, T., Kramer, P.J., Naylor, A.W., and Strain, B.R., (1984). Dynamic analysis of water stress of sunflower leaves by means of a thermal image processing. Plant Physiol. 76:266-269.

Hsiao, T.C., and Fisher, R.A., (1975). Mass flow meters. Wash. State Univ. Coll. Agric. Res. Cent. Bull. 809, 5-11.

Kramer, P.J. (1983). Water Relations of Plants. Academic Press, New York.

Michelena, V.A., and Boyer, J.S. (1982). Complete turgor maintenance at low water potentials in the elongative region of maize leaves. Plant Physiol. 69:1145-1149.

Pieters, C.A. (1975). Thermography and plant physiology. Bibl. Radiol. 6:210-217.

Rawlins, S.L. (1963). Resistance to water flow in the transpiration stream. Conn Agric. Exp. Stn. Bull. (New Haven) 664, 69-85.

Slavik, B., (1963). On the problem of the relationship
    between hydration of leaf tissue and intensity of
    photosynthesis and respiration. In "The Water Relations
    of Plants" (A.J. Rutter, F.H. Whitehead, Eds.). pp.
    225-234. John Wiley & Son, New York.
Takiuchi, and Hashimoto, Y. (1977). Measurement of leaf
    temperature by means of infrared thermometer in
    connection with plant physiological information. Trans.
    Soc. Instr. Cont. Eng. 13, 482-488 (in Japanese).

IMAGE ANALYSIS OF CHLOROPHYLL
FLUORESCENCE IN LEAVES

Kenji Omasa
Ken-ichiro Shimazaki[1]

National Institute for Environmental Studies,
Tukuba, Ibaraki 305, Japan

I. CHLOROPHYLL FLUORESCENCE

When a dark-adapted leaf was illuminated, the intensity
of chlorophyll a fluorescence shows complicated changes with
time during the illumination. These transient changes generally show the fast rise (F_O), and intermediary level (F_I)
with a small decline (F_D),then gradual increase to a maximum
(F_P), and again decline slowly with secondary maximum (F_M)
via a minimum (F_S), and reached to the steady-state level
(F_T). These changes in fluorescence intensity are called
chlorophyll fluorescence induction (CFI) or Kautsky effect
(Kautsky et al. 1960). According to Papageorgiou (1975),
these changes are termed OIDPSMT transients (Fig.1).

Since chlorophyll fluorescence is the re-emission of the
light energy which is once trapped by antenna chlorophyll
and had not been utilized for the photochemical reaction
(Duysens and Sweers 1963, Murata et al. 1966), the
fluorescence intensity virtually reflects the magnitude of
the photochemical reaction in the complementary fashion.
Fluorescence intensity is low when the primary electron
acceptor of PS II (Q_A) exsits in the oxidized state, and
high in its reduced state in fast phase of fluorescence
transient, in particular, because chlorophyll fluorescence
intensity is regulated by the availavility of the primary
electron acceptor of PSII(Q_A).

[1]Present address: College of General Education, Kyushu
University, Ropponmatsu, Fukuoka 810, Japan

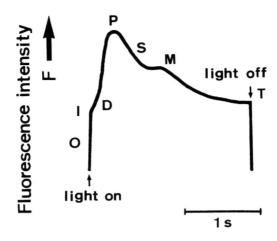

Fig.1 Typical CFI curves

The fluorescence transient from O to P takes less than several second when the leaf is illuminated with a moderate light, and the transient is called the fast phase. The fast phase of the fluorescence transient (OIDP) is closely correlated with redox reactions of photosystem II (PSII); OI represents the photoredcution of Q_A by PSII reaction center, ID dip represents rapid oxidation of Q_A by PSI (Munday and Govindjee 1969), and DP indicates the photoreduction of Q_A by the PSII reaction linked to the water-splitting enzyme system (Munday and Govindjee 1969). The fluorescence transient from P to T requires more than minutes, and is called the slow phase. The fluorescence intensity change in the slow phase is not so simple as the fast phase of fluorescence; the change include two components, photo-chemical quenching (closely correlated with the oxidation state of Q_A) and non-photochemical quenching (mainly due to trans-thylakoid pH gradient). Recently, the distinction of these two component of fluorescence quenching has been developed by modulated light fluorometers (Schreiber et al. 1986).

Chlorophyll fluorescence transient can be measured in a short time, and measured without destruction of the plant materials (e.g. isolation of chloroplasts), thereby pro-viding continuous measurement of phtotosynthetic activity which changes with time using the same leaf. The measure-ment is also able to detect the injury that had taken place in vivo but had been removed during the preparation of the chloroplasts form the leaves. Therefore, the measurement of chlorophyll fluorescecne in plant leaves in situ has been developed as a sensitive and non-destructive assay for the functional state of the photosynthetic apparatus under various kinds of stresses, such as high (Schreiber and Berry 1977), and low-temperature (Smillie and Hetherington 1983) and water stresses, photo-inhibition (Demmig and Björkman 1987), UV-irradiation (Shimazaki et al. 1988) and air pollutions (Schreiber 1978, Shimazaki et al. 1984, Omasa et al. 1987).

II. HETEROGENEITY OF THE PHOTOSYNTHETIC ACTIVITY ON THE LEAF

Recent evidence indicate that photosynthetic activity does not always show uniform distribution over the same leaf area but varies with the area to area, and sometimes forms even the "patchness" of the photosynthetic activity over the leaf under the stresses. Clear patchness of the photosynthetic activity was shown in Helianthus annus, a heterobaric leaf, as a heterogenous accumulation of starch when abscicic acid was applied to the leaves (Terashima et al 1988). This may probably be due to that a group of neighboring stomata close in localized leaf area, which limits the diffusion of CO_2 thereby inhibiting the photosynthetic CO_2 fixation in limitd leaf area. The heterogeneity of stomatal movement has been demonstrated by the different responses of the individual stomata, whose movements were measured directly by image instrumentatin system with a light microscope by Omasa et al. (1985), in addition to indirect imaging of stomatal movements using thermal image instrumentation system (Omasa et al. 1981a,b, Hashimoto et al 1984).

Moreover, photosynthetic activity itself does not respond uniformly when the plants are under environmental stresses. Heterogenous changes in photosynthetic activity were demonstrated using the technique of delayed light imaging by Björn and Forsberg (1979), and Ellenson and Amundson (1982), who measured the delayed light emission spatially over the one leaf when the leaves were treated with UV-radiation, tobacco mosaic virus (Björn and Forsberg 1979) and exposed to sulfur dioxide (Ellenson and Amundson 1982). Although the delayed light imaging is able to detect the damage to the photosynthetic apparatus and localize the damaged area before the symptom become visible, the technique does not reveal the inhibition site in photo-synthetic system because the intensity of delayed light is affected similarly by various but different factors (Morita et al. 1981; Satoh and Katoh 1983). Thus, we have developed a new system using a CCD image sensor for a spatial analysis of chlorophyll fluorescence induction. Using this system we could show both spatial differences in photosynthetic activity on whole leaf in situ and the inhibition site of the photosynthetic apparatus when the plants were exposed to sulfur dioxide (Omasa et al. 1987). In addition other examples which were applied to our system are presented.

III. IMAGE INSTRUMENTATION SYSTEM AND ITS PERFORMANCE

Ordinary TV cameras and recording systems are not suitable for a quantitative analysis of chlorophyll fluorescence induction (CFI) in attached leaves, because of their low sensitivity, large after-image, bad image quality, the presence of AGC (automatic gain control) function, and the indistinctness in timing of the playback image. Common tungsten and fluorescent lamps cannot be used as light sources for CFI imaging, because of the unevenness and flutuation in light intensity. The new image instrumentaion system (Fig. 2) was designed to overcome these problems. A highly sensitive CCD imager with uniform sensitivity and an after-image suppression was selected for a TV camera (SONY XC-47, improved type). The image quality was improved by the use of a computer-control VTR (SONY BVU-820) with a digital time base corrector (SONY BVT-800) and the pre-processing by using a high-speed TV image processor (KCR nexus 6800). The AGC function, which changes a relationship between fluorescence intensity and the gray level of the recorded image, was removed from the TV camera and the VTR. The timing of the playback image was exactly defined by the use of a shutter synchronized to the TV signal and by the time code recorded in each image. The unevenness in light source intensity was improved by the use of two xenon lamp (CERMAX LX-300F) projectors, and by attaching a special ND filter with concentric circles of different density.

After the plant had been adapted to the dark for 30 min, the CFI was provoked by irradiation of the whole leaf with two beams of blue-green light (380-620 nm) from the projectors, with filters (Corning 4-96 + two heat-absorbing filters + a special ND filter) via the shutter opening. The fluorescence image was continuously measured at a TV field interval of 1/60s by the TV camera equipped with an interference filter (Vacuum optics co. Japan IF-W, 683 nm; half-band width, 10 nm) and a red cut-off filter (Corning

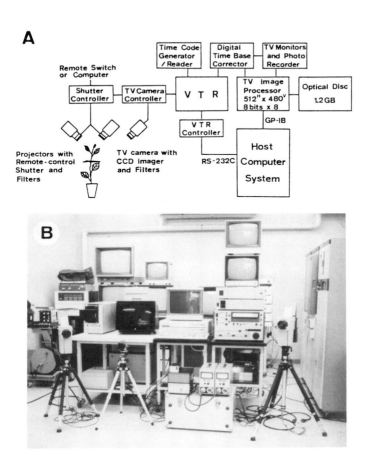

Fig.2 Image instrumentation system for quantitative analysis of CFI in attached leaves (Omasa et al. 1987). A, block diagram; B, photograph

2–64, >650 nm), and was recorded, with time code, by the VTR. The VTR image was digitized by a video A/D converter after it was played back to a still image without guard band noise through the time base corrector. A series of the digitized images (512H x 480V 8bits) was stored on an optical disc (National DU-15). A host computer system was used to control the VTR and TV image precessor.

The CFI curves, image intensity of characteristic transient levels (I,D,P,S,M,T), and image amplitude of major transient characteristics (ID, DP, PS, MT) were calculated

by the TV image processor on the basis of a series of images
after preprocessing for shading correction and noise
removal. These were represented by scales which correspond
to the A/D conversion level. The relationship between the
fluorescence intensity and the A/D conversion level of VTR
image showed a linear correlation. The image shading was
corrected by calculating the ratio of an original image to a
specific image (shading master), obtained by measuring a
uniform light of definite intensity, because the shading was
mainly caused by the lens and optical filters of the TV
camera. The noise was removed by the use of a spatial
smoothing filter, and the averaging of images digitized from
a still VTR image. For example, the image quality could be
improved to within 1% standard deviation by the shading
correction and the smoothing of 3x3 pixels after an
averaging of 10 images, when the image resolution was 280
lines. The after-image for the TV camera was about 4% at 30
ms after shutter opening, and decreased to 0.3% at 50 ms.

The CFI from a defined part of a cucumber leaf in situ
was measured by our system under different intensities of
actinic blue-green light. The CFI curves were calculated on

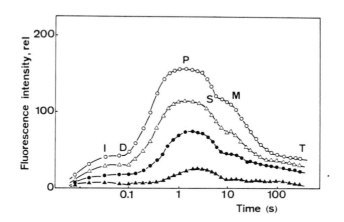

Fig.3 The CFI curves of a small area (about 1 mm^2) of a
healthy cucumber leaf in situ under different intensities of
actinic blue-green light (Omasa et al. 1987). The CFI was
obtained at the same definite area and shown by logarithmic
scale in time. Before the measurement the leaf was kept in
darkness for 30 min. Intensity of actinic light (unit= μmol
photon m^{-2}s^{-1}): (○) 200, (△) 150, (●) 100, and (▲) 50.
Light intensities were varied using ND filters. Environmental conditions: air temperature, 25.0 °C; RH, 70%

the basis of a series of fluorescence images (Fig. 3). Those clearly revealed the typical IDPSMT transients (Papageorgiou 1975) under light intensities from 50 to 200 μmol photons $m^{-2}s^{-1}$ at leaf surface. In these intensity ranges of light, we could resolve IDPSMT transients from any part of the leaf surface which has an area of at least 1 mm^2. Fluorescence intensities of I,D,P,S,M, and T, and rates in transients of DP, PS, and MT increased as the actinic light intensity increased. The appearance of peak P became more rapid with the increase of actinic light intensity.

Because the CFI is light intensity dependent, as described above, it is important to illuminate the whole leaf with uniform light when we want to compare CFI curves derived from different areas (each about 1 mm^2) of the leaf. The combination of spacial ND filters and two projector lamps, which were placed at angles of 60° and 120° to the leaf surface, achieved this. The combination kept the spatial deviation of the intensity of actinic light within 5% over a flat surface of 20 cm in diameter.

IV. DIAGNOSIS OF ENVIRONMENTAL STRESSES

1)Effect of SO_2

Sunflower (Helianthus annuus L. cv. Russian Mammoth) plants were fumigated with 1.5 ppm(v/v) SO_2 for 30 min in a growth chamber. During the fumigation, one-half of a leaf blade was covered with a thin aluminum foil to prevent the SO_2 entrance into the plant tissue. This procedure allows the comparison between a fumigated area (F) and an unfumigated area (UF) of the same leaf.

Effects of the SO_2 fumigation on CFI were analyzed by the image instrumentation system and were presented by the CFI curves and the images with gray scale (Fig. 4). In the unfumigated leaf area (UF), the CFI showed clearly the typical IDPSMT transitents (Papageorgiou 1975) and almost identical transients at any location of leaf area. Since the CFI observed upon dark-light transition of the leaf reflects the partial reactions of photosynthesis, we can detect the alteration in photosynthetic apparatus by SO_2 from the changes in CFI curves. As shown in Figure 4, B and C, both images of intensity in characteristic transient levels (I, P, M, T) and images of amplitude in major transient characteristics (ID, DP, PS, MT) in fumigated counterpart (F) strikingly differed from those in unfumigated area. Fluorescence intensity at I was raised

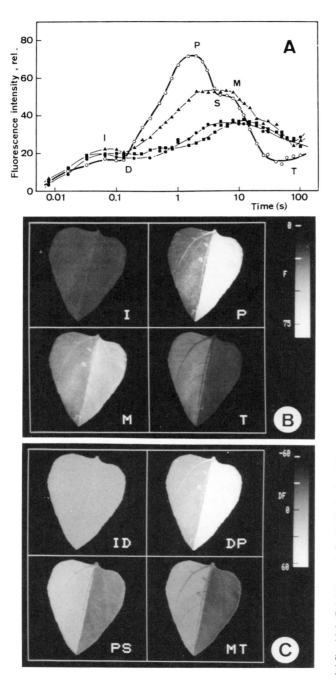

Fig.4 Effect of SO_2 on CFI in an attached sunflower leaf (Omasa <u>et</u> <u>al</u>. 1987). Sunflower plant was fumigated with 1.5 ppm(v/v) at 25.0 °C air temperature, 70%RH, and 350 μmol photons m^{-2} s^{-1} light intensity for 30 min. After dark adaptation for 30 min, CFI of an attached leaf was measured under 125 μmol photons m^{-2} s^{-1} actinic light. A, CFI curves at different sites in fumigated area (F: ●, interveinal site 1; ▲, site 2 near a large vein; ■, site 3 near a veinlet) and unfumigated area (UF; ○, interveinal site 4). B, images of intensity in characteristic transient levels (I,P,M,T). C, images of amplitude in major transient characteristics (ID, DP, PS, MT).

and at P reduced markedly, and that at T was increased in the fumigated area. Amplitude of fluorescence transients of DP-rise and PS- and MT-decline, indicating photosynthetic activity was reduced in the fumigated leaf. The changes in the intensity and amplitude varied with the location on the leaf surface; the effect of SO_2 was severe in the vicinites of interveins and veinlets than in those near large veins. Contrary to the perturbation in photosynthetic apparatus shown above, there was no visible injury in the whole surface of leaf at the end of SO_2 treatment and 2 days later.

The significance of the changes in CFI induced by SO_2 fumigation was as follows. Because fluorescence intensity in the early induction phenomena is regulated by the redox state of Q_A, a primary electron acceptor of PSII(Duysens and Sweers 1963, Murata et al 1966, Papageorgiou 1975), the elevated I level suggest that some portion of Q_A was brought to reduced state by the SO_2-fumigation. Since the DP rise in CFI reflects photoreduction of Q_A through reductant from H_2O (Papageorgiou 1975, Schreiber 1978), a diminished rise of DP was consistent with the inactivation of the water-splitting enzyme system (Shimazaki et al 1984). Since PS decline involves energy-dependent quenching (Krause et al 1981), the suppression of PS decline suggested the depression of formation of trans-thylakoid proton gradient probably due to the inactivation of the water-splitting enzyme system. However, the possibility that the PS decline was affected by the inhibition of electron flow Q_A to PSI cannot be excluded because PS decline partly reflects the oxidation of Q_A by PSI(Bradbury and Baker 1981). Suppression of MT decline was probably due to the inhibition of the trans-thylakoid proton gradient formation in addition to un-identified reactions in chloroplasts (Bradbury and Baker 1981, Papageorgiou 1975). Although the extent of the SO_2-effect on CFI differed from area to area on a single leaf, the mode of SO_2 action was essentially the same.

2) Effect of PAN

Petunia (Petunia hybrida cv. Tytan white) plants were fumigated with 0.06 ppm (v/v) peroxyacetyl nitrate (PAN) for 3 hrs in the growth chamber (Takagi et al. 1989). When the CFI of an attached leaf was measured within one hour after the fumigation, the image of the chlorophyll fluorescence at the levels of fluorescence in I, P, M and T was not affected by PAN, indicating that photosynthetic activity was not damaged (Fig 5).

After the measurement of fluorescence image, one-half of the leaf blade was covered with aluminium foil to prevent

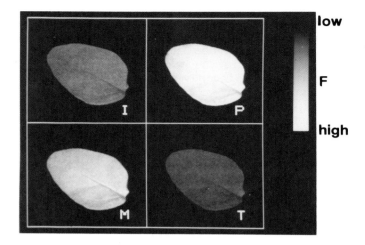

Fig.5 Immediate imaging of of CFI in an attached petunia leaf after the PAN fumigation. Petunia plant was fumigated with 0.06 ppm (v/v) PAN for 3 hrs at 25.0 °C air temperature, ca. 60% RH, and 270 µmol photons $m^{-2}s^{-1}$ light intensity. After dark adaptation for 30 min, the CFI image of an attached leaf was measured under 130 µmol photons $m^{-2}s^{-1}$ actinic light.

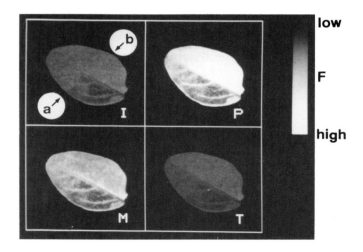

Fig.6 Role of light in PAN injuries after the PAN treatment. A lower-half (a) of leaf depicted in Fig.5 was illuminated for 12 hrs under 130 µmol photons $m^{-2}s^{-1}$ and the upper-half (b) was kept in darkness after the PAN exposure. The CFI image was measured every 3 hrs.

light illumination and allowed to stand in the light (130 μmol photons m^{-2} s^{-1}). The CFI was affected strongly when the leaf was illuminated after the PAN treatment. Fig. 6 shows images of intensity in characteristic transient levels when the leaf was illuminated for 12 hrs. The fluorescence intensities at P and M were depressed markedly in illuminated area (a), but the fluorescence was not affected virtually in unilluminated leaf area (b). This indicates that the light played an important role in the phytotoxicity of PAN.

3) Effect of UV

Effect of UV-light irradiation on CFI in sunflower leaves was investigated. A circular area of the attached leaf was irradiated with UV-light (0.07 mW cm^{-2}), (central wavelength 300nm, half-band width 10 nm) for 3 hrs. After the treatment of UV-light, the DP transient was depressed in sunflower leaves, indicating the inhibition of water-splitting enzyme system as reported previously (Smillie 1982/1983). The depression of the fluorescence intensity at the maximum level (F_p) was observed in a circular form in the leaf on which UV-light had been irradiated (Fig 7). The circular form became obscure in 6 hrs when the plants were allowed to stand under visible-light , suggseting the recovery of photosynthetic activity.

4) Effect of chilling

Attached leaves of <u>Cordyline stricta</u> plants were kept under various low temperatures for 10 hrs at night, then the plants were transferred to the growth chamber whose

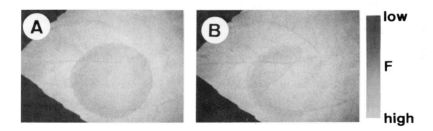

Fig.7 Effect of UV-light on F_p in attached sunflower leaf. A, 30 min after the UV-light irradiation. B, 6 hrs later. A circular area of the leaf was irradiated with 300 nm UV-light (0.07 mW cm^{-2}) for 3 hrs at 25.0 °C air temperature, 70% RH, and 350 μmol photons $m^{-2}s^{-1}$ visible-light intensity.

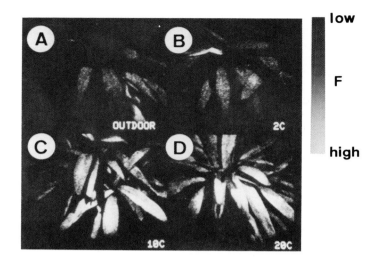

Fig.8 Effects of low temperature at night on F_p image. A, outdoor conditions at night (the lowest temperature, ca. 0 °C); B, night temperature, 2 °C; C, 10 °C; D, 20 °C.

temperature was 20 °C and F_p was measured after an equilibration of the plant with this temperature (Fig.8). The F_p depressions caused by the chilling treatment under outdoor conditions (A, the lowest temperature, ca. 0 °C) at night and 2 °C night temperature (B) were severe. Since the leaves were crooked, some spatial difference in F_p was occured. The slope of DP transient was smaller in the chilling plants than in the untreated plants, being consistent with the inhibition of the water-splittng enzyme system (Smillie and Nott 1979). The depression of the fluorescence intensity was gradually restored and became almost normal in 12 hrs when plants were allowed to stand at 20 °C.

V.   CONCLUSIONS

The image instrumentation system of chlorophyll fluorescence makes it possible to resolve the IDPSMT change at any leaf area over the whole single leaf. When the fluorescence induction was affected in the fast phase of fluorescence induction (IDP) by various environmental factors, it is easy to determine the inhibition site of photosynthetic apparatus. However, it will be difficult when the marked change was observed in the slow phase of fluorescence induction because the slow phase consist of two different components, photochemical and non-photochemical quenching.

Recently, Schreiber et al.(1986) developed a new fluorescence measuring system and was able to discriminate the photochemical and non-photochemical quenching in the slow fluorescence change by a pulse modulated technique. More recently, Daley et al. (1989) obtained the fluorescence images of the slow phase of fluorescence induction in combination with the technique of Schreiber et al.(1986) and estimated the photosynthetic electron transport over the leaf area of Xanthium strumarium. They found a heterogenous distribution of photosynthetic electron transport activity when stomata closed in group by application of abscicic acid to the plant leaves. In combination with our image instrumentation system (Omasa et al.1987) and the pulse modulated system, the image analysis of chlorophyll fluorescence will provide an integrative information of the functional state of photosynthetic apparatus over the whole leaf.

REFERENCES

Björn, L.O., and Forsberg, A.S.(1979) Imaging by delayed light emission (photoluminography) as a method for detecting damage to the photosynthetic system. Physiol. Plant. 47: 215.

Bradbury, M., and Baker N.R (1981) Analysis of the slow phases of the in vivo chlorophyll fluorescence induction curve. Changes in the redox state of photosystem II electron acceptors and fluorescence emission from photosystems I and II. Biochem. Biophys. Acta. 635:542

Daley, P. F., Raschke, K., Ball, J. T., and Berry, J. A. (1989) Topography of photosynthetic activity of leaves obtained from video images of chlorophyll fluorescence. Plant Physiol. 90: 1233

Demmig, B., and Björkmann, O. (1987) Comparison of the effect of excessive light on chlorophyll fluorescence (77K) and photon yield of O_2 evolution in leaves of higher plants. Planta 171: 171.

Duysens, L. M. N., and Sweers, H. E. (1963) Mechanism of two photochemical reactions in algae as studied by means of fluorescence. In Studies on Microalgae and Photosynthetic Bacteria. Edited by Japanese Soc. Plant

Physiol. pp. 353–372. University of Tokyo Press, Tokyo.

Ellenson, J.L., and Amundson, R.G. (1982) Delayed light imaging for the early detection of plant stress. Science 215: 1104.

Hashimoto, Y. Ino, T., Kramer, P.J., Naylor, A.W., and Strain B.R. (1984) Dynamic analysis of water stress of sunflower leaves by means of a thermal image processing system. Plant Physiol. 76:266

Kautsky, H., Appel, W., and Amann, H.(1960) Die Fluoreszenz-kurve und die Photochemie der Pflanze. Biochem. Z. 332: 277.

Krause, G.H., Brisntais, J-M., and Vernotte C. (1981) Two mechanisms of reversible fluorescence quenching in chloroplasts. In Photosynthesis. I. Photophysical Processes- Membrane Energization. Edited by G. Akoyunoglou. pp. 575. Balaban International Science Services, Philadelphia

Morita, S., Itoh, S., and Nishimura, M. (1981) Acceleration of the decay of membrane potential after energization of chloroplasts in intact Zea mays leaves. Plant Cell Physiol. 22: 205.

Munday, C. J., and Govindjee (1969) Light-induced changes in the fluorescence yield of chlorophyll a in vivo III. The dip and peak in Chlorella pyrenoidosa. Biophys. J. 9: 1

Murata, N., Nishimura, M., and Takamiya, A. (1966) Fluorescence of chlorophyll in photosynthetic systems. II. Induction of fluorescence in isolated spinach chloroplasts. Biochim. Biophys. Acta 120: 23.

Omasa, K., Abo, F., Aiga I., and Hashimoto Y. (1981a) Image instrumentation of plants exposed to air pollutants-quantification of physiological information included in thermal infrared images. Trans. Soc. Instrum. Control Eng. 17:657 (in Japanese and English summary): Res. Rep. Natl. Inst. Environ. Stud. JPN 66:69 (1984)(in English translation).

Omasa, K., Hashimoto Y., and Aiga I. (1981b) A quantitative analysis of the relationships between O_3 sorption and its acute effects on plant leaves using image instrumentation. Environ. Control Biol. 19:85

Omasa, K., Hashimoto, Y., Kramer, P. J., Strain, B. R., Aiga, I., and Kondo, J. (1985) Direct observation of reversible and irreversible stomatal responses of attached sunflower leaves to SO_2. Plant Physiol. 79: 153.

Omasa, K., Shimazaki, K., Aiga, I., Larcher, W., and Onoe, M. (1987) Imaging analysis of chlorophyll fluorescence transients for diagnosing the photosynthetic system of attached leaves. Plant Physiol. 84: 748.

Papageorgiou, G. (1975) Chlorophyll fluorescence: An intrinsic probe of photosynthesis. In Bioenergetics of Photosynthesis. Edited by Govindjee. pp. 320. Academic Press, New York.

Satoh, K., and Katoh, S. (1983) Induction kinetics of millisecond-delayed luminescence intact Bryopsis chloroplasts. Plant Cell Physiol. 24: 953.

Schreiber, U. (1978) Chlorophyll fluorescence assay for ozone injury in intact plants. Plant Physiol. 61: 80.

Schreiber, U., and Berry, J. (1977) Heat-induced chlorophyll fluorescence changes in intact leaves correlated with damage of the photosynthetic apparatus. Planta 136: 233

Schreiber, U., Schliwa, U., and Bilger, W. (1986) Continuous recording of photochemical and non-photochemical chlorophyll fluorescence quenching with a new type of modulation fluorometer. Photosynth. Res. 10: 51

Shimazaki, K., Igarashi,T., and Kondo, N. (1988) Protection by the epidermis of photosynthesis against UV-C radiation estimated by chlorophyll fluorescence. Physiol. Planta. 74: 34.

Shimazaki, K., Ito, K., Kondo, N., and Sugahara, K. (1984) Reversible inhibition of the photosynthetic water-splitting enzyme system by SO_2-fumigation assayed by chlorophyll fluorescence and EPR signal in vivo. Plant Cell Physiol. 25: 795.

Smillie, R. M. (1982/1983) Chlorophyll fluorescence in vivo as a probe for rapid measurement of tolerance to ultraviolet radiation. Plant Sci. Lett. 28: 283.

Smillie, R.M., and Hetherington S.E. (1983) Stress tolerance and stress-induced injury in crop plants measured by chlorophyll fluorescence in vivo. Plant Physiol. 72:1043-1050.

Smillie, R. M., and Nott, R. (1979) Assay of chilling injury in wild and domestic tomatoes based on photosystem activity of chilled leaves. Plant Physiol. 63: 796.

Takagi, H., Tobe, K., Takeshita, S., and Omasa K. (1989) Development of PAN exposure system for plants. J. Agri. Meteorol. 45:39

Terashima, I., Wong, S-C., Osmond, C. B., and Farquhar, G. D.(1988) Characterization of non-uniform photosynthesis induced by abscisic acid in leaves having different mesophyll anatomies. Plant Cell Physiol. 29: 385.

NUCLEAR MAGNETIC RESONANCE
RESEARCH ON PLANTS

Paul J. Kramer
James N. Siedow

Department of Botany
Duke University

Janet S. MacFall

School of Forestry and
Environmental Studies
Duke University

I.   INTRODUCTION

Nuclear magnetic resonance technology is based on discoveries made independently by Felix Bloch and Edward Purcell and their colleagues in 1946. Its earliest use was in spectroscopic studies of the structure of molecules containing hydrogen. However, increase in the strength of electromagnets and the availability of large computers capable of performing Fourier transformations and storing and processing large amounts of data increased the sensitivity and usefulness of NMR many fold. These developments made NMR spectroscopy applicable to low concentrations of phosphorus (^{31}P), carbon (^{13}C), and other elements in addition to ^1H, containing unpaired protons or neutrons in their nuclei. Thus NMR spectroscopic studies of metabolism in living plant tissues became possible and will be discussed in detail later in this paper.

Measurement Techniques in Plant Science

The same technological advances that increased
the usefulness of NMR spectroscopy, combined with
Lauterbur's introduction of magnetic field
gradients (Lauterbur, 1973), made high definition
imaging of living tissue possible. At first, NMR
imaging was used chiefly in clinical medicine
where it is superior to X-rays for some purposes
because it does not use ionizing radiation and
usually gives better images of soft tissues and
blood. Now the potential of NMR imaging for
studies on plant water content, water binding, and
water flow, and for observations of root-soil
relationships is being explored with encouraging
results. NMR imaging is of particular interest
because it is noninvasive and nondestructive,
allowing repeated observations to be made on the
same organ or tissue without injury.

Because of the public's fear of the term
"nuclear" the process is known in medical circles
as magnetic resonance imaging or MRI, somewhat to
the dismay of physicists. Imaging will be
discussed in detail in a later section, but first
we will explain how NMR signals are obtained.

It is difficult to explain the origin of NMR
signals to nonphysicists, but by using analogies
and glossing over details an approximate
explanation can be provided. Production of NMR
signals depends on the fact that the nuclei of
some kinds of atoms possess unpaired protons or
neutrons which cause them to behave like tiny
spinning bar magnets. They are surrounded by
magnetic fields, the strength of which is a
characteristic of each kind of nucleus and is
known as its magnetic moment. When placed in a
static external magnetic field the otherwise
randomly oriented, spinning nuclei become oriented
in the applied field and precess like spinning
tops in a gravitational field. The frequency of
precession, known as the Larmor frequency, is
related to the kind of nucleus and the strength of
the magnetic field, as shown in the following
equation:

$$\omega = \gamma B_O \qquad\qquad\qquad (Eq. 1)$$

where omega (ω) is the Larmor frequency or
frequency of precession, gamma (γ) is the
gyromagnetic ratio, and B_O is the strength of the

applied magnetic field. The gyromagnetic ratio is
a constant for each kind of nucleus, expressed in
MHz/Tesla. Its value is 42.57 for ^{1}H, 10.71 for
^{13}C, and 17.23 for ^{31}P.

Slightly over half the nuclei are aligned in a
low energy state, parallel to the applied static
magnetic field, and the remainder in a high energy
state, antiparallel to the applied field. When
subjected to radiation at their Larmor frequency,
applied perpendicular to the static magnetic
field, all of the nuclei precess synchronously,
perpendicular to the magnetic field. When the
radiation is turned off the nuclei return to their
original orientation. The time required for
return of the nuclei to their original orientation
is termed the relaxation time or decay time, and
this depends on the rate at which energy is
transferred to the "lattice" or environment in
which the nuclei exist. For example, the
relaxation time of ^{1}H protons is much shorter in
ice than in liquid water. It also is shorter for
water bound to membranes and other cell structures
than for vacuolar water.

Changes in energy level produce resonance
resulting in radiofrequency signals that can be
detected in an appropriate coil located in the
immediate vicinity of the tissue under study.
Sometimes the coil used to generate the
radiofrequency radiation also is used to receive
the signals generated by the resonating nuclei,
and functions as a transceiver or combined
transmitter and receiver. The signals are
amplified and collected in a computer where they
are processed by Fourier transform and stored for
future use, either to produce spectrograms as in
NMR spectroscopy or images as in NMR imaging. As
the individual signals are very weak, a large
number usually must be collected and averaged by
the computer to provide usable data.

The strength of the signals depends on the kind
and concentration of nuclei, and their
environment, including the strength of the applied
magnetic field, the timing of the radiofrequency
pulses used to generate resonance, and conditions
that affect relaxation time. Use of gradient
coils to maintain a linear gradient in strength of
the magnetic field across the object under study
makes it possible to locate the source of signals

from individual voxels or volumes of tissue and produce two or three dimensional images. This is feasible because, as mentioned earlier, the Larmor frequency increases with increasing strength of the applied magnetic field (eq.1). Thus signals from the stronger part of the field can be distinguished from those originating in a weaker part of the field. All of this requires very careful adjustment of the gradient coil strength and pulse timing, especially when a resolution measured in millimeters or micrometers is desired.

The hardware essential for NMR imaging includes a strong static magnet, coils to produce a gradient of magnetic field strength necessary to locate signals in space, and coils to generate the radiofrequency radiation necessary to excite the magnetized nuclei and detect the resulting signals. Large computer capacity is required to process and store the signals in a memory from which they can be received and processed to produce images or spectra. By the use of strong super-conducting magnets and strong field gradients, combined with appropriate pulse sequences, the potential exists for study of a variety of phenomena in plants. These include (1) dehydration and rehydration of various tissues, (2) differentiation between free water and bound water on various plant structures or in ice, (3) hopefully, measurement of flow and diffusion, (4) study of root-soil water relations, (5) study of developmental morphology and stress tolerance of specific tissues such as anthers, ovules, and meristematic regions, and (6) possible improvement of our understanding of how injury is caused by insects and pathogens. All of this is in addition to the study of plant metabolism by NMR spectroscopy discussed in the next section of this chapter.

Most NMR images of plant tissue are based on the concentration and degree of binding of ^1H protons in water. This is possible because of the high concentration of water in living tissue and the high gyromagnetic ratio of ^1H. Sometimes images are affected by signals from lignified, suberized or cutinized tissue, creating problems that are discussed in the section on imaging. As mentioned earlier, the intensity of the signals and the characteristics of the resulting images usually depend on the concentration of protons and their relaxation time, which is related to their environment, also on operator-controlled timing of pulse sequences. By management of pulse sequence timing images can be produced that emphasize either spin density, N(^{1}H), spin-lattice relaxation time (T1), or spin- spin relaxation time (T2).

Definitions

A few frequently used terms will be defined.

Tesla. One Tesla = 10,000 gauss, the gauss being the primary unit of strength of a magnetic field. The magnetic field of the earth is only about 1 gauss, that of a clinical magnet used for whole body imaging about 2 Tesla, (20,000 gauss), while the magnets used for research on small objects such as plant organs may exceed 10 Tesla in strength.

Pixels. The individual elements or sources of signals that collectively constitute an image. Increasing signal strength results in brighter pixels and brighter areas on an image.

Voxels. The volume of tissue represented by each pixel. This varies considerably depending on the objectives, but might be a volume of plant tissue with a "slice thickness" of 1.0 or 1.2mm and a width of 0.1mm or less on the other sides (see Fig. 2).

Fourier Transform. This refers to the mathematical treatment used to convert data from one form to another, as from frequency versus time to frequency versus space, the latter being required to produce images. It is done by a program in the computer.

Spin Density (N^{1}H). refers to the concentration or density of nuclei under study per unit of volume or voxel. In work on plants most of the protons are the hydrogen nuclei of water and spin density can sometimes be used to indicate the concentration of water. However, this simplistic assumption sometimes is invalidated by signals derived from lignified tissue, suberin, and wax, creating problems of interpretation. This will be dealt with in the section on imaging.

Spin-Lattice Relaxation Time (T1). This is the
time required for protons excited to their higher
energy state by radiofrequency pulses to return to
their lower energy level by loss of energy to
their surroundings or lattice. It can be used to
differentiate between bound and free water because
it is longest in free water (about 3 seconds) and
decreases with increase in intensity of binding,
to less than a second in most plant tissue.
    Spin-Spin Relaxation Time (T2). This is the
time during which the transverse magnetization
acts sufficiently coherently to yield a signal.
    Pulse Repetition Time (TR). Refers to the
timing of successive radiofrequency pulses. A
short TR yields an image dependent on T1, a longer
TR removes this dependence.

II.   NUCLEAR MAGNETIC RESONANCE SPECTROSCOPY

James N. Siedow

    While the first report of the use of in vivo
NMR spectroscopy in plants involved the uptake of
$^{13}CO_2$ and its incorporation into metabolites
(Schaefer et al., 1975), the most successful in
vivo NMR studies of plant metabolism to date have
utilized the isotope ^{31}P (Roberts, 1984). There
are several reasons for this. Phosphorus-31 is
the lone naturally-occurring isotope of
phosphorus, so no enrichment is required to
enhance the levels present in the plant.
Phosphorus-containing metabolites exist in
relatively high concentrations in plant tissues
(Pradet and Raymond, 1983), and they play a
crucial role in both photosynthesis and
respiration. The chemical shift of inorganic
phosphate (Pi) is sensitive to pH over the range
6.0-8.0 and, as such, can provide a monitor of
intracellular pH. In addition to inorganic
phosphate the major phosphorus-containing species
which can be monitored in plants include sugar-
phosphates, ADP, ATP, PPi, and UDP-glucose. ^{31}P
NMR has been applied to a wide range of plant
tissues including root tips (Roberts et al., 1980,
1984), root nodules (Mitsumori et al., 1985),

tubers (Kime et al., 1982), stems (Mitsumori et al., 1985), leaves and leaf discs (Foyer et al., 1982; Waterton et al., 1983; Foyer and Spencer, 1984), plant tissue culture (Martin et al., 1982; Rebeille et al., 1983), pollen (Mitsumori et al., 1985), and unicellular algae (Mitsumori and Ito, 1984).

Roberts et al. (1980) first utilized ^{31}P NMR in plants to measure cytoplasmic and vacuolar pHs in vivo in maize root tips. This seminal study identified two features of plant cells which must be taken into account when doing NMR on plant tissues. First, in mature plant cells the vacuole generally occupies such a large percentage of the total cell volume (> 80% in most herbaceous tissues), that the Pi signal associated with the vacuole can swamp the much smaller Pi signal associated with the cytoplasm. As a result, Roberts et al. (1980), limited their study to the less highly vacuolate cells in the first 2-3 mm of the root tip. Second, the concentration of phosphorus-containing metabolites in plant cells is well below that found in more metabolically active animal tissues. As a result the number of repetitive scans needed to obtain a usable spectrum with plant tissues is often large and even with modern, high-field spectrometers (360-500 MHz), acquisition times will range from a few minutes (Roberts et al., 1980) to well over an hour (Foyer and Spencer, 1984). In spite of these limitations, ^{31}P NMR has been successfully used to measure cytoplasmic and vacuolar pH in a wide variety of nonphotosynthetic tissues since the initial work of Roberts and coworkers (Mitsumori et al., 1985; Kime et al., 1982; Martin et al., 1982).

Initial attempts to distinguish cytoplasmic and vacuolar Pi pools in photosynthetic tissues showed a single, broad resonance which peaked at the chemical shift value expected of vacuolar Pi. In wheat leaves, this large linewidth appears to be due to "inhomogeneous broadening" associated with bulk magnetic susceptibility differences between the aqueous tissue phase and intercellular air spaces (Waterton et al., 1983). This problem could be overcome in wheat by vacuum infiltration of water into the air spaces. However, Foyer and Spencer (1984) reported that this solution may be

limited to wheat and doesn't work with the leaves
of dicots. They found that a more general
solution to this broadening problem involved
infiltrating the leaves with a 10 mM EDTA
solution. This indicated that in addition to the
inhomogeneous broadening, free paramagnetic ions
in the leaf were contributing to linewidth
broadening in noninfiltrated leaves. The
combination of complexing ions with EDTA and
saturating air spaces with H_2O served to overcome
this problem and should be generally applicable to
any leaf type.

    Using intact <u>Chlorella</u> cells, Mitsumori and Ito
(1984) observed a single phosphate resonance (pH
7.3) in cells incubated in the dark. Upon
illumination, the pH of one compartment in the
cell rose to a value of 8.1 while that of a second
compartment dropped to pH 6.8. Because no
appreciable vacuolar volume exists in <u>Chlorella</u>,
the compartment whose pH increased upon
illumination was assigned to the chloroplast
stroma and the compartment whose pH decreased was
assumed to be the cytoplasm. The light-induced
increase in stroma pH, but not the decrease in
cytoplasmic pH, could be eliminated by addition of
DCMU. In higher plants, Waterton <u>et al</u>. (1983)
observed a light-induced increase in the pH of the
"cytoplasmic" Pi pool in wheat leaves upon
illumination. The stromal origin of this change
was suggested, but not demonstrated
experimentally.

    Using nonphotosynthetic plant tissues, Roberts
and coworkers have been quite successful in the
application of ^{31}P NMR to study the energetics of
metabolic responses in root tips; particularly the
changes that take place when roots are subjected
to hypoxic conditions. The earlier work was
reviewed by Roberts (1984) and will not be
considered here. More recently, Roberts <u>et al</u>.
(1984a) used NMR (^{31}P and some ^{13}C) to establish
that lactic acid formation in roots placed under
hypoxic conditions acts to initially lower the
cytoplasmic pH which, in turn, serves as a signal
to stimulate ethanol production. The ability to
carry out ethanolic instead of lactic acid
fermentation allows the plant to maintain the
cytoplasmic pH near neutrality and, as such,
tolerate longer periods under hypoxic conditions.

In subsequent studies (Roberts et al.; 1984b, 1985), it was shown that leakage of acid from the vacuole to the cytoplasm was responsible for the acidosis which ultimately leads to cell death in roots under prolonged exposure to hypoxic conditions. Further, this leakage takes place more rapidly in roots which are less tolerant of hypoxic conditions (Roberts et al., 1985). These experiments point to the possibility that properties of the tonoplast membrane may distinguish flooding-tolerant from flooding-intolerant plant species, and demonstrate further the potential of NMR to measure directly the in vivo responses of plant metabolism to environmental stimuli. In this example, a theory put forth ten years earlier (Davies et al., 1974) was tested directly for the first time.

^{13}C has not been widely applied to the study of higher plant metabolism due, in large part, to the fact of its natural abundance being only 1.1% (Martin, 1985). This severely limits sensitivity in unenriched systems. However, given the importance of carbon-containing compounds in both photosynthetic and respiratory metabolism, there is reason to expect that in plant systems which can be enriched with ^{13}C (via $^{13}CO_2$), the potential for obtaining insights into carbon metabolism is quite large. One area where in vivo NMR may be useful is in the study of Crassulacean acid metabolism (CAM).

Theories of the regulation of carbon metabolism in CAM plants center around the early morning period when PEP carboxylase activity is shut down and malate decarboxylation begins. Cytoplasmic pH and/or malate levels are felt to play a major role in this regulation (Osmond, 1978; Wu and Wedding, 1985). If changes in cytoplasmic pH (malate) lie at the heart of the regulation of CAM, NMR currently provides the only method of directly measuring those changes in vivo.

In an attempt to address the question of CAM regulation, using NMR, Stidham et al. (1983) took advantage of the fact that the leaves of CAM plants take up large amounts of CO_2 at night, accumulating it in the form of malate to levels as high as 0.1-0.2 M. By exposing leaves of Kalanchoe tubiflora to $^{13}CO_2$ during a night period, the malate which accumulated was highly

enriched for ^{13}C allowing easy measurement of the
malate NMR. The goal was to take advantage of the
pH dependence of the malate chemical shifts in an
attempt to monitor cytoplasmic and vacuolar pHs _in
vivo_. It was found that, after labeling, both C-1
and C-4 of malate were enriched in [13]C. This is
in keeping with previous observations regarding
the randomization of the initially C-4 labeled
malate by the action of fumarase within the
mitochondria (Bradbeer et al. 1958). The malate
resonances measured were all associated with a low
pH (<4.0) region which presumably represented the
large vacuolar pool of malate. During the course
of the day, the size of this malate pool decreased
and its pH rose, in keeping with current concepts
of the flow of carbon in CAM plants (Osmond,
1978).

An additional feature came out of this study
(Stidham et al., 1983). The malate labeled at C-4
appeared to be utilized preferentially during the
light period relative to malate which had
undergone randomization (and was thus labeled at
C-1). The significance of this observation was
not clear at the time, but it was suggested that
malate which gets shunted through the mitochondria
may not be stored in the same compartment as the
bulk malate pool. More recent work (N. Schmitt
and J.N. Siedow, unpublished observations) has
refined this earlier observation using a system
for illuminating leaves directly in the NMR
cavity. This allows real-time measurements to be
taken of the loss of malate during the
decarboxylation phase of photosynthesis. This
work has confirmed the results of Stidham et al.
(1983) with respect to the apparent preferential
loss of malate labeled in the C-4 position
relative to that labeled at C-1. However,
extraction of malate from the leaf tissue and
subsequent measurement of label distribution
showed that the ratio of C-4 to C-1 labeled malate
was identical to that seen at the beginning of the
illumination period (approximately 2:1). The
presence of a different ratio in intact leaves can
be attributed to the appearance of a significant
amount of cytoplasmic malate about 90 minutes
after the onset of illumination. The chemical
shift of the C-4 of malate at pH greater than 7.0
coincides with the chemical shift of the C-1 at a

pH of around 4.0 (i.e., the vacuole) (Stidham <u>et al</u>., 1983). These results suggest that a significant amount of malate does appear in the cytoplasm, but some time after the onset of illumination. Further studies are being carried out to establish the validity of this suggestion.

The utility of enriching the leaves not only of CAM plants, but also of C_3 and C_4 plants by exposure for extended periods to $^{13}CO_2$ can overcome the problem of the low natural abundance of ^{13}C and points the way to future applications of ^{13}C NMR in the study of photosynthetic carbon metabolism. With the various Calvin cycle intermediates labeled, ^{13}C NMR can be carried out using intact leaves to see how the Calvin cycle responds (+/- illumination) in leaves which have been subjected to a variety of different environmental treatments (heat, chilling, salt). Under these conditions, sites of disruption of the operation of the Calvin cycle can be ascertained with greater precision and a clearer picture can be obtained of whether different environmental stresses act on a few common, susceptible sites of action or there exist a wide range of stress-specific sites of inhibition of photosynthesis.

III.   MAGNETIC RESONANCE IMAGING

Janet S. MacFall

Magnetic resonance imaging (MRI) is an extremely versatile technology for non-destructive plant studies. To date, it has been used chiefly to show the possibilities of non-destructive study of plant form and function, especially with respect to water relations. Realistic expectations of the type of information which can be provided by this technique and correct interpretation of acquired images rests on a basic understanding of the fundamental processes of the technique. Some of its limitations will become evident later in this chapter.

A. Equipment

There are many components in a magnetic
resonance imager. The source of the primary,
static magnetic field usually is a superconducting
magnet, with a bore diameter that can vary in size
from a few cm to nearly 50 cm - large enough to
accommodate a person, as for clinical
applications. The field strength of the magnet
may vary from a fraction of a Tesla (=10,000
Gauss) to 7 Tesla or more. Magnet size and
strength usually are determined by the
technologies of magnet construction and the
desired application. For small volume samples in
which the highest resolution is desired, a small
bore, high field magnet usually is most
appropriate. If the specimen is large and the
highest resolution may not be necessary, as for
example, in the imaging of logs, large volumes of
soil, or in large animal studies, larger bore
magnets are used. Due to the relationship
established by the Larmor equation, magnetic field
homogeneity is of importance in standardizing
frequency of precession. Electromagnetic "shim"
coils are fitted to the magnet, allowing careful
adjustment of the magnetic field until acceptable
homogeneity is achieved.
Three magnetic gradients in the primary
magnetic field placed along three perpendicular
spatial axes are critical to image production.
The use of gradients within the magnetic field is
the primary difference between magnetic resonance
imaging and nuclear magnetic resonance
spectroscopy. The magnetic gradient is formed
from coils of wire placed on each side of the
center of the magnetic field. These coils are
supplied with electrical current flowing in
opposite directions, allowing addition to and
subtraction from the magnetic field created by the
primary magnet. When current is flowing through
these coils, a magnetic field gradient is
produced, expressed in Gauss/cm, which should be
as linear as possible to reduce image distortion.
These gradients are switched on and off at
intervals and power levels determined by the
chosen pulse sequence for data acquisition. Use
of these gradients and the Larmor equation make

magnetic resonance imaging possible. They also serve to determine and limit the slice thickness and the field of view used in image acquisition.

A coil capable of generating pulses of radiofrequency waves (RF) is placed either around or adjacent to the specimen. This is tuned to broadcast at the Larmor frequency (see Introduction), of the nuclei under study in the primary field. The RF coil is used to flip the spinning nuclei in different timings and sequences, depending on the application and pulse program. It is pulsed many times for periods measured in milliseconds, typically in multiples of 128 or 256, repeatedly flipping the spinning nuclei down in a 90° angle from the perpendicular, then flipping them 180°, then allowing relaxation into alignment with the primary magnetic field.

As the spins return to a relaxed condition, signal is emitted from the sample, creating a transverse magnetization in a receiver coil placed either around or next to the specimen. This induces an AC current several magnitudes less than the initial RF pulse. This weak signal is amplified and filtered, and the signal amplitude, frequency and period recorded and digitized.

Once a data set is acquired on a specimen, several sets of calculations are required to produce the final image. Many acquired signals are averaged to increase the signal-to-noise ratio. The data are then recalculated through a Fourier transformation. This allows conversion of the signal from a plot of signal frequency and amplitude over time to signal frequency and amplitude over space.

Most imaging units are composed of several computers linked by a communications network. For example, the commercial system used in our research has four computers – a host computer, an array processor, a status control module and a pulse control module. These components allow acquisition, storage and rapid processing of large data sets. Long term data storage can be achieved in many ways, including archiving on streaming tape or optical disks. Figure 1 illustrates the components of an MRI system.

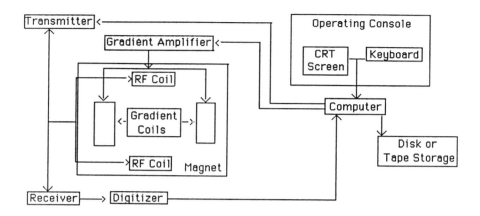

Figure 1. Schematic drawing of the components of
a magnetic resonance imaging system.

B. Factors Affecting Image Quality

    Four factors affecting image quality which are
of interest in the acquisition of an MR image are
(1) field of view, (2) spatial resolution, (3)
slice thickness, and (4) contrast. The field of
view is the area of the specimen which is imaged.
For plant studies, a typical field of view might
be 10 mm x 10 mm. Spatial resolution is defined
as the area within the field of view contained in
each pixel. For example, in most applications an
image is composed of 256 x 256 pixels. For a
field of view of 10 mm, the spatial resolution is
equal to 10 mm/256 = .039 mm. Data are actually
acquired on a three-dimensional volume, as the
specimen is three-dimensional in nature.
Therefore, the gradients have been designed to
allow acquisition of information on a
predetermined "slice" of the specimen. This can
vary from fractions of a mm to several cm through
the entire specimen. The information obtained in
the slice is then projected as a 2D image, defined
by the field of view and spatial resolution. A
diagram illustrating this procedure is shown in
Figure 2. In general for plant work, as in light

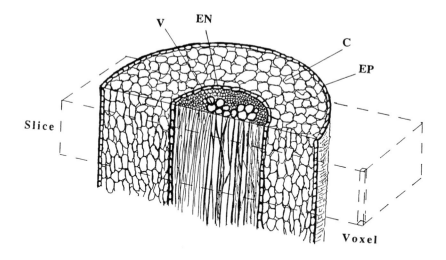

Figure 2. Outline of the tissue volume of a
primary root which is scanned during the data
acquisition of a single slice. Each side of the
slice is 256 (or some multiple of 128) pixels,
distributed over the chosen field of view (ex.
10mm x 10mm). EP - epidermis, C - Cortex, EN -
endodermis, V - vascular tissue.

microscopy, the thinner the slice the better the
definition in the image of structural features.
Contrast is the difference in signal intensity or
brightness between adjacent pixels.
     In biological research there are numerous
protocols which make use of colorimetric assays.
In these methods, the Beer Lambert Law is followed
within a given range in which the concentration of
a substance is directly proportional through a
linear relationship to absorbance of a specific
wavelength of light. In MR imaging, however, the
relationship between signal intensity, instrument
settings and the specimen is much more complex.
For a simple, spin-echo pulse sequence the amount
of signal (brightness of image) obtained is
proportional to acquisition parameters and the
specimen in the following way:

$$S(^1H) = N(^1H)(1-exp^{(-TR/T1)})(exp^{(-TE/T2)})$$

$S(^1H)$ = signal intensity, TR = time of pulse
sequence repetition, TE = time between initiation
of the pulse sequence and maximum echo amplitude.
The terms $N(^1H)$, T1, and T2 have been defined in
the introduction.

In MRI, image contrast depends not only on the
water content of the tissue and surrounding matrix
(soil, sand, air, etc.), but also on the degree of
water binding which influences relaxation times.
It is the heterogeneous nature of the tissue under
study with respect to $N(^1H)$, T1, and T2 which are
the primary contrast determining factors in MR
imaging (Wehrli, MacFall and Newton,1983). Machine
parameter settings of TR and TE can be used to
enhance visualization, emphasizing either water
content or degree of water binding within tissue.

C.   Animal Versus Plant Imaging

Most of the early work with MR imaging was
clinical work on humans and experimental animals.
It is a routinely used technique for diagnosis of
diseases involving the brain, the spinal cord,
pelvis, and musculoskeleton, and is an excellent
technique for studying the neck, liver, and some
other organs. By appropriate programming, healthy
tissue can be distinguished from diseased or
aberrant tissue such as aseptic necrosis (Budinger
and Margulis, 1988), chiefly by differences in T1
and T2 relaxation times. It is particularly
useful in research on effects of carcinogens on
small animals because the living animals can be
imaged repeatedly to observe development of
lesions, where traditional methods require killing
the animals after various exposure periods and
performing autopsies.

The development of MRI techniques to describe
flow has generated a great deal of interest both
in the clinical and research settings. Again, by
tailoring the pulse sequence and machine parameter
settings to the specific application and specimen
under study, rapid laminar flow within large
vessels can be detected as either a bright region
or a dark region. An example might be in imaging

the path of flow of a bolus of blood following
arterial flow into the leg. Observation of water
flow in plants will be discussed later.

Plants are potentially excellent study subjects
for MR imaging compared with animals for two
reasons. The first is that they remain immobile
in the apparatus for long periods without
complaints or injury, permitting long data
acquisition times. The heartbeat, breathing,
impatience with confinement, and twitching of
animals requires short exposure times or special
techniques to coordinate machine pulsing with
movement. The second advantage is the
heterogeneous distribution of water in plants,
where both content and degree of binding vary
among tissues (Johnson, et al., 1987). Plant
anatomical features can often be distinguished
simply from images acquired with long repetition
and short echo times, which weights signal
strength on the water content of the associated
tissue. Lignified woody tissue, such as that in
the xylem and supportive tissue usually has a
lower water content than unlignified and
parenchyma tissue, but has shorter relaxation
times due to the molecular interactions and state
of molecular hydration associated with
lignification. For example, water is distributed
so homogeneously in animal tissues that image
contrast usually depends on differences in T1 and
T2 relaxation times.

D.   Uses Of NMR Imaging In Plant Research

Although the potential usefulness of NMR
imaging is widely recognized, relatively few
studies have been made on plants due to the
limited availability of the equipment. Not only
are suitable magnets and computers expensive, but
their effective operation requires a highly
trained staff, and the cost of equipment and
operation exceeds the budget of most plant science
departments. Imaging seems particularly useful
for research in dehydration and rehydration of
various tissues, to differentiate between free and
bound water and to study differences in stress
tolerance of various tissues such as stamens and
pistils, study of root-soil interrelationships,

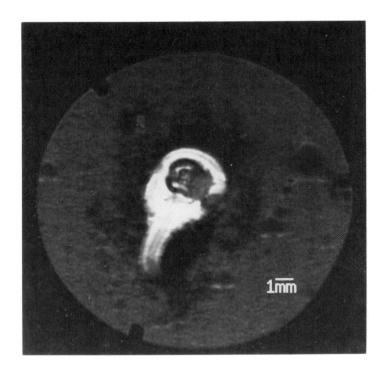

Figure 3. MR image of the taproot and lateral
root of a loblolly pine seedling planted into
moist sand. The dark region surrounding the root
is the region of water depletion from the sand.

study of developmental morphology, and a variety
of problems in pathology and insect injury. A few
examples will be discussed in more detail.
    Quantitative study of water potential and water
content is an important potential application of
MRI. As mentioned previously in this chapter,
there is not necessarily a direct relationship
between water content and signal intensity or
brightness of image. However, under some
conditions, MacFall and coworkers (1990) have
demonstrated a linear relationship between signal
intensity and water content of a fine, acid-washed
sand. This presents the possibility of
quantitative water determinations from images
acquired with MRI. Work is ongoing at present in
our lab to determine a suitable reference material

to place in the field of view which may be used to
standardize the signal intensity in the image,
potentially allowing determination of water
content within the sand. It was confirmed that
areas of sand images which were bright contained
comparatively greater water contents than dark
regions. This observation allows the acquisition
of images of water depletion around a root, as
seen in Figure 3.

As the technique develops, there is the
possibility of determining intact plant water
potential and/or content from MR images. At
present, MR spectroscopic studies of macerated or
excised plant tissue suggest multiple T1
relaxation times (for example within a leaf).
Fitting techniques for T1 determinations make
calculations of more than 2 T1 values uncertain.
When one considers the diversity of cell types
within most plant parts (e.g., mesophyll,
epidermis, xylem, phloem) and the
compartmentalization of water within a cell (bound
to walls, bound to membranes, cytoplasmic,
vacuolar, etc.), it is likely that there are many
degrees of water binding. Additionally, if the
tissue is damaged cell rupture is likely to
influence the state of water binding and therefore
will affect relaxation measurements. Veres and
Johnson, in work in our lab (personal
communication) have shown with MR images of fern
roots different T1 values between the xylem,
cortex and root margin and the values change with
different states of hydration and plant water
potential. Further study is needed to relate
signal intensity and relaxation time to water
content and plant water potential.

Water distribution and exchange within intact
tissue have been studied to a limited extent. In
an elegant work, rapid water movement into intact
and excised wheat grains has been imaged (Jenner,
et al., 1988). Water binding, distribution, and
compartmentalization also was described in leaves
of Mesembryanthemum crystallinum grown under high
salinity conditions. Water of differing
mobilities and relaxation times was found in
bladder cells, water-conducting vessels, and two
types of mesophyll cells. It was speculated that
the different types of mesophyll cells contribute
to the asymmetric water exchange described in the

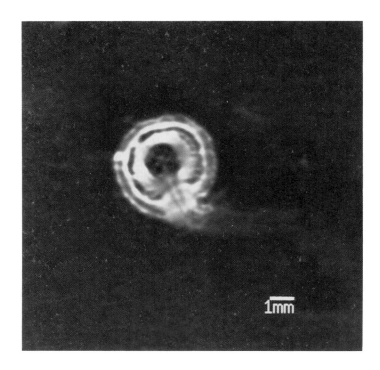

Figure 4. Taproot and attached lateral root of
loblolly pine seedling. The vascular connections
between the two roots can be seen clearly.

literature of Crassulacean acid metabolism
(Walter, et al., 1989). MR imaging may also
contribute to investigation of the theory that
water in leaves exists in at least three
isotopically distinct groups (Yakir, et al.,
1989).
     Images of intact roots at a spatial resolution
of .03 - .1 mm have shown intricate and distinct
patterns of water distribution within the tissue
(Brown, et al., 1986; MacFall, et al., 1990; Veres
and Kramer, personal communication). This is well
illustrated in Figure 4. Signal intensity and T1
measurements within the root tissue decreased as
the plant dehydrated, demonstrating a dependence
on water content for signal intensity and T1
relaxation times. Connelly and coworkers (1987)

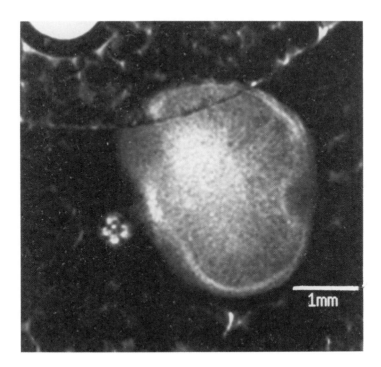

Figure 5.   High resolution MR image of a soybean nodule.

obtained images of maize roots that clearly show structural detail through the use of contrast agents (Mn^{+2} and D_2O) taken up by the root.   They were able to observe the flow of these contrast agents into the vascular system through a change in signal intensity.   By use of techniques similar to clinical techniques used to measure blood flow, Van As and Schaafsma (1984) measured rates of flow through a transpiring cucumber stem with MRI. Flow through capillary tubes of similar diameter to the cucumber vascular elements was used as a model to calibrate the technique.   Size of the vascular elements is limiting, however, because only large diameter vessels are suitable for flow measurement by this technique.
        High resolution images can also be made of other plant tissues, as seen in Figure 5 illustrating a soybean nodule and root. Physiological processes on an individual organ

such as this can be studied by repeated
acquisition of images of the same plant undergoing
any selected treatment or experiment.

   There are many questions in plant pathology
which might be addressed from the unique viewpoint
of MRI. Core breakdown of Bartlett pears has been
observed, allowing nondestructive comparison of
different storage techniques (Wang and Want,
1989). Nematode infections causing formation of
root galls has been studied with MRI (Matyac, et
al., 1989). Wood structure, including growth
rings and knots in aspen have been seen in images
acquired by MRI (Hall, et al., 1986). Regions of
root necrosis in loblolly pine seedlings have been
observed in our laboratory with MRI, while the
surface of the root showed only slight
discoloration (Figure 6). Although this technique

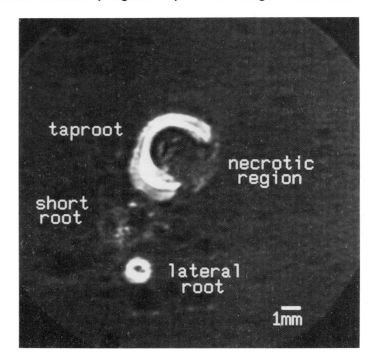

Figure 6. Cross-sectional view through a loblolly
pine seedling tap root. A necrotic region can be
clearly distinguished.

offers a unique potential in allowing study of
individual subjects, providing an opportunity to
directly and non-destructively observe disease
progression, limited use has been made of MRI for
this application to date.

Imaging is proving particularly useful for the
study of root-soil relations which usually cannot
be observed without destroying the natural
relationship of the roots to the surrounding
substrate by excavation. Removal of roots from
the substrate usually disrupts the root hairs,
root cap and the layer of mucigel at the root-soil
interface (Curl and Truelove, 1986). Several
studies have demonstrated the feasibility of
imaging roots in situ in sand, soil, and other
media (Bottomley, et al., 1986; Rogers and
Bottomley, 1987). A high content of ferromagnetic
particles (> 4%) causes magnetic field
inhomogeneities and reduces image quality. High
water content in some synthetic potting mixes such
as perlite, Ottawa sand, and peat give a strong
signal that obscures the root image (Rogers and
Bottomley, 1987) and some sands produce field
inhomogeneity, termed magnetic susceptibility,
that distort the root images. However, excellent
images of roots have been obtained both in cross
and longitudinal section, including what appears
to be the rhizosphere.

Regions of water depletion also have been
observed in images of woody loblolly roots in
moist sand (MacFall, et al., 1990). Such studies
will provide important information concerning the
role of root density in water uptake. They also
verify the absorption of water through suberized
roots (Chung and Kramer, 1975).

E.   Developmental and Morphological Studies

NMR imaging is valuable for developmental
studies because it is nondestructive and repeated
observations can be made over time on the same
tissue. For example, stem tips could be imaged
repeatedly over time to show the development of
leaf and flower primordia and their vascular
connections. It might be possible to follow the
transformation of a stem tip from the vegetative
to the flowering condition. Development of seeds

in ovaries and changes in water content of
maturing seeds can be followed, as well as water
uptake by germinating seeds. Root growth and
branching also can be followed in some media, and
excellent images of cross-sections of intact,
growing roots and stems can be obtained at
microscopic definition. The potential usefulness
of NMR imaging for developmental studies was
discussed by Lohman and Ratcliffe (1988) and its
use in morphological studies by Brown, et al.
(1988).

Another exciting and unique potential
application of MRI to plant studies is the
possibility of developing three-dimensional
reconstructions of the root tissues and
surrounding soil. Images of adjacent slices can
be rapidly acquired in succession and processed
into a three-dimensional image of the specimen.
Such image processing techniques present the
possibility of making volumetric measurements and
comparisons over time, for example, of the surface
area of roots and the volume of soil comprising
the water depletion zone. These reconstruction
techniques are at present in the early stages of
development, but as their use becomes more
widespread in the clinical setting, the hardware
and software required will likely become
increasingly available to researchers in other
disciplines.

F.   Summary

Although magnetic resonance imaging has great
potential in the study of plant physiology, few
applications have been made to date of the
technique. The possibility of the non-destructive
study of plants with resolutions approaching the
cellular level will provide unique opportunities
to study process physiology and form in a manner
which has not previously been possible. As
magnetic resonance imaging units become more
common, and it is possible to devote more machine
time to research, it is likely that this technique
will be more widely used in agricultural and
physiological research.

ACKNOWLEDGMENTS

We wish to acknowledge the invaluable
assistance provided by Dr. G. Allan Johnson and
his staff in the Diagnostic Physics Laboratory of
the Duke University Medical Center in the research
on imaging plant material.

REFERENCES

Bottomley, P.A., Rogers, H.H., and Foster, T.H.
(1986). NMR imaging shows water distribution
and transport in plant root systems in situ.
Proc. Natl. Acad. Sci. 83, 87-89
Bradbeer, J.W., Stinson, S.L., and Stiller, M.J.
(1958). Malate synthesis in Crassulacean
leaves I. The distribution of ^{14}C in malate
of leaves exposed to $^{14}CO_2$ in the dark.
Plant Physiol. 33, 66-70.
Brown, J.M., Johnson, G.A., and Kramer, P.J.
(1986). In vivo magnetic resonance
microscopy of changing water content in
Pelargonium hortorum roots. Plant Physiol.
82, 1158-1160.
Brown, J.M., Thomas, J.F., Cofer, G.P., and
Johnson, G.A. (1988). Magnetic resonance
microscopy of stem tissues of Pelargonium
hortorum. Bot. Gaz. 149, 253-259.
Budinger, T.F., and Margulis, A.R., (eds.) (1988).
Medical Magnetic Resonance: A Primer. Soc.
of Magnetic Resonance in Medicine, Berkely,
Calif.
Chung, H.H., and Kramer, P.J. (1975). Absorption
of water and ^{32}P through suberized and
unsuberized roots of loblolly pine. Can. J.
For. Res. 5, 229-235.
Connelly, A., Lohman, J.A.B., Loughman, B.C.,
Quiquampoix, H., and Ratcliffe, R.G. (1987).
High resolution imaging of plant tissues by
NMR. J. of Exp. Bot. 195, 1713-1723.
Curl, E.A., and Truelove, B. (1986). The
Rhizosphere. Springer-Verlag. New York.
Davies, D.D., Grego, S., and Kenworthy, P. (1974).
The control of the production of lactate and
ethanol by higher plants. Planta 118, 297-
310.

Foyer, C. and Spencer, C. (1984). The application of ^{31}P NMR to the study of phosphate metabolism in whole leaves. Ann. Report of Res. Inst. for Photosyn., Univ. Sheffield, pp. 23-25.

Foyer, C., Walker, D.A., Spencer, C. and Mann, B. (1982). Observations on the phosphate status and intracellular pH of intact cells, protoplasts and chloroplasts fromphotosynthetic tissue using ^{31}P NMR. Biochem. J. 202, 429-434.

Hall, L.D., Rajanayagam, V., Stewart, W., and Steiner, P.R. (1986). Magnetic resonance imaging of wood. Can. J. For. Res. 16, 423-426.

Jenner, C.F., Xia, Y., Eccles, C.D., and Callaghan, P.T. (1988). Circulation of water within wheat grain revealed by nuclear magnetic resonance micro-imaging. Nature 336, 399-402.

Johnson, G.A., Brown, J., and Kramer, P.J. (1987). Magnetic resonance microscopy of changes in water content in stems of transpiring plants. Proc. Natl. Acad. Sci. USA. 84, 2752-2755.

Kime, M.J., Ratcliffe, R.G., Williams, R.J.P., and Loughman, B.C. (1982). The application of ^{31}P NMR to higher plant tissue I. Detection of spectra. J. Exptl. Bot. 33, 656-669.

Lauterbur, P.C. (1973). Image formation by induced local interactions-examples employing nuclear magnetic resonance. Nature 242, 190-191.

Lohmman, J.A.B. and Ratcliffe, R.G. (1988). Prospects for NMR imaging in the study of biological morphogenesis. Experentia 44, 666-679.

MacFall, J.S., Johnson, G.A., and Kramer, P.J. (1990). Observation of a water depletion region surrounding loblolly pine roots by magnetic resonance imaging. Proc. Natl. Acad. Sci. USA. 87, 1203-1207.

Martin, F. (1985). Monitoring plant metabolism by ^{13}C, ^{15}N and ^{14}N NMR spectroscopy: A review of the applications to algae, fungi and higher plants. Physiol. Veg. 23, 463-490.

Martin, J-B., Bligny, R., Rebeille, F., Douce, R.,
    Leguay, J.J., Mathieu, Y., and Guern, J.
    (1982). ^{31}P NMR study of intracellular pH of
    plant cells cultivated in liquid medium.
    Plant Physiol 70, 1156-1161.
Matyac, C.A., Cofer, G.P., Bailey, J.E., and
    Johnson, G.A., (1989). In situ observations
    of root-gall formation using NMR imaging. J.
    Nematol. 21, 131-134
Mitsumori, F., and Ito, O. (1984). Phosphorus-31
    nuclear magnetic resonance studies of
    photosynthesizing Chlorella. FEBS Letters
    174, 248-252.
Mitsumori, F., Yoneyama, T., and Ito, O. (1985).
    ^{31}P NMR studies on higher plant tissue.
    Plant Sci. Lett. 38, 87-92.
Osmond, C.B. (1978). Crassulacean acid
    metabolism: A curiosity in context. Ann.
    Rev. Plant Physiol. 29, 379-414.
Pradet, A., and Raymond, P. (1983). Adenine
    nucleotide ratios and adenylate energy charge
    in energy metabolism. Ann. Rev. Plant
    Physiol. 34, 199-224.
Rebeille, F., Bligny, R., Martin, J-B., and Douce,
    R. (1983). Relationship between the
    cytoplasm and the vacuole phosphate pool in
    Acer pseudoplatanus cells. Archiv.
    Biochem.Biophys. 225, 143-148.
Roberts, J.K.M. (1984). Study of plant
    metabolism in vivo using NMR spectroscopy.
    Ann. Rev. Plant Physiol. 35, 375-386.
Roberts, J.K.M., Ray, P.M., Wade-Jardetzky, N.
    and Jardetzky, O. (1980). Estimation of
    cytoplasmic and vacuolar pH in higher plant
    cells by ^{31}p NMR. Nature 283, 870-872.
Roberts, J.K.M., Callis, J., Wemmer, P., Walbot,
    V., and Jardetzky, O. (1984a). The
    mechanism of cytoplasmic pH regulation in
    hypoxyic maize root tips and its role in
    survival under hypoxia. Proc. Natl. Acad. of
    Sci. USA 81, 3379-3383.
Roberts, J.K.M., Callis, J., Jardetzky, O., and
    Walbot, V. (1984b). Cytoplasmic acidosis as
    a determinant of flooding intolerance in
    plants. Proc. Natl. Acad. Sci. USA 81, 6029-
    6033.

Roberts, J.K.M., Andrade, F.H., and Anderson, I.C.
    (1985). Further evidence that cytoplasmic
    acidosis is a determinant of flooding
    intolerance in plants. Plant Physiol. 77,
    492-494.
Rogers, H.H., and Bottomley, P.A. (1987). In situ
    nuclear magnetic resonance imaging of roots:
    Influence of soil type, ferromagnetic
    particle content, and soil water. Agron. J.
    79, 957-965.
Schaefer, J., Stejskal, E.O., and Beard, C.F.
    (1975). ^{13}C NMR analysis of metabolism in
    soybeans labeled by $^{13}CO_2$. Plant Physiol 55,
    1048-1053.
Stidham, M.A., Moreland, D.E., and Siedow, J.N.
    (1983). ^{13}C NMR studies of crassulacean acid
    metabolism in intact leaves of Kalanchoe
    tubiflora. Plant Physiol. 73, 517-520.
Van As, H., and Schaafsma, T.J. (1984).
    Novinvasivs measurement of plant water flow
    by nuclear magnetic resonance. Biophysical
    J. 45, 469-472.
Walter, L., Balling, A., Zimmermann, U., Haase,
    A., and Kuhn, W. (1989). NMR imaging of
    leaves of Mesembryanthemum crystallinum L.
    plants grown at high salinity. Planta 178,
    524-530.
Wang, C.Y., and Want, P.C. (1989). Nondestructive
    detection of core breakdown in bartlett pears
    with NMR imaging. Hortscience 24, 106-109.
Waterton, J.C., Bridges, I.G., and Irving, M.P.
    (1983). Intracellular compartmentation
    detected by ^{31}P NMR in intact photosynthetic
    wheat-leaf tissue. Biochim. Biophys. Acta
    763, 315-320.
Wehrli, R.W., MacFall, J.R., and Newton, T.H.
    (1983). Parameters determining the
    appearance of NMR images. In: T.H. Newton
    and D.G. Potts (eds.), Advanced Imaging
    Techniques, Vol 2 in Modern Neuroradiology.
    Calvadel Press, San Anselmo, CA pp 81-117.
Wu, M-X., and Wedding, R.T. (1985). Diurnal
    regulation of phosphoenolpyruvate carboxylase
    from Crassula. Plant Physiol. 77, 667-675.

Yakir, D., DeNiro, M.J., and Gat, J.R. (1990).
    Natural deuterium and oxygen-18 enrichment in
    leaf water of cotton plants grown under wet
    and dry conditions: evidence for water
    compartmentation and its dynamics. Plant,
    Cell and Environ. 13, 49-56.